Mass Spectrometry and Nutrition Research

RSC Food Analysis Monographs

Series Editor:
P.S. Belton, *School of Chemical Sciences, University of East Anglia, Norwich, UK*

How to obtain future titles on publication:
A standing order plan is available for this series. A standing order will bring delivery of each new volume immediately on publication.

For further information please contact:
Book Sales Department, Royal Society of Chemistry,
Thomas Graham House, Science Park, Milton Road, Cambridge,
CB4 0WF, UK
Telephone: +44 (0)1223 420066, Fax: +44 (0)1223 420247, Email: books@rsc.org
Visit our website at http://www.rsc.org/Shop/Books/

Mass Spectrometry and Nutrition Research

Edited by

Laurent B. Fay and Martin Kussmann
Nestlé Research Centre, Lausanne, Switzerland

RSCPublishing

RSC Food Analysis Monographs No: 9

ISBN: 978-1-84973-036-5
ISSN: 1757-7098

A catalogue record for this book is available from the British Library

Published by The Royal Society of Chemistry,
Thomas Graham House, Science Park, Milton Road,
Cambridge CB4 0WF, UK

Registered Charity Number 207890

For further information see our web site at www.rsc.org

Introduction to Mass Spectrometry in Nutrition and Health

LAURENT B. FAY[a] AND MARTIN KUSSMANN[b, c]

[a] Nestlé Research Centre, Nutrition and Health Department, Vers-chez-les-Blanc, PO Box 44, CH-1000, Lausanne 26, Switzerland; [b] Nestlé Research Centre, Vers-chez-les-Blanc, PO Box 44, CH-1000, Lausanne 26, Switzerland; [c] Faculty of Science, Aarhus University, Ny Munkegade, Building 1521, DK-8000, Aarhus C, Denmark

Like all animals, human beings must feed everyday from the cradle to the grave in order to supply their body with the right amount of water, nutrients and energy. From the early hunter-gatherers to the myriad of different cuisines today and all over the world, human beings have developed a wide range of preparations and cooking practices rooted in their environment, cultural mores and preferences. Food is sourced from plants, animals or other categories such as fungus or fermented products and provides nutrients grouped into two categories: macronutrients encompassing fat, protein, and carbohydrates and micronutrients that are minerals and vitamins. In addition food contains water and fibres. Moreover, food is not only consumed for sustaining body functions, but also for pleasure and enjoyment, and is an essential component of social interaction.

Worldwide food production has grown rapidly over the past 30 years, managing to exceed population growth. Consequently, the world can today

RSC Food Analysis Monographs No. 9
Mass Spectrometry and Nutrition Research
Edited by Laurent B. Fay and Martin Kussmann
© The Royal Society of Chemistry 2010
Published by the Royal Society of Chemistry, www.rsc.org

produce enough food to provide every person with more than 2700 calories per day, *i.e.* a level normally sufficient to ensure that all have access to adequate food, provided distribution is not too unequal. However, as of 2009, the Food and Agricultural Organization (FAO) estimates that 1.02 billion people are undernourished worldwide. This is the highest number since 1970, the earliest year for which comparable statistics are available (source: www.fao.org). Lack of essential energy and protein stunts the bodies, minds and hopes of some 200 million children. In parallel and ironically, the world is facing a serious obesity crisis. Indeed, 1.6 billion people worldwide are overweight with at least 400 million being obese (data for 2015 are projected to be over 700 million). Worldwide, 20 million children under five are estimated to be overweight. Between those two extremes, there are many disease states that can be either caused or alleviated by changes in diet. Diet imbalances or micronutrient deficiencies can negatively impact health, leading to diseases such as scurvy or osteoporosis, as well as psychological and behavioural problems.

It is well recognized that nutrition exerts one of the strongest life-long environmental impact on human health and the interplay between nutrition and health has been known for centuries: the Greek doctor Hippocrates (4th century BC) recommended using food as medicine and vice versa; Sun Si-Miao, a famous doctor of the Tang Dynasty (7th century AD) stated that "When a person is sick, the doctor should first regulate the patient's diet and lifestyle". Today nutrition is one of the hottest topics debated in the press and there is not a single day without the advertisement of nutritional advices, magical recipes to for example lose fat and reshape our body, or new food components that for example strengthen our arteries or delay Alzheimer's disease.

Today's nutrition science attempts to understand how and why specific dietary aspects influence health. Historically, nutrition research had defined macro- and micronutrient requirements at population level and has thereby provided recommendations to avoid mal- and undernutrition of the general public. As such, nutrition science began just prior to the dawn of the 20th century. At that time, Christiaan Eijkman, a Dutch military doctor, found that rice skin contained some factors counteracting beri-beri—a disease that debilitates and kills people by attacking their hearts and nervous systems. This led to further research and the conclusion that a diet of over-milled rice was the main cause of beri-beri in man. In 1901, Grijns extracted an essential nutrient— later to be identified as the B vitamin thiamine—in the outer layers of rice that protected against beri-beri. A whole new era in nutrition and medicine was started, largely based on observational trial-and-error experiments.

Most of our current knowledge of health and food is still based on such observations and correlations. Indeed, when assessing the relationship between diet and health, a distinction must be made between causality and correlation. A causal relationship between diet and health can be defined as one in which the chain of events, which link the ingestion of a given food with a defined bio-logical endpoint, is established and the mechanism of action is known. In a correlative relationship, a health/disease endpoint appears to be related in some way to a given food but the nature and extent of that relationship is not established. Most of our current knowledge concerning diet and health only facilitates correlations. Assigning dietary health benefits or disease risks to a particular dietary pattern is difficult for a number of reasons. These include the complexity of diet, the co-occurrence of many nutrients in foods, the possible interactions between diet and the genetic background of the individual and also environmental factors.

The new era of nutrition research aims to move from empirical knowledge to evidence-based molecular science. This is because the myriad of food components interact with our body at system, organ, cellular and molecular level. Modern nutrition and health research focuses on promoting health, reducing the risk of disease and optimizing physical and mental performance. Understanding the interplay between food and the human body requires holistic approaches because nutritional improvement of one health aspect must not be compromised by deterioration of others. Nowadays, holistic technologies—and mass spectrometry with its outstanding analytical capabilities is considered as one of those—have on the one hand to deliver a holistic view of metabolism and on the other to understand the inter-individual differences in human metabolism. Moreover, the concept of biologically functional foods requires the understanding of the disease aetiology and, on that basis, the mechanisms of prevention followed by the identification of active food constituents and the demonstration of their bioavailability and efficacy. Biomarkers have been proven as being an important tool to understand the latter. When combined with biomarkers of susceptibility, they should help understand and explain the responsiveness of human to diet (whole food or individual compounds), and ultimately reveal the relationship between nutrition and health. Based on the strong correlation between bioavailability, efficacy and susceptibility, sets of biomarkers at protein, nutrient and metabolite level will ultimately be required to describe the pathway from an exposure, via internal dose, target tissue dose, biologically effective dose to early biological effect, altered structure and/or function and the final outcome.

Mass spectrometry has played a fundamental role in the advancement of all aspects of life sciences in general and nutrition science in particular. From early development aiming at coupling a gas chromatograph to a mass spectrometer in the mid-1950s to modern UPLC-MS, the compound range accessible by mass spectrometers has been extended from mostly volatile and rather low molecular weight organic compounds to non-volatile and fragile high molecular weight biomolecules, and even beyond that to large supra-molecular assemblies. This virtually unlimited mass range plus the information-richness enabled by controlled fragmentation and structural analysis of these entities has rendered mass spectrometry one of the most powerful analytical platforms in modern health and life sciences.

Mass spectrometry is *per se* not quantitative, meaning that the response of a given analyte typically depends on its chemical nature. However, thanks to developments such as isotopically labelled standards, high-resolution based direct comparison of liquid-chromatography mass spectrometry traces ("label-free") or targeted parallel quantification based on specific fragmentation behaviours of selected analytes ("multiple-reaction monitoring"), relative and absolute quantification of bioactives in food and body as well as their bioefficacy markers is now possible.

As food ingredients and their sensorial and physiological benefits come rather as an ensemble than as individual, purified active principles, the analysis of both the food matter and its effects conferred on the body requires sensitive methods that can be tuned to be once compound-specific and once holistic depending on whether the interest is targeted characterization or global analysis. Mass spectrometry is therefore the ideal tool for modern nutrition research as it can deliver sensitivity, specificity and comprehensiveness.

Markers for nutritional bioavailability and bioefficacy can be addressed in a hypothesis-driven, targeted fashion or at global scale without any pre-assumption. Today, nutritional research is taking advantage of mass spectrometry to comprehensively decipher biologically active food components and to assess their bioavailability at different organ sites, as well as their site-specific and eventually overall health benefits.

Mass spectrometers are also utilized to address other aspects of food and nutritional research than health. Practically all modern mass spectrometric techniques have fed into analytical platforms for nutritional research. The instruments are for example used to investigate the pleasure aspects of food, *i.e.* its sensory qualities. Flavour is the sum of taste and olfaction. LC-MS/MS techniques typically serve to characterize taste molecules, whereas GC-MS instruments analyze the volatile molecules that make up for olfaction of food and drinks. Safety and quality assessment employ a large range of mass spectrometric instrumentation from isotope ratio MS (IR-MS) for authenticity checks to hyphenated MS techniques for ensuring quality, safety and compliance of food products.

This book has two objectives: introducing and inspiring nutritionists to the use of mass spectrometry and convincing mass spectrometric scientists to develop and apply their tools to questions of nutrition and health. By addressing the major modern mass spectrometric techniques, reviewing their application to the analysis of food macro- and micronutrients and assessing the utility and potential of mass spectrometry in addressing nutritional health effects, we hope to achieve this ambitious goal.

Contents

RSC Food Analysis Monographs No. 9
Mass Spectrometry and Nutrition Research
Edited by Laurent B. Fay and Martin Kussmann
© The Royal Society of Chemistry 2010
Published by the Royal Society of Chemistry, www.rsc.org

SECTION 2: MASS SPECTROMETRY ANALYSIS OF FOOD INGREDIENTS

Abbreviations

AC	affinity chromatography
AMS	accelerator mass spectrometry
APCI	atmospheric pressure chemical ionization
APPI	atmospheric pressure photoionization
AQUA	absolute quantitative analysis
CAD	collisionally activated dissociation
CAE	chemically assisted fragmentation
CE	capillary electrophoresis
CEC	electrochromatography
CI	chemical ionization
CID	collision induced dissociation
cIEF	capillary isoelectric focusing
CIS	co-ordination ion-spray
COSY	correlation spectroscopy
CSIA	compound-specific isotope analysis
DESI	desorption electrospray ionization
DIGE	differential gel electrophoresis
EA	elemental analyzer
EC	electron capture
ECD	electron capture dissociation
ECNCI	electron capture negative chemical ionization
EI	electron ionization
ELSD	evaporative light scattering detection
ESI	electrospray ionization
ETD	electron transfer dissociation
FAB	fast atom bombardment
FIA	flow injection analysis
FPLC	fast protein liquid chromatography
FTICR	Fourier transform ion cyclotron resonance
FWHM	full width at half height maximum
GC	gas chromatography

GC-MS	gas chromatography-mass spectrometry
GLC	gas–liquid chromatography
HMBC	heteronuclear multiple bond correlation
HPLC	high performance liquid chromatography
HSQC	heteronuclear single quantum correlation
ICAT	isotope-coded affinity technology
ICP-MS	inductively coupled plasma mass spectrometry
IE	ionization energy
IEF	isoelectric focusing
IRMS	isotope ratio mass spectrometry
IT	ion trap
iTRAG	isobaric tag for relative and absolute quantitation
LC	liquid chromatograph
LC-MS	liquid chromatography-mass spectrometry
LOD	limit of detection
MALDI	matrix-assisted laser desorption ionization
MRM	multiple reaction monitoring
MS	mass spectrometry
MS^n	multi stage mass spectrometry
MS/MS	tandem mass spectrometry
m/z	mass-to-charge
NI	negative ionization
NCI	negative chemical ionization
NMR	nuclear magnetic resonance
PAGE	polyacrylamide gel electrophoresis
PDA	photodiode array
PGC	porous graphitized carbon
PI	positive ionization
ppm	part per million
PSD	post-source decay
PTM	post-translational modification
QqQ	triple quadrupole
QqToF	quadrupole/time-of-flight
rf	radio frequency
RPLC	reversed phase liquid chromatography
SCX	strong cation exchange [chromatography]
SDMS	synthesis/degradation ratio mass spectrometry
SEC	size exclusion chromatography
SIM	single ion mass
S/N	signal-to-noise [ratio]
SPE	solid phase extraction
SRM	single reaction monitoring
Th	Thompson [unit]
TIC	total ion current
TIMS	thermal ionization mass spectrometry
TLC	thin layer chromatography

TMT	tandem mass tag
TOCSY	total correlation spectroscopy
ToF	time-of-flight
ToF/ToF	time-of-flight/time-of-flight
TSI	thermospray ionization
UPLC	ultra-high performance liquid chromatography

About the Editors

Dr. Laurent B. Fay is currently the Head of the Nutrition and Health Department at the Nestlé Research Center in Switzerland. After a Master in Nutrition in 1981 he completed his PhD in Biochemistry at Dijon University, France, in 1985, and joined Nestlé in 1989 after three years in the pharmaceutical industry conducting metabolic and pharmacokinetic studies.

Dr. Fay's current research aims at developing nutritional approaches that promote health and wellness throughout life. His research team focuses on providing practical food and beverage solutions for optimal physical and cognitive performance, growth and development, protection, and weight management. Composed of nutritionists and biologists from diverse fields including microbiology, molecular biology, immunology, allergy, gut physiology and cognitive science, they apply the scientific advances of today to meet the nutrition needs of tomorrow. They bridge nutrition, metabolism and health through an integrated approach, transforming fundamental knowledge into product concepts.

Dr. Fay is the author of more than 160 peer-reviewed publications. He is honorary member of the Swiss Group for Mass Spectrometry.

Prof. Dr. Martin Kussmann leads the Functional Genomics Group at the Nestlé Research Centre and is Honorary Professor at the Faculty of Science, Aarhus University, Denmark. His team at Nestlé is responsible for Nutrigenomics and Nutrigenetics. They develop and integrate gene and protein expression profiling with bioinformatics and computational molecular science. Major application fields are immunology, diabesity, and digestive health. Being educated as an analytical biochemist, Prof. Kussmann has acquired research experience in the pharmaceutical (Pharmacia-Upjohn/

Sweden, Cerbios-Pharma/Switzerland), biotechnological (Prof. Dr. D. Hochstrasser, GeneProt/ Switzerland) and nutritional (Nestlé) industry.

Prof. Kussmann holds a M.Sc. and a Ph.D. in Chemistry, obtained at the Universities of Aachen and Konstanz, Germany, and at the University of California, San Francisco, USA (Prof. Dr. A. L. Burlingame). During his doctorate and post-doctorate (Prof. P. Roepstorff, University of Southern Denmark, Odense), he has specialised in analytical biochemistry, proteomics and genomics.

Prof. Kussmann has (co-) authored 50 peer-reviewed publications, has edited books and journal issues, and is an internationally requested author and speaker. He is a member of the *ASBMB* (Am. Soc. Biochem. Mol. Biol.), *DGMS* (German Soc. Mass Spectrom.), *NuGO* (European Nutrigenomics Organization), *SGMS* (Swiss Soc. Mass Spectrom.) and *SPS* (Swiss Proteomics Soc.). He is an Editorial Board Member of the *Journal of Proteomics; The Open Journal of Proteomics; Proteomics Insights; and Endocrine, Metabolic and Immune Disorders-Drug Targets (EMID-DT)*.

Section 1
Mass Spectrometry Technologies

CHAPTER 1

Mass Spectrometry Technologies

LAURENT B. FAY[a] AND MARTIN KUSSMANN[a,b]

[a] Nestlé Research Centre, Vers-chez-les-Blanc, PO Box 44, CH-1000, Lausanne 26, Switzerland; [b] Faculty of Science, Aarhus University, Ny Munkegade, Building 1521, DK-8000, Aarhus C, Denmark

1.1 Introduction

Mass spectrometers are molecular balances. They can determine the size, quantity and structure of inorganic and organic compounds. Traditionally, these measures had been limited to volatile organic compounds, but for 20 years mass spectrometers have generated such information also on large, fragile and non-volatile molecules such as vitamins, peptides, proteins, oligo- and poly-saccharides, and even DNA and RNA.

Mass spectrometry (MS) has become an essential analytical tool in modern life sciences, not only thanks to its sensitivity but also to the large amount of information delivered by this technique from a structural and a quantitative viewpoint. Typically, mass spectrometers enable structure elucidation of organic molecules *via* the determination of molecular weight and the study of fragmentation patterns. Moreover, and increasingly importantly, such instruments can quantify these organic molecules. Additionally, in scientific areas other than the life sciences (geochemistry, ecology, food chemistry, forensic and sport science), mass spectrometry is widely deployed to precisely determine stable isotope ratios of exogenous or endogenous molecules. ^{13}C labelled compounds are mainly used in mass spectrometry as internal standards or to

RSC Food Analysis Monographs No. 9
Mass Spectrometry and Nutrition Research
Edited by Laurent B. Fay and Martin Kussmann
© The Royal Society of Chemistry 2010
Published by the Royal Society of Chemistry, www.rsc.org

generate metabolite information. Whereas hydrogen/deuterium exchange experiments are used to distinguish between isomeric structures of analytes.[1]

In 1910, J. J. Thomson was the first to build a so-called "parabola spectrograph" meant for the determination of mass-to-charge (m/z) ratios of ions. Following this pioneering work, A. J. Dempster and F. W. Aston developed the first mass spectrometric instrument. Dempster constructed a magnetic analyzer that focused ions into an electrical collector, while Aston utilized both electrostatic and magnetic fields to focus ions onto a photographic plate. From the late 1930s to the early 1950s, A. Nier in collaboration with J. H. E. Mattauch, R. F. K. Herzog and K. T. Bainbridge (amongst others) incorporated many developments in vacuum technologies and electronics for power supplies and ion detection. Their work significantly improved magnetic focusing instruments leading to better performance, convenience and lower costs. Double-focusing machines, attaining greater precision by adding an electrostatic analyzer, were also greatly refined. These instruments were built for the purpose of accurately determining the exact atomic weights of the elements and their isotopes; they made use of Faraday cups to convert particle impacts into an electric current for signal recording. Tremendous progress has been made since this pioneering work in, for example, ion generation, ion transmission, ion detection, signal amplification and, last but not least, computing technologies to control the instrument and record the data.

Since the 1980s, mass spectrometry has become one of the most popular analytical platforms for the identification and/or quantification of organic molecules in complex samples such as body fluids, tissues and food matrices. During less than two decades, we have witnessed the transformation of mass spectrometers from multi-purpose research grade instruments operated only by instrumental experts into user-friendly computer-embedded solutions dedicated to specific measurements such as nutrient/metabolite and peptide/protein identification and quantification. Even the detailed analysis of genetic and genomic material is now being addressed by mass spectrometry. A Google® search with the term "mass spectrometry" results in more than six million entries. In 2007, the mass spectrometry market was estimated at \$2 billion with an expected 8% annual growth rate through 2010 (www.allbusiness.com/instrument-business-outlook/1179913-1.html).

The performance of mass spectrometers in combination with ionization techniques can be defined by several intrinsic parameters, *i.e.* mass resolving power (or resolution), mass accuracy, sensitivity and linear dynamic range (Figure 1.1).

The **mass resolving power or resolution** is defined as the ratio $m/\Delta m$, with the mass (m) at the apex of the mass signal and Δm the width at x% height (typically 50%) of this mass signal, designated by the full width at half height maximum (FWHM). The **mass accuracy** is described by the ratio between the mass error (difference between measured and real mass) and the theoretical mass, often represented as parts per million (ppm), *e.g.* a mass accuracy of 100 ppm corresponds to a theoretical mass of 1000 with a measured mass at 999.9. The **sensitivity** is described by the ratio between the intensity level of the mass signal and the intensity level of the noise. The **linear dynamic range** is

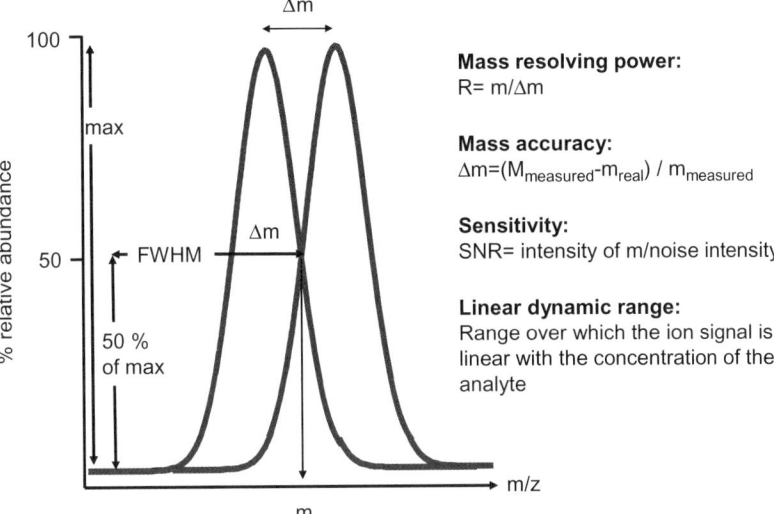

Figure 1.1 The performance of a mass spectrometer is characterized by: (1) mass resolving power (the minimum mass difference that can still be separated from a given mass); (2) mass accuracy (the accuracy of determining the real mass of any compound); (3) sensitivity (signal-to-noise for a given absolute amount and type of analyte); and (4) linear dynamic range.

described as the range of linearity of the ion signal measured as a function of the analyte concentration.

The enormously broad scope of nutritional research (*e.g.* organic and inorganic nature of the analytes, their volatility or thermal instability, the wide polarity range from water soluble compounds to lipophilic molecules), and the need to measure isotopic abundance to investigate the metabolic fate of nutrients require the deployment of virtually all the mass spectrometric instrumentations available today.

There are many books describing in great detail various mass spectrometric technologies either from a purely instrumental perspective[2,3] or from a more applicative viewpoint.[4,5] Below we briefly describe each instrumentation starting with the ionization sources, and then describe the various mass analyzers and ion detection devices commonly available on the market. The hyphenation with different chromatographic systems is covered at the end of the chapter.

1.2 Ionization Sources

Any species—be it an organic molecule or an inorganic element—to be analyzed in a mass spectrometer needs to be ionized unless already present in an ionic form. Ionisation of an analyte (M) takes place by removing or adding an electron to yield an $M^{+\bullet}$ or $M^{-\bullet}$, respectively. Both these species have the same mass as the original molecule, with the mass of the electron being negligibly small. The analyte may also be ionized by addition or subtraction of charged species (*e.g.* H$^+$) to give [M + H]$^+$ or [M − H]$^-$ with masses that are different from that of the starting analyte.

Ionization is a key process in mass spectrometry and much instrumental and theoretical work has been devoted to understanding the processes that convert a neutral molecule into an ionized species (cation or anion). This process takes place in the so-called ionization source, which is also responsible for the transfer of the newly produced ions into the gas phase prior to their introduction into the analyzer of the mass spectrometer. There are several ionization sources namely:

- electron ionization;
- chemical ionization;
- electrospray ionization;
- atmospheric pressure chemical ionization;
- atmospheric pressure photoionization; and
- matrix-assisted laser desorption ionization.

1.2.1   Electron Ionization

Electron ionization (EI)—formerly also called electron impact ionization—is the oldest method of ionization. It originates from the work of J. J. Thompson (1856–1940) who won the Nobel Prize in physics in 1906 for having generated ionized species following a discharge of electricity into gases. The unit for the mass-to-charge (*m/z*) values in mass spectrometry, the Thompson (Th), is the recognition of his groundbreaking work.

Energetic electrons from a heated filament are accelerated by an electric field through the high vacuum ion chamber containing the gaseous sample. Such energetic electrons will transfer part of their energy to the neutral volatilized analyte, causing an electron to be ejected from the molecule forming a molecular ion through the reaction:

$$M + {}^\bullet e^- \rightarrow M^+ + 2\,e^-$$

In most electron ionization conditions, only one molecule out of a million will be ionized. The energy of the electron beam is approximately 70 eV; keeping this energy constant between instruments allows standardized spectra to be obtained and the assembly of EI spectra libraries for identification of compounds with so far unknown EI spectra.

Most of the organic analytes have an ionization potential (*i.e.* energy required to liberate an electron from the molecule) in the range 6–10 eV. Therefore, the excess of energy remaining in the molecular ion produces fragmentation, which is often pronounced enough to decompose the molecular ion. Indeed, the process of electron ionization is considered a "hard" ionization

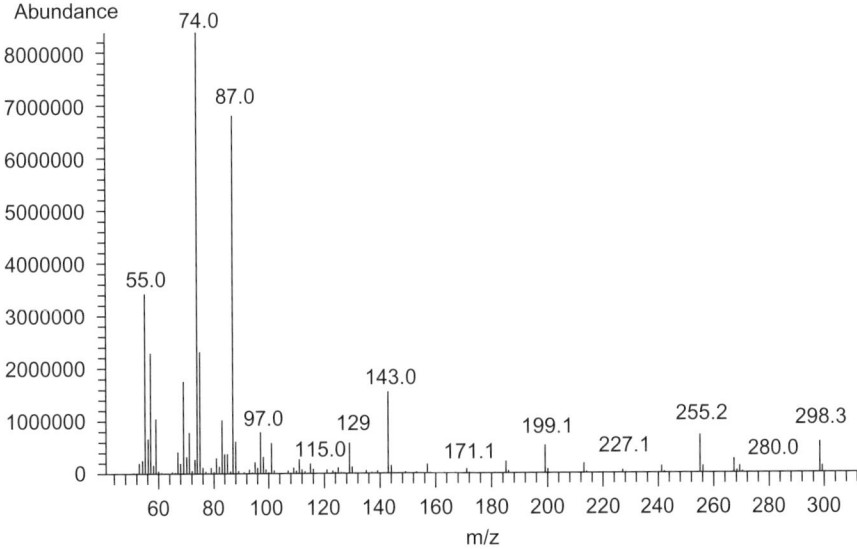

Figure 1.2 Electron ionization (EI) mass spectrum of methyl stearate. The molecular ion ($M^{+\bullet}$) is at m/z 298.

process resulting in considerable fragmentation of the molecular ion due to the excess of energy imparted to the analyte (see, for example, Figure 1.2). Fragmentation of the molecular ion produces both even and odd electron fragment ions.

1.2.2 Chemical Ionization

Chemical ionization (CI) relies on gas phase ion-molecule reactions to ionize the neutral analyte. First, the collision of an ion with a reactant gas such as methane gives rise to $CH_4^{+\bullet}$, which reacts with another methane molecule yielding a stable CH_5^+ ions.[6] This so-called reactant gas ion collides with the analyte, ionizing the latter usually by proton transfer and giving a protonated molecule:

$$M + CH_5^+ \rightarrow [M + H]^+ + CH_4$$

These ion-molecule reactions need to take place at a source pressure of about 0.1–1.0 Torr, which is a much higher pressure than the one used for EI (10^{-5} to 10^{-6} Torr). In the CI ion source, the pressure of 0.1–1.0 Torr favours ion-molecule reactions before the ions are repelled from the ion chamber. Moreover, under this high pressure, electron capture is very efficient making the generation of negative ions just as likely as generation of positive ions. Negative CI is, therefore, a frequently used technique. The sensitivity of positive CI is similar to the sensitivity of EI. However, for highly electrophilic analytes,

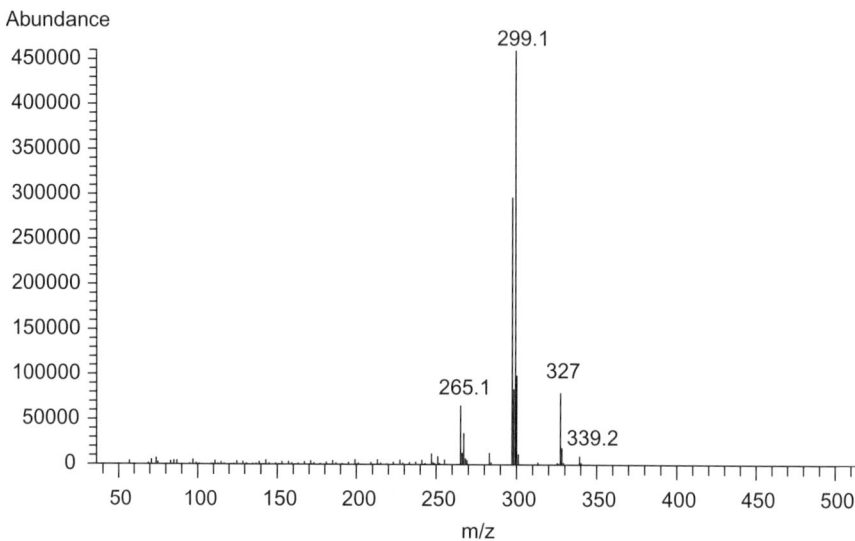

Figure 1.3 Positive-ion chemical ionization (CI) mass spectrum of methyl stearate. The protonated molecular ion $(M + H)^+$ is at m/z 299. The signal at m/z 327 corresponds to the adduct ion $[M + C_2H_5]^+$.

negative CI is many times more sensitive than positive CI enabling, for example, the detection of 200 zeptomol (10^{-21} mol) of methyl uracil as its pentafluoro derivative.[7]

Chemical ionization is a softer ionization method compared with electron ionization as a greater abundance of ionized analyte can be detected (see, for example, Figure 1.3), making CI useful for determining the molecular weight of unknown analytes. The most popular reactant gases for chemical ionization are methane, isobutane and ammonia.

1.2.3 Atmospheric Pressure Chemical Ionization

Atmospheric pressure chemical ionization (APCI) resembles CI in the sense that ionization is conferred by a reactive reagent. In APCI, the ions are formed at atmospheric pressure using electrons emitted either by a ^{63}Ni foil or, more commonly today, by a corona discharge needle. In such a corona discharge mode, a needle is located near the ion source and a very high negative voltage is applied to it. The analyte solution is sprayed into a heated nebulizer and converted into a fine mist which is carried by a flow of nitrogen and passes through the corona discharge. Primary ions ($N_2^{+\cdot}$, $O_2^{+\cdot}$ or $H_2O^{+\cdot}$) are formed by electron ionization *via* electrons from the corona. They collide with the solvent molecules forming secondary clusters of reactant gas ions, *e.g.* $H_3O^+(H_2O)_n$. Analyte ionization takes place *via* gas phase ion-molecule reactions such as proton transfer and charge exchange. Proton transfer occurs if the analyte has a high proton affinity, whereas for charge exchange, the analyte should possess low ionization energy. Only relatively small and stable compounds up to about 1000–1500 Da can be analyzed using APCI because of the heat required in the vaporization process. In addition, APCI often suffers from a high background due to efficient ionization of gases, solvents and impurities that have high proton affinities.

1.2.4 Atmospheric Pressure Photoionization (APPI)

Atmospheric pressure photoionization (APPI) involves the absorption of energy from an ultraviolet (UV) source. Ionization is induced with vacuum-ultraviolet 10 eV photons emitted by a krypton discharge lamp. The photons can ionize compounds that possess ionization energy (IE) below their own energy (10 eV); this includes most analytes, but excludes most of the typically used gases and solvents. Single-photon ionization occurs according to following the reaction:

$$M + hy \rightarrow M^{+\cdot} + e^-$$

An alternative route may also take place if a carrier gas such as nitrogen is used which strongly absorbs the UV radiation:

$$N_2 + hv \rightarrow N_{2^*}$$

$$N_{2^*} + M \rightarrow N_2 + M^{+\cdot} + e^-$$

Typical analytes possess ionization energies in the range 7–10 eV, whereas the common carrier gases have higher IE values (Table 1.1). Therefore, the analytes can be selectively ionized without interference from ionized carrier gas molecules. Moreover, APPI allows the ionization of less polar analytes because the ionization depends on the ionization energy of the analyte rather than its proton affinity, as in the case of ESI (see below) and APCI.

1.2.5 Electrospray Ionization

The development of electrospray ionization (ESI) for the analysis of biological macromolecules was rewarded with the attribution in 2002 of the Nobel Prize in chemistry to J. B. Fenn (together with Koichi Tanaka for "the

Table 1.1 Ionization energies (eV) of the common carrier gases used for mass spectrometric analysis.

Name	Symbol	Ionization potential (eV)
Hydrogen	H	13.6
Nitrogen	N	14.5
Argon	Ar	15.7
Helium	He	24.5

development of methods for identification and structure analyses of biological macromolecules" and K. Wüthrich for "his development of nuclear magnetic resonance spectroscopy for determining the three-dimensional structure of biological macromolecules in solution"). Electrospray ionization revolutionized the way ions were transferred from a solution to the gas phase.

There is ample literature debating the mechanism of electrospray ionization.[8] In a nutshell, analytes are introduced to the source in an ionic solution at a low flow rate (a few µl min^{-1}). This solution passes through an electrospray needle with a high potential difference between its tip and a nearby counter electrode (typically in the range 2.5–4 kV). The positive charges in the solution are repelled by each other and by the positively charged needle walls, so that a cone-shaped flow is formed on the tip of the needle. In a similar way, negatively charged ions can be formed by setting a negative voltage on the needle wall.

As the solvent evaporates, droplets shrink with a resulting increase of the charge density on their surface. Eventually, the droplets reach a state called "Rayleigh's instability limit", which causes them to undergo a series of coulombic fissions until gas-phase ions are left.[9] There is another theory, by Iribarne and Thomson,[10] stating that the droplets start emitting ions directly to the gas phase when the repulsion forces on the droplet surface become high enough to break the surface tension ("ion evaporation").

ESI is a very soft ionisation method as the analytes retain very little residual energy upon ionization. It is applicable to small polar molecules for which almost no fragmentations are observed, and to large macromolecules (*e.g.* proteins or polysaccharides) as multicharged ions are produced (see, for example, Figure 1.4). Multicharged ions have proved extremely useful in determining the molecular weight of large macromolecules more precisely. Mathematical deconvolution of the series of peaks allows calculating their molecular weights with an accuracy of 10 ppm.[11]

As a recent variant of the ESI method, the so-termed DESI technique—standing for "desorption electrospray ionization"—has been developed by Cooks *et al.*[12,13] In this approach, the cone-shaped electrospray is targeted directly onto solid surfaces under ambient pressure in order to directly desorb ions, which are then sampled into a vacuum. Hence, DESI has the ability to record mass spectra on ordinary samples in their native environment, without sample preparation or pre-separation—all this by creating ions outside the instrument. Extremely rapid analysis is coupled with high sensitivity and high chemical specificity. The technique is appealing for direct, minimally invasive analysis of surfaces and can be even extended to biological applications such as skin analysis. These characteristics are applied to, for example, high-throughput metabolomics, natural product discovery, biological tissue imaging or detection of explosives. DESI may be of future use for *in vivo* clinical analysis and its adaptation to portable mass spectrometers.

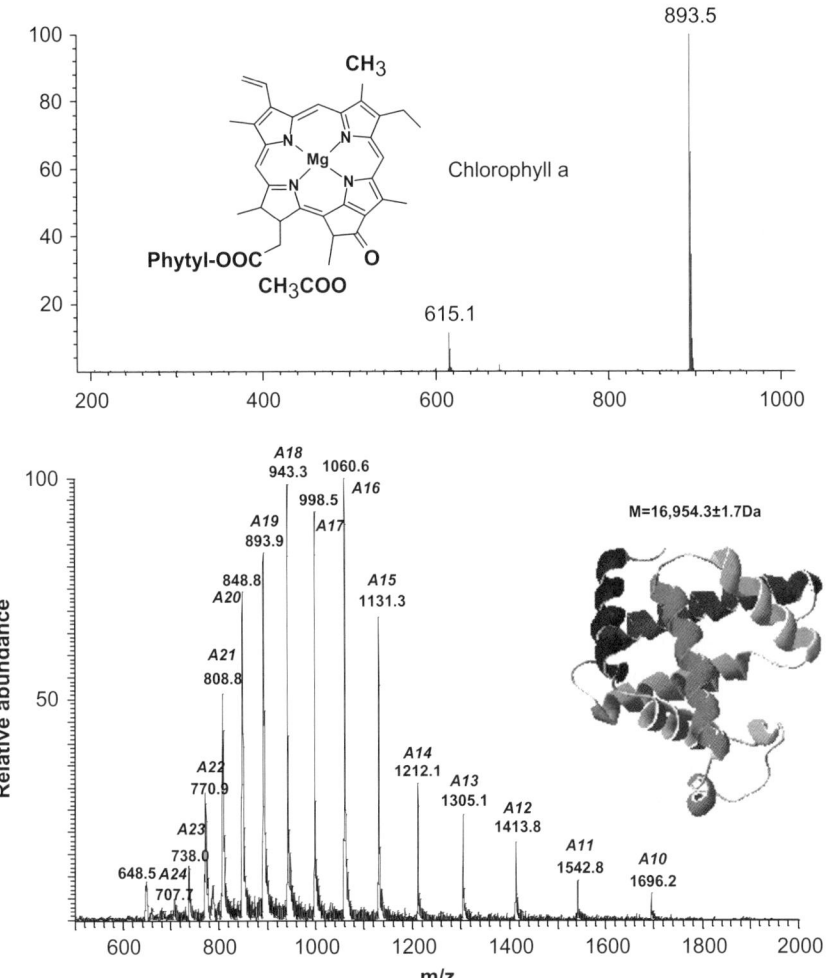

Figure 1.4 Positive electrospray ionization (ESI) mass spectrum of chlorophyll a (top panel) and myoglobin (bottom panel). The protonated molecular ion $(M + H)^+$ of chlorophyll is at m/z 893. The insert in the bottom panel shows the three-dimensional structure of the protein and denotes its molecular mass.

1.2.6 Matrix-assisted Laser Desorption Ionization

In matrix-assisted laser desorption ionization (MALDI), a low concentration of analyte is mixed with a large excess of a matrix on a metal plate and exposed after drying to a pulsed nitrogen laser beam (337 nm) or an Nd-YAG laser (266 nm). The matrix has a strong absorbance at the laser wavelength and is highly sublimable. Following each laser pulse, analyte ions are produced. Organic matrix compounds typically used are 2,5-dihydroxybenzoic acid,

3,5-dimethoxy-2-hydroxy-*trans*-cinnamic acid (sinapinic acid) and α-cyano-4-hydroxy-*trans*-cinnamic acid.

MALDI ionization is mediated by the absorption of the laser energy by the strongly UV-absorbing matrix. This causes rapid vibrational excitation of the matrix, leading to localized disintegration of the "solid solution". The clusters ejected from the surface consist of analyte molecules surrounded by matrix and salt ions. Stabilization of the photo-excited matrix molecules takes place through proton transfer to the analyte. Cation attachment to the analyte is also favoured by this process, forming the characteristic $[M + X]^+$ analyte ions with $X = H$, Na, K, *etc.*

1.3 Mass Analyzers

The mass analyzer is central to mass spectrometric technology. Its role is to separate ions—molecular ions (M^+), adduct ions such as $[M + H]^+$ and $[M + Na]^+$, and fragment ions—with a defined mass accuracy and resolution. In this chapter, we consider several basic types of mass analyzers currently used in life sciences and nutrition, namely:

- electromagnetic sector;
- quadrupole mass filter;
- ion trap;
- time-of-flight analyzer;
- Fourier transform ion cyclotron resonance analyzer; and
- the Orbitrap system.

The various analyzers are different in conception and performance, each with its own strengths and weaknesses. Often, they work as a standalone mass analyzer, but the current trend points towards hyphenated systems in order to combine the advantages of different analyzers in one mass spectrometer:

- triple-quadrupole;
- quadrupole-ion trap;
- quadrupole-time of flight;
- ion trap-time of flight;
- time of flight-time of flight;
- ion trap-Fourier transform ion cyclotron resonance; or
- ion trap-Orbitrap tandem mass spectrometers.

The last three systems are strongly dedicated to the analysis of macromolecules.

1.3.1 Electromagnetic Sector Analyzer

The magnetic sector analyzer is a momentum separator. Accelerated ions from the ionization source enter a flight tube and travel into the magnetic field. The

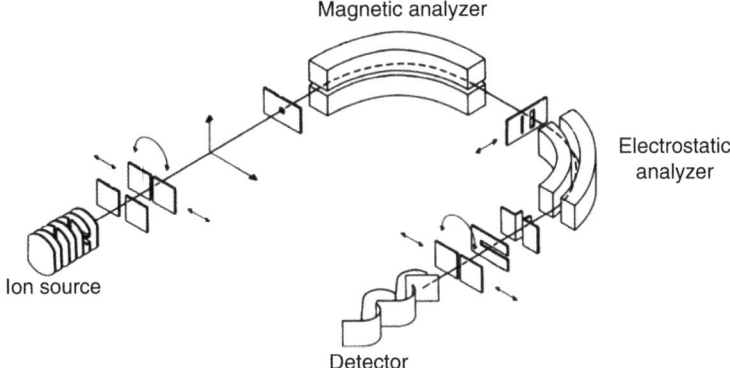

Figure 1.5 Electromagnetic sector analyzer.

accelerating voltage from the source determines the kinetic energy (zeV) imparted to the ions. Once accelerated, the ions are focussed with slit lenses to reduce their energetic spread before entering the magnetic field (Figure 1.5). The ions experience a centrifugal force (mv^2/r, with r being the radius of curvature of the magnetic sector and v the velocity of the ions) due to the constraint ($Bzev$) imparted by the magnetic field (B). From these, the following equations allow calculation of m/z values:

$$zeV = 1/2\,mv^2 \text{ and } Bzev = mv^2/r$$

$$m/z = (B^2R^2e)/2V$$

At constant values of the magnetic field (B) and the accelerating potential (V), each value of m/z will correspond to a certain radius of curvature. In scanning operation, the radius of curvature is fixed and the accelerating voltage is held constant while the magnetic field is scanned. A certain value of the magnetic field will bring ions at a certain m/z to a focus at the detector placed at the exit of the flight tube. A scan of the magnetic field from lower to higher strength enables the detection of ions from low to high m/z values.

In a double focussing analyzer, after the ion beam has been accelerated through a potential of 2–8 kV, it passes a curved electric sector where an electric potential forces the ions to follow the curve and to separate according to their translational energy. Ions having the same energy are therefore focussed on a line. The electric sector is not a mass analyzer but rather an energy filter. The narrow energy-focussed ions then enter the magnetic field to be separated according to their m/z ratio. Combining the magnetic sector with the electric sector allows a double focussing effect to be obtained in which the magnetic sector focuses directionally while the electric sector focuses energetically. Two types of double focussing geometries can be obtained depending on whether the different m/z values are focussed at different points on the detector (Mattauch–Herzog geometry with no magnetic scanning), or the ions are focussed onto a single point of the detector following magnetic scanning (Nier–Johnson double focussing mass spectrometer).

The typical resolution obtained under normal scan mode is below 5000, meaning that the separation limit is reached between ions at m/z 5000 and 5001, or 500 and 500.1, or 50 and 50.01. In exact mass measurement (and during calibration of the instrument), the magnetic field is fixed and the accelerating voltage is scanned. This is done to enhance measurement precision and to avoid the so-called magnetic hysteresis phenomenon: when a field is applied to the magnet and is then removed, it retains a residual magnetization changing its original strength. Double focussing mass spectrometers can easily reach a resolution power of up to 50 000.

1.3.2 Quadrupole Mass Filter

Quadrupole filters are composed of four cylindrical rods placed in parallel to each other as depicted in Figure 1.6.

A dc voltage (U) is applied to one set of opposite and electrically connected rods while a radio frequency (rf) voltage is applied to the other pair of rods. After a low voltage acceleration (few eV), ions follow a complex trajectory in the quadrupole mass filter. For a given pair of dc and rf voltages, ions oscillate with a finite and stable amplitude and pass through the quadrupole to reach the detector. If the oscillations are infinite, they will be unstable and the ions will collide with the rods and not reach the detector. For a specific combination of dc and rf voltages, the ions have therefore a stable trajectory and a stability diagram of the dc voltage *versus* the rf voltage can be drawn (Figure 1.7). The rf voltage is generally kept constant while the polarity of the dc voltage is

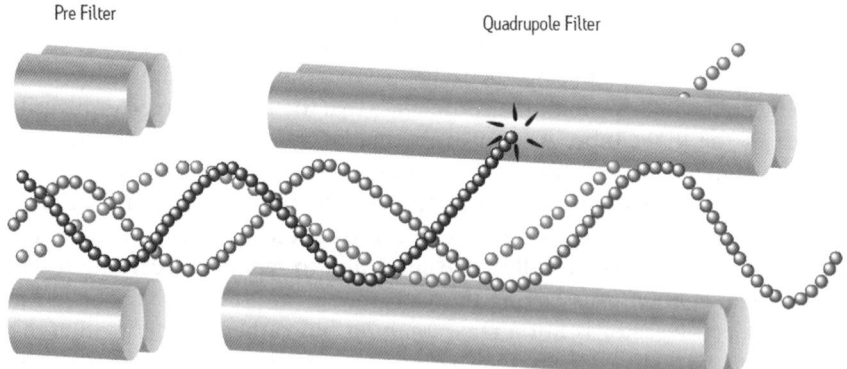

Pre Filter Quadrupole Filter

Figure 1.6   A quadrupole mass filter consists of four parallel metal rods. A (+)dc and a (−)dc voltage (U) is applied to each pair of rods. An ac potential [Vcos(Ωt)] is superimposed and applied 180 °C out of phase to each pair of rods. For given dc and ac voltages, only ions of a certain mass-to-charge ratio pass through the quadrupole filter and all other ions are ejected of their original path (image courtesy Waters Corp.).

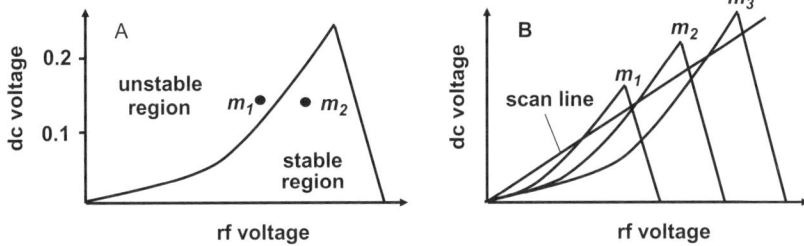

Figure 1.7 Stability of masses in a quadrupole: dc voltage *versus* rf voltage for masses m_1 (unstable) and m_2 (stable and passes through the quadrupole) for one (A) or multiple (B) sets of dc and rf voltages.

switched. Each ion with its specific m/z ratio has a different stable region. Figure 1.7 depicts the capability of the quadrupole to scan m/z ratios (m_1, m_2, m_3...m_n). The slope of the scan line corresponds to a fixed ratio of the dc and rf voltages. Moving along this line corresponds to scanning m/z ratios of the ions of interest.

The quadrupole mass filter is a more robust and cheaper instrument than magnetic sector machines. Its ability to scan the entire mass range 20–2000 Da very rapidly, acquiring several spectra per second, makes quadrupoles well suited for gas chromatography-mass spectrometry (GC-MS) with capillary columns. Moreover, as the rod voltages can be switched in a couple of milliseconds and because the dynamic range of quadrupoles reaches 10^6, quadrupole instrumentation is very powerful for quantitative mass spectrometry.

1.3.3 Ion Trap Analyzer

Extension of the quadrupole technology led to the development of ion traps. Instead of four rods, these use a ring electrode and two end cap electrodes arranged in a cylindrical symmetry. The two end caps have slits for ions entering into the trap and for the detection of the ions exiting the trap. Originally, ionization was performed within the trap but today's instruments use external sources for the ionization (Figure 1.8). The two end cap electrodes are grounded and a radio frequency voltage is applied to the ring electrode generating a quadrupolar field inside the trap. Ions therefore follow trajectories confined in a well-defined space.

In operation, the ions introduced into the trap are cooled by a dampening gas such as helium at a pressure of about 1 mTorr. The dampening gas induces low-energy collisions that thermally cool the ions and reduce their kinetic energy, shrinking their trajectories toward the centre of the ion trap and thus enabling trapping of injected ions. Trapped ions are further focused toward the centre of the trap by means of an oscillating rf potential applied to the ring electrode. The radio frequency voltage causes ions to oscillate stably with a so-termed

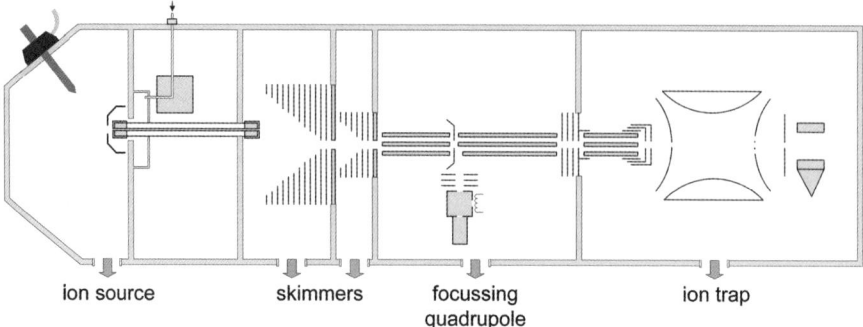

ion source   skimmers   focussing   ion trap
                                      quadrupole

Figure 1.8   Schematic drawing of an ion trap mass spectrometer (image courtesy
                 Bruker Corp.).

secular frequency. The amplitude of the oscillation determines the m/z range of
ions that will be trapped.

Ion stability depends on the mass and charge of the ion, the size of the ion
trap (r), the oscillating frequency of the radio frequency (w), and the amplitude
of the voltage on the ring electrode (V). The dependence of ion motion on these
parameters is described by the parameter qz:

$$qz = 4eV/mr^2w^2$$

Linear ramping of the radio frequency voltage will create unstable ion trajectories. Ions with lower m/z values will be ejected before ions with higher m/z
values. This scanning method is called **mass-selective axial instability mode** and
limits higher m/z values to about 1500 due to the limitation of the radio frequency voltage applied (7500 V).

Another scanning method known as **resonance ejection** has been developed to
extend the m/z range measurable with the trap. In this mode, a low amplitude
ac voltage is applied to the end caps and the qz value of an ion of interest is
changed until the secular frequency of the ion matches the frequency of the
applied ac voltage. When resonance occurs, the amplitudes of ion trajectories
linearly increase with time. A high-amplitude ac voltage will cause resonance
ejection of the ion.

The deployment of a linear or two-dimensional quadrupole ion trap as a high
performance mass spectrometer has been reported.[14] The mass analysis is
performed by ejecting ions out of a slot in one of the rods using the mass-
selective instability mode of operation. Resonance ejection and excitation are
utilized to enhance mass analysis and to isolate and activate ions for multi stage
mass spectrometry (MS[n]) capability. Improved trapping efficiency and
increased ion capacity are observed relative to a three-dimensional (3D) ion
trap with similar mass range. Mass resolution is comparable to 3D traps,
including high resolution at slower scan rates, although adequate mechanical
tolerance of the trap structure is a requirement.

1.3.4 Time-of-flight Analyzer

Time-of-flight (ToF) analyzers separate accelerated ions based on their different velocities. According to the equation of kinetic energy, the velocity (v) of an ion of mass (m) accelerated through a potential (V) is:

$$v = (2zeV/m)^{1/2}$$

As the velocity of an ion is mass dependent, ions with different masses (m_1, m_2, m_3 ...) accelerated and passing through a field-free region will arrive at the detector at different time depending on their velocities (v_1, v_2, v_3 ...). If the field-free region has a length l, this means for the ion velocity v:

$$t = 1/v \text{ and } v = (2zeV/m)^{1/2}$$

$$t = (m/z)/(2eV)^{1/2}$$

Consequently, the time that an ion takes to reach the detector is proportional to the square root of its m/z value. The larger the mass of a singly charged ion, the longer it takes to travel through the field-free region.

Early time-of-flight analyzers suffered from a low mass resolving power. This was due to the spread of spatial and energy distribution imparted to the ions in the ionization source. Pulsed ion extraction systems have been developed to compensate for this phenomenon.[15] After desorption/ionisation, the ions are kept in the ion source under a field-free condition for a short period of time (*e.g.* 100–400 ns), before they are extracted with a high electrical field and accelerated towards the detector.

Delayed extraction has also been used with orthogonal ion extraction geometry.[16] In the more recent orthogonal acceleration time of flight (oaTOFMS) mode, ions are formed in a continuous ion source, accelerated and then focussed into a very thin ion beam. As the ions traverse an orthogonal sampling region, a sudden voltage pulse is applied ejecting a portion of the beam orthogonally. This packet of ions is then accelerated into the time-of-flight drift region. Ions of different m/z values have different velocities and hence arrive at the detector at different times relative to the orthogonal acceleration pulse. By precisely recording these arrival times, a time-of-flight mass spectrum is produced. The axial ion beam is typically sampled at between 10 000 and 100 000 times per second; individual time-of-flight spectra are generally summed before storing to file. As the initial energy spread of the ions is generally very low in the orthogonal direction compared with the axial direction, the spread of ion arrivals for a particular m/z value is minimized and high mass resolution can be obtained.

Additional time focussing can also be achieved through an electrostatic mirror (reflectron) at the end of the flight tube. The role of this electrostatic mirror is to focus ions with the same m/z ratio but slightly different kinetic energies. Ions with a higher kinetic energy penetrate this mirror more deeply before reflection compared with ions having a lower kinetic energy (Figure 1.9).

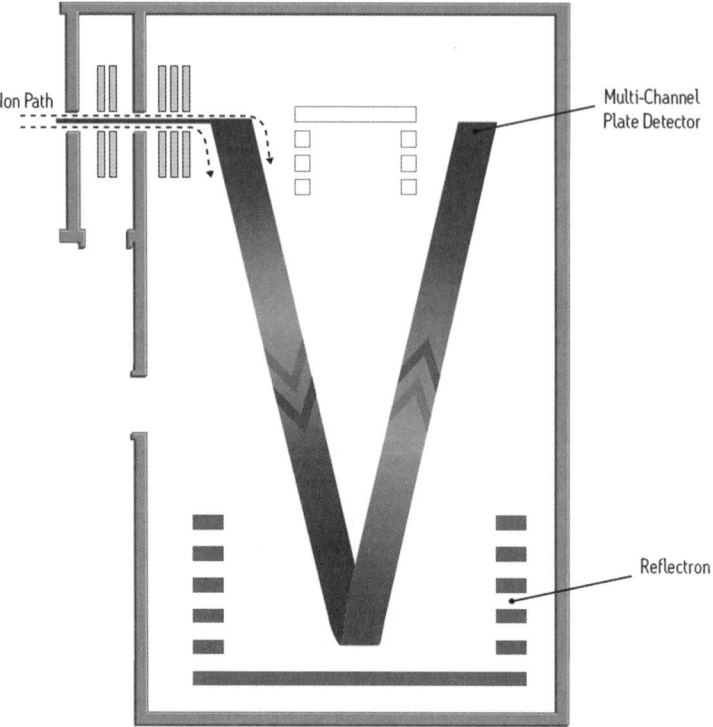

Ion Path

Multi-Channel
Plate Detector

Reflectron

Figure 1.9   Schematic drawing of a time-of-flight mass spectrometer (image courtesy
Waters Corp.).

This double focussing effect increases mass resolution by one order of magnitude and resolution greater than 10 000 can be reached.

1.3.5   Fourier Transform Ion Cyclotron Resonance Analyzer

The Fourier transform ion cyclotron resonance mass spectrometer (FTICR-MS) is the unsurpassed instrument in terms of mass resolution and mass accuracy. It can routinely deliver mass accuracies below five parts per million (5 ppm) with mass resolution above 10^6. Like the ion trap, FT-ICR-MS stores ions in a cavity but storage is achieved in a high-vacuum chamber using ion cyclotron resonance in a strong and static magnetic field generated by a superconducting magnet of 3.5–9.4 Tesla, or even today, 11.5 T.

The ion storage of FTICR-MS is based on the principle that an ion entering a strong magnetic field will undergo a circular motion (cyclotron motion) that is perpendicular to the magnetic field lines. The cyclotron motion has a resonance frequency specific to the ion's m/z ratio. Mass analysis is based on the resonance frequencies of the trapped ions. Trapped ions are excited by an oscillating electric field (rf pulse) having the same frequency as the ion motions. Ions will then absorb kinetic energy, gradually dislocate from the centre of the ICR cell, approach its plate boundary and undergo dissociative collisions and ion-molecule reactions. This generates image currents that decay as the ions are destroyed through collisions (Figure 1.10). This electrical signal is converted into a mass spectrum by Fourier transformation.

In FTICR-MS, all resonating ions are measured simultaneously which leads to high sensitivity as there is no need for a scanning process. Moreover, modern FTICR-MS uses external ionization sources in order to maintain very low pressure in the ICR cell, ensuring very high resolution power. Moreover, similarly to ion traps, ion chemistry experiments and different time-dependant mass spectrometric stages can be performed.

1.3.6 Orbitraps

Orbitraps[17] can reach almost as high a mass resolution and accuracy as FT-ICR analyzers. In such an instrument, a linear ion trap mass spectrometer is coupled to an Orbitrap mass analyzer *via* an rf-only trapping quadrupole with a curved axis (Figure 1.11). The latter injects pulsed ion beams into a rapidly changing electric field in the orbitrap wherein the ions are trapped at high kinetic energies around an inner electrode. Image current detection is subsequently performed after a stable electrostatic field is achieved. Fourier transformation of the acquired transient allows wide mass range detection with high resolving power, mass accuracy and dynamic range. The entire instrument operates in liquid chromatography-mass spectrometry (LC-MS) mode (one spectrum per second) with a nominal mass resolving power of 60 000, and uses automatic gain control to provide high accuracy mass measurements—within 2 ppm using internal standards and within 5 ppm with external calibration. The maximum resolving power exceeds 100 000 (FWHM). Rapid, automated data-dependent capabilities enable real-time acquisition of up to three high mass accuracy MS/MS spectra per second.

1.4 Ion Detectors

Ion detectors are devices able to generate an electrical current with its intensity being proportional to the abundance of the ions. Ions leaving the mass analyzer are either directed sequentially to a single-channel detector or dispersed simultaneously to a focal-plate (array) detector. Quadrupole or ion trap analyzers use single-channel detectors whereas time-of-flight instruments are equipped with focal-plate detectors. All of these devices generate an electrical current that is subsequently digitalized and transferred to a computer for data management and storage.

The most common single-channel detector is the electron multiplier which is composed of a series of 10–20 electrodes or dynodes. When an ion hits the first electrode, it releases electrons that are accelerated by an electric potential to the

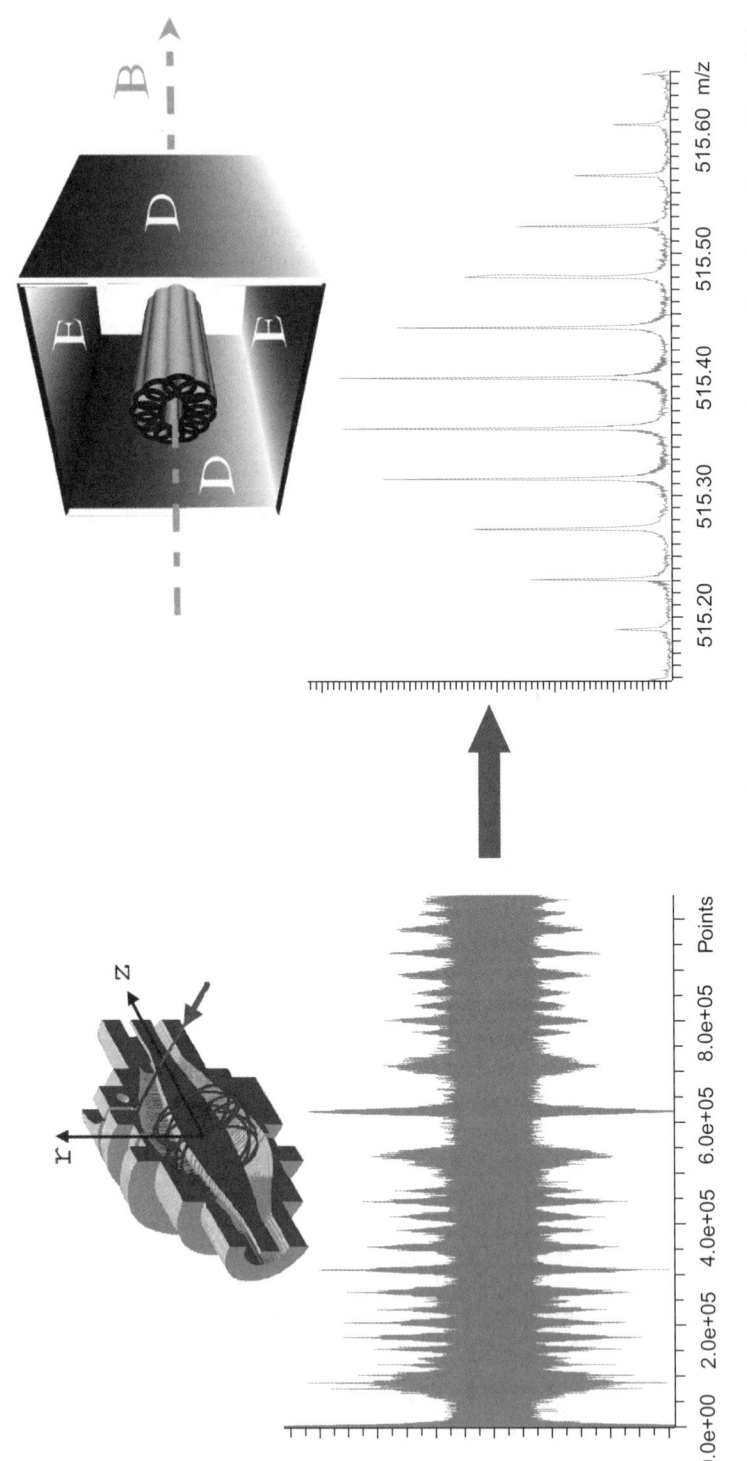

Figure 1.10	Schematic drawing of a Fourier transform ion cyclotron resonance mass spectrometer (top panel) and transformation of the typical free-induction decay signal into a mass spectrum (bottom panel).

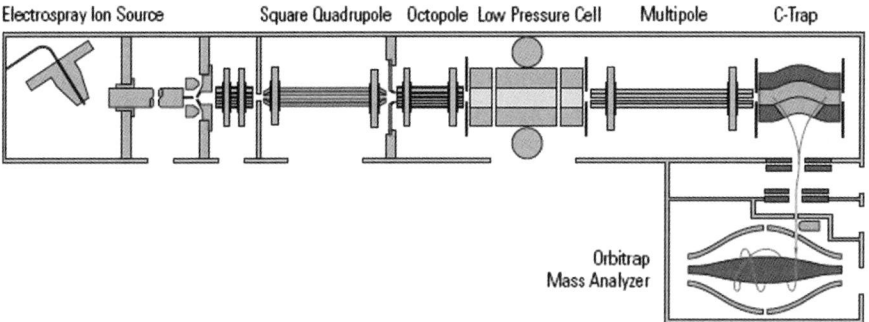

Electrospray Ion Source   Square Quadrupole   Octopole   Low Pressure Cell   Multipole   C-Trap

Orbitrap
Mass Analyzer

Figure 1.11   Schematic drawing of an orbitrap mass spectrometer (image courtesy Thermo Fischer Scientific).

second dynode. Each electron hitting the dynode causes the release of a bunch of new electrons that are again accelerated by the electric potential to the third dynode. This cascading effect continues through all dynodes and provides an electron multiplying effect with a gain of 10^6 to 10^8.

An alternative design is the channel electron multiplier, in which the series of dynodes is replaced by one continuous electrode shaped like a conical tube. Ions hit the inner surface of the tune and generate electrons that are multiplied through a cascade of collisions with the wall of the detector. The gain obtained with a channel electron multiplier is similar to the one obtained with electron multiplier made of a series of dynodes.

The second widely used single-channel detector is the Daly detector in which the ions impact a plate and release electrons. Those electrons are subsequently accelerated onto a scintillator emitting photons which are detected by a photomultiplier producing an electric current.

The second category of ion detection devices corresponds to focal-plane array detectors allowing the simultaneous detection of all ions exiting the mass analyzer. Modern devices are composed of 10^4 to 10^7 micro-channel electron multipliers arranged side-by-side as an array. Each micro-channel electron multiplier works independently and emits electrons when ions hit it. Focal-plane array detectors monitor a certain window of m/z values simultaneously. More ions can be recorded per unit of time leading to greater sensitivity compared with single-channel detectors.

1.5   Tandem Mass Spectrometry

Tandem mass spectrometry or MS/MS involves two stages of mass analysis separated by a reaction or fragmentation step.[18] MS/MS instruments drastically reduce the "chemical noise" due to their high specificity. Nevertheless, ions selected for fragmentation in the first MS (precursor ions) and fragments ions produced in the second MS must be stable to be detected.

The fragmentation of the selected ions is the essential step in any MS/MS experiment. Many methods are deployed to fragment ions resulting in different fragmentation types and different structural information about the analyte. First, metastable ions can be generated if the ionization process such as electron impact is strong enough to produce ions with sufficient internal energy. Such metastable ions will remain in non-equilibrium for a short period of time before auto-dissociation occurs within the mass spectrometer. Second and very common in MS/MS experiments, post-source fragmentation can be performed by adding energy to the selected precursor ions. In such situations, energy is added through post-source collisions with neutral atoms or molecules, absorption of radiation, or the transfer or capture of an electron by a multiply charged ion. Collision-induced dissociation (CID), also referred to as collisionally activated dissociation (CAD), involves the collision of an ion with a neutral atom or molecule in the gas phase and subsequent dissociation of the ion.

There are two fundamentally different approaches to MS/MS:[19]

- tandem-in-space;
- tandem-in-time.

Tandem in-space instruments have two independent mass spectrometers in physically different locations in the instrument. In a **tandem in-time** instrument, the separation is accomplished with ions trapped in the same place, with multiple separation steps taking place over time. In these instruments, multiple analyses can be performed, which are referred to as MS^n.

Triple quadrupole (QqQ) and quadrupole/time-of-flight (QqToF) instruments are typical representatives of tandem in-space instrumentation. There are four possible scanning experiments:

- **Product ion scan.** A precursor ion is selected by the first mass analyzer and then subjected to collision-activated dissociation in a collision cell. All resulting masses are detected in the second mass analyzer.
- **Precursor ion scan.** The product ion is selected in the second mass analyzer and the precursor masses are scanned in the first mass analyzer.
- **Neutral loss scan.** The first mass analyzer scans all masses. The second mass analyzer also scans, but at a defined offset from the first mass analyzer. This offset corresponds to a neutral loss that is commonly observed for the class of compounds (*e.g.* water loss for alcohols).
- **Selected reaction monitoring.** Both mass analyzers are set to a selected mass (*e.g.* precursor ion and the corresponding fragment ions). Largely employed for quantification, this mode maximizes the instrument duty cycle and delivers high selectivity and high sensitivity.

Triple quadrupole instrumentation (QqQ) was developed in the early 1980s and is today considered the workhorse for targeted, quantitative analysis of complex samples. Recent developments led to higher resolution QqQ instruments capable of peak widths in the order of 0.1 amu (FWHM) and mass assignment accuracies of <10 milli amu. QqTOF instruments further enhance mass resolution (*ca.* 5000–20 000 FWHM) and deliver, in addition, excellent mass accuracy (<5 ppm) and the ability to record a complete mass spectrum for each pulse of ions injected into analyzer. However, a major weakness of QqToF-type instruments is their inefficient precursor ion scanning capabilities compared to triple quadrupoles, since this mode does not benefit from simultaneous time-of-flight detection.

In addition to QqQ and QqTOF machines, several other instrumental set-ups are today commercially available such as quadrupole-ion trap, time-of-flight/time-of-flight (ToF/ToF)[20] or magnetic sector-ion trap—even if this latter configuration tended to disappear following the boom in the time-of-flight market.

Tandem-in-time instruments are, in general, ion-trapping mass spectrometers such as two-dimensional or three-dimensional quadrupole ion traps and Fourier transform ion cyclotron resonance (FTICR) analyzers. The various stages of mass spectrometry are conducted within the same physical trapping volume but at different times during the experiment. This approach allows for:

- transfer of the ions from the ion source into the trapping volume;
- isolation of the precursor ions of interest by ejecting the undesired ions;
- fragmentation of these precursor ions; and
- performance of a final mass analysis.

In principle these devices are capable of many stages of mass spectrometry (MS^n, see above). Ion traps are inherently scanning instruments and limited to product ion scanning. However, since they can record a complete mass spectrum of each pulse of ions introduced into the trapping volume, they are very sensitive instruments.

1.6   Isotope Ratio Mass Spectrometry (IRMS)

Isotope ratio mass spectrometry is dedicated to the very accurate determination of isotope ratios (mostly $^2H/^1H$, $^{13}C/^{12}C$, $^{15}N/^{14}N$, $^{18}O/^{16}O$) in different biological samples.[21] The theory and practice of IRMS has been reviewed in detail elsewhere[22–24] and are only outlined in this chapter.

The IRMS device (or gas IRMS, Figure 1.12) is meant only for the analysis of light stable isotopes such as ^{13}C, ^{15}N, ^{18}O, ^{34}S, and ^2H, which are then transformed into CO_2, N_2, CO, SO_2, and H_2 gases, respectively, and finally introduced into the ion source. In order to measure isotopic ratios with "high-precision", two inlet systems provide fast introduction of a reference gas into the ion source as well as the gas produced by the gas chromatograph (GC), elemental analyzer (EA) and liquid chromatograph (LC). Subsequently, these gaseous analytes collide with an electron beam in a high vacuum; these molecules then lose electrons to produce positive ions. The IRMS sensitivity for CO_2

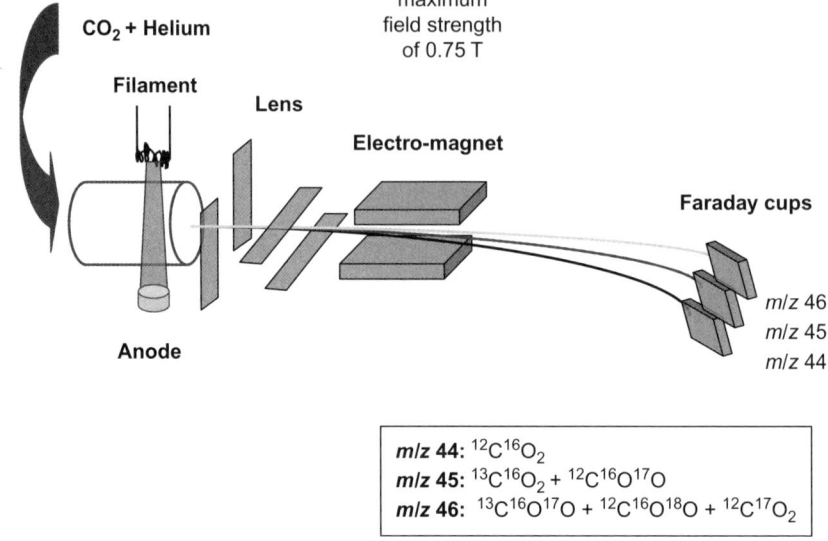

Figure 1.12   Schematic drawing of an isotope ratio mass spectrometer and example of a carbon dioxide (CO_2) molecule separation based on different element isotope compositions.

can be described by the number of ions collected per input of molecule; the absolute sensitivity is around one ion per 800–1200 molecules in dual inlet mode. These ions are accelerated through a flight tube in an electromagnetic field. Finally they are separated according to their m/z ratio and counted by a multiple Faraday cup collector to eliminate fluctuations in the ion beam intensity. For ^{13}C measurements, the Faraday cups allow simultaneous measurement of three ions at m/z 44, 45, and 46. These ions represent the different isotopomers $^{12}C^{16}O^{16}O$ (m/z 44), $^{13}C^{16}O^{16}O$, and $^{12}C^{17}O^{16}O$ (m/z 45), and $^{12}C^{16}O^{18}O$ (m/z 46).[24]

The peak area of each isotopomer is measured and finally transformed into a $^{13}C/^{12}C$ ratio using m/z 44, 45 and 46. The m/z 46 isotopomer is measured to assess the ^{18}O isotope content, from which the ^{17}O content is calculated to obtain the true ^{13}C value measured for m/z 45/.[25]

The term "compound-specific isotope analysis" (CSIA) was coined by Schoell in the late 1980s and refers exclusively to GC coupled to IRMS for high-precision isotopic analysis. CSIA became feasible following the work of Matthews and Hayes[26] with the introduction of the combustion furnace with Pt and CuO as the oxygen source based on the work of Sano and colleagues. They used a GC-MS equipped with a catalytic combustion unit to monitor urinary metabolites after ingestion of ^{13}C aspirin.[27]

Different interfaces are currently commercially available to measure isotopic ratios of elements other than carbon.[28] The minimum sample size or limit of detection (LOD) in IRMS is usually reported in nanograms of element (carbon atom) or amplitude of CO_2 signal associated to an isotopic precision (standard deviation). The IRMS precision is in theory statistically limited by the number of counts from the minor ion beam, indicating that a certain signal amplitude is needed to measure the isotopic ratio of carbon. Therefore, with an efficiency of 10^{-4} (ions/molecules), 0.31 nanomoles of carbon are theoretically needed to achieve a standard deviation of 0.1%.[29]

IRMS can be considered as complementary rather than a competitor to conventional MS methods. For example, using GC-MS or LC-MS, the measurable range of different isotopomers spans from 0.1 to 100% while, with IR-MS, the isotopic ratio determination complementarily covers the range 0.001–0.1%.[23,28] Stable isotope abundances are expressed, using the "delta" (δ) notation, as the ratio of the two isotopes of interest in the sample compared with the same ratio in an international reference standard. Since the differences in ratios between the sample and standard are usually very small, they are expressed as parts per thousand (‰) or per million deviation from the standard.

1.7 Hyphenated Mass Spectrometry

1.7.1 Gas Chromatography-Mass Spectrometry

In 1955, Gohlke and McLafferty (cited in ref. 30) established the first coupling between a gas chromatograph and a mass spectrometer, offering to the analytical community one of the most powerful spectroscopic techniques. Mass spectrometry coupled to gas chromatography (GC-MS) has a long and successful standing in analyzing volatile fractions of food and drinks,[31,32] but also in quantifying non-volatile nutrients such as amino acids,[33,34] fatty acids[35,36] and sugars.[37,38]

Gas chromatography instrumentation comprises a gas control unit, a sample introduction system, a column placed into a temperature-programmable oven and a transfer line (or interface) to the mass spectrometer. The gas control unit ensures flow-rate or pressure control of the gas flow passing through the injector, the column and the transfer line.

The sample introduction is critical as it should allow the sample to enter as a narrow band at the very beginning of the column without thermal degradation and analyte discrimination. For GC-MS analysis, the most frequently used techniques are split-less injection and on-column injection. In the split-less mode, the sample is introduced into a high-temperature chamber in order to be volatilized. The gaseous sample is then totally (no-split) transferred to the column, which is heated to a temperature below the boiling point of the sample solvent for the sample solution to be condensed at the top of the column. After rapid heating of the column to remove the solvent, the temperature is slowly programmed to ensure the separation of the analytes of interest. In the on-column mode, the sample is directly introduced as a liquid into the top of the column. On-column injection generates less sample discrimination than

Table 1.2   Chemically bonded general purpose stationary phases for capillary
                 GC (adapted from ref. 54).

Common stationary phases[a]	Composition	Maximum operating temperature (°C)
CP-Sil-5 DB-1 HP-1 SPB-1	dimethylpolysiloxane	325
CP-Sil 8; DB-5; HP-5; SPB-5	5% phenyl, 95% dimethylpolysiloxane	325
CP-Sil 19; DB-1701; HP-1701; SPB-1701	14% cyanopropylphenyl, 86% dimethylpolysiloxane	275
CP-Sil 24; DB-17; HP-50; SPB-50	50% phenyl, 50% dimethylpolysiloxane	280
CP-Sil 43; DB-225; HP-225	25% cyanopropyl, 25% phenyl, 50% dimethylpolysiloxane	200
CP-Wax 52; DB-Wax; Supelcowax	Polyethylene glycol	250
CP-Wax 58; DB-FFAP; HP-FFAP; Nukol	Polyethylene glycol nitroterephthalic acid ester	250

[a]Stationary phases from Chrompack (CP-Sil), J&W Sci (DB), Agilent Technologies (HP) and Supelco (SPB). Similar phases are available from numerous other manufacturers.

split-less injection, but the risk of column contamination is higher. Carrier gases are commonly helium or hydrogen at a flow of 1 mL min^{-1} when working with open capillary columns.

The GC column is placed in a temperature-programmable oven. During chromatography, the oven temperature is linearly increased at a rate of 4–20 °C per minute until the column's maximum operating temperature is reached. This temperature depends on the stationary phases chemically bonded to the column. Table 1.2 lists the most commonly used stationary phases for open tubular capillary gas chromatography. Column lengths range from 10–100 m with an internal diameter of 0.25–0.50 mm. Stationary phase film thicknesses are in the range 0.1–2 µm. The separation of the analytes is based on their partitioning coefficient with the stationary phase. Compounds with a higher boiling point and/or stronger solubility in the stationary phase will be retained and therefore elute later.

Last but not least, the transfer line in most of today's instruments is a heated direct line from the end of the capillary column to the ion source of the mass spectrometer. This direct line ensures that the totality of the sample will be transferred to the ion source. Any dead volume must be avoided to preserve the chromatographic resolution and no metallic piece must be in contact with the sample, which would otherwise decompose. The carrier gas flow enters into the high vacuum region of the source, necessitating the use of an efficient pumping system and a highly diffusible carrier gas (*e.g.* helium).

GC-MS is a straightforward analytical technique for the analysis of volatile and non-polar molecules that are thermally stable (*e.g.* flavours, hydrocarbons, short chain fatty acids). Thermally unstable molecules (sugars, amino acids, organic acids) need to be chemically modified (derivatisation) in order to increase their thermal stability and render them amenable to a gaseous state. Tables 1.3 to 1.5 summarize the most commonly used derivatives for the GC-MS analysis of biologically relevant analytes.

1.7.2 High Performance Liquid Chromatography-Mass Spectrometry

Coupling a liquid chromatograph to a mass spectrometer is not straightforward. Indeed, a flow of $1\,mL\,min^{-1}$ of water corresponds to more than $1000\,mL\,min^{-1}$ of vapour. In addition, most mobile phases used for the high pressure liquid chromatography (HPLC) separation contain inorganic salts and buffers, many of which are non-volatile. Moreover, the analytes themselves are often non-volatile and thermally unstable.

However, despite the technical hurdles to be overcome when coupling an HPLC to an MS, the application potential and range of HPLC-MS are wider than those of GC-MS. HPLC-MS is perfectly suited for the analysis of polar, ionic, thermally unstable and non-volatile molecules and requires less sample preparation than GC-MS.

Among the various interfaces designed to couple an HPLC to a mass spectrometer, two are currently widely used—namely electrospray and APCI (see Section 1.2.3 and 1.2.5, respectively). In terms of application, (Figure 1.13) electrospray may cover a more limited range of polarity than APCI, but its potential to analyse thermally labile compounds is wider; in particular, a virtually unlimited range of biologically important macromolecules can be addressed including (glyco-)proteins, oligo- and poly-saccharides, and DNA/RNA. Thus, today's HPLC-MS based techniques find broader applications in nutrition research than GC-MS methods as they provide a means to investigate nutrient stability, absorption, metabolism and metabolite excretion. Hence, they can bridge the gap between the content of a particular ingredient in a food matter and its biological endpoint.[39,40]

1.8 Accelerator Mass Spectrometry

Among the arsenal of techniques available, accelerator mass spectrometry (AMS) is an emerging approach for the investigation of the fate of nutrients, especially micronutrients in the body over a longer period of time, including their bioconversion and tissue distribution.[41–43] This ultra-sensitive analytical instrument[44,45] measures the amount of naturally rare, radioactive nuclides such as ^3H, ^{14}C, ^{41}Ca, *etc.* with a sensitivity of about 1 in 10^{15} for carbon.

Table 1.3 Synoptic chart on derivative formation with oxygenated groups.

Functional group	Procedure	Examples of products
—OH (Primary, secondary and tertiary alcohols; phenols; carbohydrates)	Silylation	—O—Si(CH$_3$)$_3$
	Acylation	—O—CO—CH$_3$; —O—CO—CF$_3$
	Benzoylation	—O—CO—C$_6$H$_5$; —O—CO—C$_6$F$_5$
	Alkylation	—O—CH$_3$; —O—CH2—C$_6$F$_5$
	Oxidation	—O—CHO; —COOH
	Dansylation	Ar—O—Dns
	Reaction with Dis-Cl	—O—Dis
	Reaction with FDNB	7-Nitrobenzofurazan
	Reaction with NBD-Cl	
	Ion-pair formation	Ar—O⁻M⁺
C=O (Aldehydes and ketones)	Oxime formation	=NOH ; =NOCH$_3$
	Oxime formation and silylation	=NOSI(CH$_3$)$_3$

Reaction	Product
Ketal/acetal formation	(dioxolane structure)
Hydrazone formation	C_6H_5—NH—N=C(CH$_3$)$_2$
Schiff's base formation	R—N=C(CH$_3$)$_2$
Silylation	$(H_3C)_3SiO$—... —C(=O)—$OSi(CH_3)_3$
Oxidation	—COOH
Reduction	—OH
Aldonitrile formation	—CH=N—OH → —CN
—COOH (Carboxylic acids)	
Esterification (alkyl)	—CO_2—CH_3; —CO_2—CH_2CF_3
Esterification (aryl)	—CO_2—$CH_2C_6H_5$; —CO_2—$CH_2C_6F_5$
Silylation	—CO_2—$Si(CH_3)_3$
Reduction	—CH_2—OH
Decarboxylation	R—COOH → R—H
Cyclization	Depends on parent compound
Ion-pair formation	R—COO^- M^+

Table 1.3 Continued.

Functional group	Procedure	Examples of products
(vicinal diol, —C(OH)—C(OH)—)	As for —OH, but also cyclic boronate formation	cyclic boronate ester (O—B—R on dioxaborolane ring)
	Acetal or ketal formation	cyclic acetal/ketal (R, R' on dioxolane ring)
H—C(OH)—COOH (α-Hydroxy acids)	Oxidative cleavage	R—COOH + R'—COOH
	As for the individual groupings, but also boronation	—CH—C=O cyclic boronate with C$_4$H$_9$
	Reduction	H—C(OH)—CH$_2$OH

Group	Treatment	Product
—CO—COOH (α-Keto acids)	Oxidative cleavage (periodate)	—COOH
	Oxidation (mild)	—CO—COOH
	Simultaneous acylation and esterification	(acylated/esterified derivative containing —O—C_2H_5 and —C(=O)—O—C(=O)—C_3F_7 groups)
	As for the individual groupings, but also cyclization with 1,2-di-aminobenzene followed by silylation	(quinoxalinone derivative with R substituent and N—Si$(CH_3)_3$)
	Reduction	R—CH(OH)—CH$_2$—OH (diol)
R—CO—OR′ (Esters)	Oxidative cleavage	—COOH
	Esters may be analyzed chromato-graphically without derivation, but where R′ is involatile: ester interchange (trans-esterification)	R—CO—OCH_3
	Reduction	R—CH_2—OH + R′—OH
	Alkaline hydrolysis	R—COOH + R′—OH

Table 1.3 Continued.

Functional group	Procedure	Examples of products
R—CO \O R'—CO (Acid anhydrides)	Direct chromatographic analysis	
	Reduction	$R-CH_2-OH + R'-CH_2-OH$
	Hydrolysis	$R-COOH + R'-COOH$
	Esterification	$R-CO-OR'' + R'-CO-OR''$
R—O—R' (Ethers)	Ethers may be chromatographed without derivatization Cleavage with hydriodic acid	$R-I + R'-I$

Table 1.4 Synoptic chart on the formation of derivatives from nitrogen-containing groups.

Functional group	Procedure	Examples of products
—NH$_2$ (Primary amines; amino acids; amino sugars)	Acylation	—NH—CO—CH$_3$; —NH—CO—CF$_3$
	Benzoylation	—NH—CO—C$_6$H$_5$; —NH—CO—C$_6$F$_5$
	Silylation (mild)	—NH—Si(CH$_3$)$_3$
	Silylation (vigorous)	—N[Si(CH$_3$)$_3$]$_2$
	Treatment with CS$_2$	—N=C=S
	Thiourea formation	
	Schiff's base formation	—N=CH—C$_6$F$_5$
	2,4-Dinitrophenylation	
	Sulfonamide formation	—NH—Dns

Table 1.4 Continued.

Functional group	Procedure	Examples of products
	Carbamate formation	$-NH-CO_2-CH_3$
	Carbylamine reaction	$-N{\equiv}C$
	Treatment with nitrous acid	$-OH$
	Treatment with fluorescamine	*N*-substituted 2-phenyl-pyrrolin-4-ones
	Treatment with pyridoxal	Pyridoxylidine derivative
	Treatment with NBD-Cl	7-Nitrobenzo-furazan
	Alkylation	$-N\begin{smallmatrix}CH_3\\CH_3\end{smallmatrix}$
	Ion-pair formation	$R-NH_3{}^+\,X^-$
$-NH-R$ (Secondary amines, imino acids, substituted amino sugars)	Acylation	$-N\overset{R}{\underset{}{}}-CO-CH_3;\quad -N\overset{R}{\underset{}{}}-CO-CF_3$
	Benzoylation	$-N\overset{R}{\underset{}{}}-CO-C_6H_5;\quad -N-CO-C_6F_5$
	Silylation	$-N\overset{R}{\underset{R}{}}-Si(CH_3)_3$
	2,4-Dinitro-phenylation	

Group	Reaction	Product
R—N—R' \\ | R" (Tertiary amines)	Sulfonamide formation	
	Treatment with NBD-Cl	
	Ion-pair formation	$R\text{—}NH_2^+X^-$ (with R')
	Hofmann degradation	→ alkenes
	Carbamate formation	$R\text{—}N(R')\text{—}CO\text{—}CH_2\text{—}C_6F_5$
	Ion-pair formation	$R_3NH^+X^-$
Quaternary ammonium salts	Thermal decomposition	Tertiary amines
	Dealkylation with sodium benzene thiolate	Tertiary amines
	Ion-pair formation	$R_4N^+X^-$
—CO—NH$_2$ (Amides)	Silylation (vigorous)	$\text{—}C{=}N\text{—}Si(CH_3)_3$ $\quad\quad\;\;\text{|}$ $\quad\quad\; O\text{—}Si(CH_3)_3$
	Acylation (vigorous)	$\text{—}CO\text{—}NH\text{—}CO\text{—}C_6F_5$
	Alkaline hydrolysis	$\text{—}COOH$
	Reduction	$\text{—}CH_2\text{—}NH_2$
	Dehydration	$\text{—}C{\equiv}N$
	Alkylation	$\text{—}CO\text{—}N(CH_3)_2$

Table 1.4 Continued.

Functional group	Procedure	Examples of products
—CO—NH—R (Alkylamides)	Acylation (vigorous)	—CO—N—CO—C$_6$F$_5$ $\quad\quad\mid$ $\quad\quad$R
	Silylation	—CO—N—R $\quad\quad\mid$ $\quad\quad$Si(CH$_3$)$_3$
	Alkylation	—CO—N—R $\quad\quad\mid$ $\quad\quad$CH$_3$
	Reduction	CH$_2$—NH—R
	Hydrolysis	—COOH + R—NH$_2$
—CO—N$\big\langle^{R}_{R'}$ (Dialkylamides)	Reduction	—CH$_2$—N$\big\langle^{R}_{R'}$
	Hydrolysis	—COOH + R—N$\big\langle^{}_{R'}$H $\qquad\qquad$ R
R—NH—CO—NH—R' (Substituted ureas; carbamides)	Acylation (vigorous)	$\begin{array}{c} \text{O} \quad\quad \text{R'} \quad \text{O} \\ \parallel \qquad\quad \mid \quad\ \parallel \\ \text{R—N—C—N—C—CF}_3 \\ \quad\ \mid \\ \quad\ \text{C—CF}_3 \\ \quad\ \parallel \\ \quad\ \text{O} \end{array}$
	Alkaline hydrolysis	R—NH$_2$; R'—NH$_2$

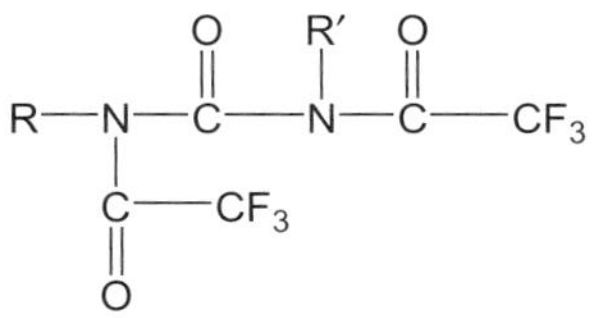

(Substituted guanidines)

Structure: R—N=C(—C(=O)—CF₃)(=O) ... N(R')—C(=O)—CF₃ ... CF₃

Acylation (vigorous) →

$$R-N-C-NR'-CO-CF_3$$ with N—CO—CF₃ and CO—CF₃ groups

(Substituted ureas)

(Amino acids)

Structure:

$$H-\underset{COOH}{\overset{NH_2}{C}}-$$

Reaction	Derivative
Hydrolysis cyclic derivatives	Substituted ureas
Esterification followed by acylation	1. *N*-TFA amino acid *n*-butyl esters
	2. *N*-TFA amino acid methyl esters
	3. *N*-HFB amino acid *n*-propyl esters
	4. *N*-Acetyl amino acid *n*-propyl esters
Acylation in aqueous solution, solvent extraction and esterification	5. *N-iso*-Butyloxycarbonyl amino acid methyl esters
Acylation followed by silylation	6. *N*-TFY amino acid TMS esters
Silylation	7. *N*-TMS amino acid TMS esters
Treatment with 2,4-dinitrofluorobenzene in aqueous solution, followed by solvent extraction and esterification	8. *N*-DNP amino acid methyl esters
Alkylation, combined with esterification	9. *N*-Dimethylaminomethyl amino acid methyl esters
Acylation and cyclization	10. Substituted oxazolin-5-ones or oxazolidin-5-ones

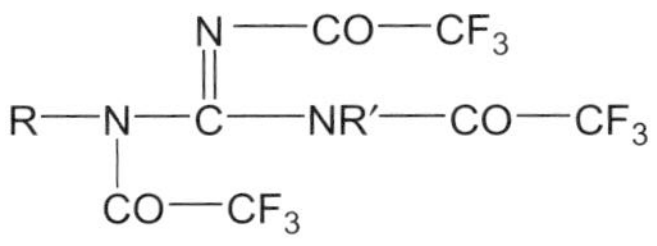

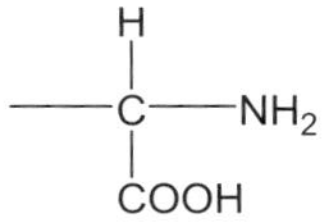

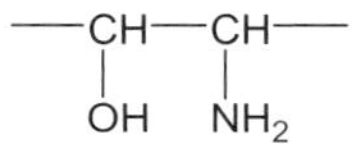

(Amino alcohols)

Structure:

$$-CH-CH-$$ with OH and NH₂

As for individual groups, but see also:

Cyclic boronate formation →

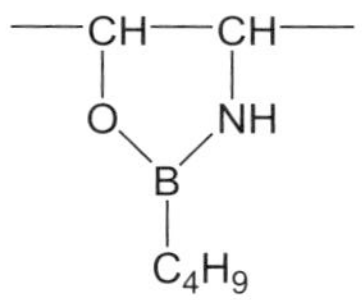

$$-CH-CH-$$ with ring: O—B(—C₄H₉)—NH

Table 1.4 Continued.

Functional group	Procedure	Examples of products
—NO$_2$ (Nitro compounds)	Simultaneous acylation and silylation	—CH—CH—NH—CO—CF$_3$ O—Si(CH$_3$)$_3$
	Chromatography without derivatization	
	Reduction	—NH$_2$
$\diagdown$C=N—OH (Oximes)	Silylation	$\diagdown$C=N—O—Si(CH$_3$)$_3$
	Acylation	$\diagdown$C=N—O—CO—CH$_3$
		$\diagdown$C=N—O—CO—CF$_3$
	Dehydration	—C≡N
$\diagdown$N—NO (Nitrosamines)	Analyzed without derivatization	
—C≡N (Nitriles)	May be analyzed without derivatization, but also:	
	Hydrolysis	—COOH
	Reduction	—CH$_2$—NH$_2$
—C≡N —CNO (Cyanates) —NCO (Isocyanates) —CNS (Thiocyanates) —NCS (Isothiocyanates)	May be analyzed directly	

Table 1.5 Synoptic chart on the formation of derivatives with miscellaneous groupings.

Functional Group	Procedure	Examples of products
—SH (Thiols, sulfydryls; sensitive to oxidation)	Acylation	—S—CO—CH₃; —S—CO—CF₃
	Silylation	—S—Si(CH₃)₃
	Alkylation	—S—CH₃
	Benzoylation	—S—CO—C₆H₅; —S—CO—C₆F₅
	Treatment of with NBD—Cl	7-Nitrobenzofurazan
	Various other fluorescent derivative formation schemes	
	Reaction with 2,3-DNFB	
	Treatment with idoacetate	—S—CH₂—COOH
	Treatment with acrylonitrile and hydrolysis	—S—CH₂—CH₂—COOH
	Reduction with Raney nickel	Hydrocarbon
	Oxidation	—S—S—
—SO₃H (Sulfonic acids)	Ion-pair formation	—SO₃H
	Halogenation	R—SO₃⁻M⁺
	Esterification	R—SO₂—Cl
		R—SO₂—OR
—S—R— (Thioethers)	May be chromatographed directly, but also:	
	Oxidation	See following two sections

Table 1.5 Continued.

Functional Group	Procedure	Examples of products
(S→O)	May be chromatographed directly	
(O→S→O)	May be chromatographed directly	
C=C (Alkenes also applicable to alikynes)	Hydrogenation	
	Epoxide formation	
	Epoxide formation and hydrolysis	
	Oxidation	
	Oxidative cleavage	
	Ozonolysis	

—CH—CH—
   $\|$
  CH$_2$
(Cyclopropyl derivatives)
	Hydrogenolysis	—CH—CH$_2$—
		   $\|$
		  CH$_3$
		
		—CH$_2$—CH—
		      $\|$
		     CH$_3$
		
		—CH$_2$—CH$_2$—

⟨phenyl⟩—CH$_2$—X (Benzyl derivatives)	Hydrogenolysis	Toluene; X—H
R—S—S—R' (Disulfide)	May be chromatographed directly, but also Reduction	R—SH + R'—SH
R—X (Halogenated compounds)	May be chromatographed directly	

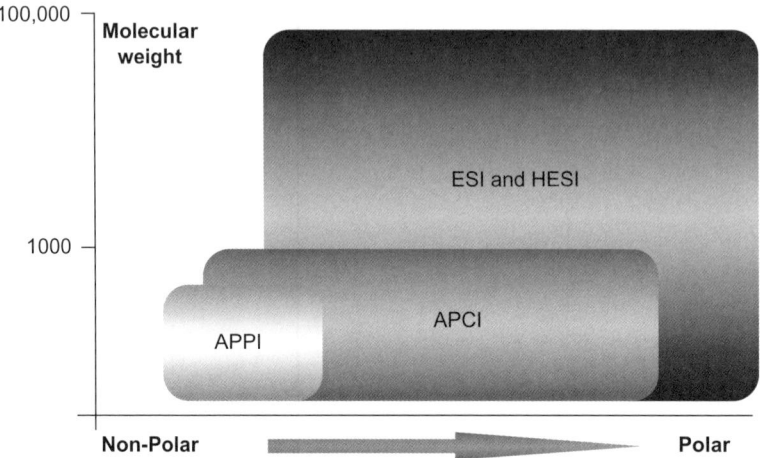

Figure 1.13 Relative applicability of LC-MS ionization techniques with regard to polarity and molecular mass of the analyte (image courtesy Thermo Fischer Scientific).

The advantage of measuring naturally rare radioactive nuclides is that the background of such atoms in biological systems is low and thus very small concentration changes can be detected providing very good sensitivity and dynamic range. Accordingly, low but physiologically relevant doses of micronutrients can be administered and investigated. These doses cause minimal radiation exposures below safety thresholds. In addition, the nutrient metabolites can be traced for prolonged periods of time. Major drawbacks of AMS are the often problematic purchase of radiolabelled compounds and the fact that the measured information is not compound-specific since AMS pools all different metabolites during the combustion step and therefore abandons selectivity.

An accelerator mass spectrometer counts individual nuclei. The sample is put into a negative ion source (Figure 1.14). Using a caesium ion gun focussed on a small spot on the sample, negative ions are produced and then selected by an injection magnet depending on their m/z ratio. Those negative ions are injected into the positive terminal of a Van der Graaf particle accelerator where all molecular ions are broken up and positive ions are formed.

The Van der Graaf particle accelerator consists of two accelerating gaps with a large positive voltage in the middle. The centre of the accelerator is charged to a voltage of up to $1–5 \times 10^6$ V. The negative ions travelling down the beam tube are accelerated towards the positive terminal. At the terminal they pass through an electron stripper and emerge as positive ions. These are repelled from the positive terminal and are accelerating to ground potential. The final velocity of the ions reaches a few percent of the speed of light (about 50 million miles per hour). The ion beam passes into a high-energy analysing magnet that separates ^{12}C, ^{13}C and ^{14}C, and their respective currents

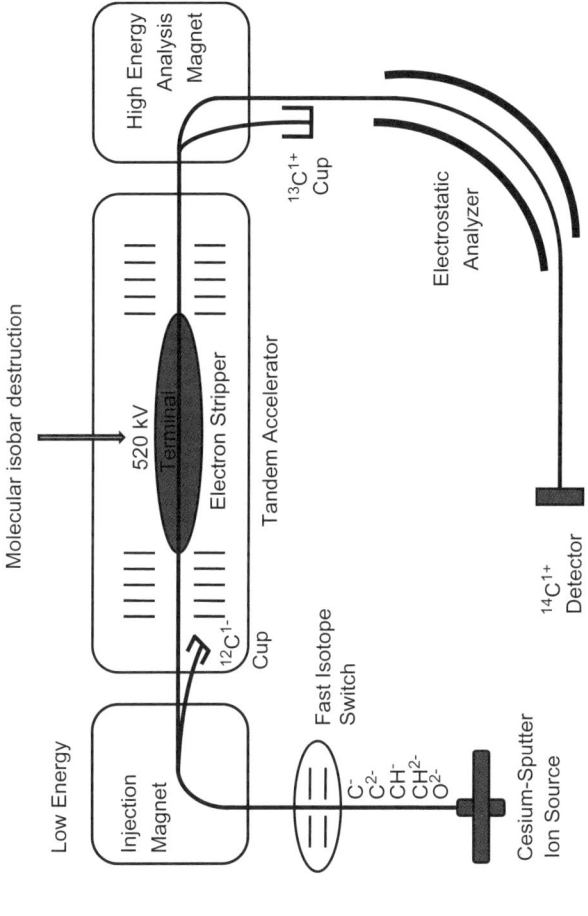

Figure 1.14 Schematic drawing of an accelerator mass spectrometer (AMS) (reproduced from ref. 55).

are measured with Faraday cups. The ^{14}C beam is then focused by a quad-rupole and electrostatic cylindrical analyzer, and the atoms are counted in a gas ionization detector.[46]

1.9 Inductively Coupled Plasma Mass Spectrometry

Mass spectrometry is also much used for the analysis of inorganic materials in addition to its application to organic molecules. Given the nutritional impor-tance of micronutrients such as calcium, magnesium, iron, selenium and the safety issue that the presence of some metals (*e.g.* lead, arsenic, mercury) in foodstuffs may represent, several mass spectrometric technologies were devel-oped to trace and quantify those elements.[47,48] Among them, many are volatile (natively or after derivatization) and can be analysed by conventional electron ionisation techniques.[49]

A large number of inorganic nutrients are, however, non-volatile and require special ionization methods. Traditionally, thermal ionization mass spectro-metry (TIMS) has been the preferred technique for the measurement of these inorganic materials—primarily for elemental analysis and accurate measure-ment of isotopic ratios. Briefly, a thermal ionization source consists of two heated filaments. The sample is deposited on the first filament and evaporated in the ion source through heating. The evaporated analytes transferred on the second filament are ionized through the loss of an electron. Ionization is therefore a two-step process: evaporation of the sample from a heated filament and removal of an electron from the evaporated analyte. Separation of the ions is subsequently performed by either a quadrupole mass filter or an electro-magnetic sector analyzer.

Inductively coupled plasma mass spectrometry (ICP-MS) was launched as a new tool for elemental analysis in the 1980s.[50,51] The development of ICP-MS techniques with better isotope-ratio measurement and interference removal capabilities (*e.g.* single and multi-detector ICP-MS and reaction/ collision cell ICP-MS) has conferred a more prominent role on ICP-MS in the determination of elemental composition and stable isotope tracers in nutri-tional research.[52]

ICP ionization is applicable both to solid-state and solution-phase samples with a high ionization efficiency ($>90\%$). Solid samples are introduced through laser ablation, whereas aqueous samples are introduced using a nebulizer producing a fine mist using high velocity argon. The aerosol then passes into a spray chamber to produce small droplets to be vaporized in the plasma torch, which uses an auxiliary flow of argon. The torch is located in the centre of a radio frequency coil, through which rf energy is passed. The intense rf field causes collisions between the Ar atoms, generating a high-energy plasma. The sample aerosol is instantaneously decomposed in the plasma (plasma tem-perature is in the order of 6000–10 000 K) to form analyte atoms which are simultaneously ionized. The resulting ions are sampled from the plasma and transferred to the mass spectrometer.[53]

1.10 Conclusions

Remarkable technological developments during the last two decades have transformed mass spectrometry into a tool that today plays a pivotal role in accelerating the pace of discoveries in life sciences. Many fields of science require strong analytical chemistry and, in virtually all areas of life sciences, advances in instrumentation drive discovery.

A widespread need for mass spectrometric applications in nutrition supports its further distribution in food research laboratories. However, a major barrier in the past for broad application was the high costs of the instrumentation. Today, the availability of instruments with different performance capabilities from table-top to high-end devices, an increased user friendliness, and high-throughput potential are major development areas in mass spectrometry instrumentations for food and nutrition research.

To achieve even higher selectivity and thus to minimize interferences, new mass analyzers with increased resolution and reduced prices have been developed and introduced. A typical example for this trend was the introduction of Q-ToF instruments 10–15 years ago, introduction of high-resolution triple-quadrupoles five years ago and the recent introduction of Orbitraps. Beside increased resolution, emphasis is also put on improved mass accuracy during development, for example, of Q-ToF (5–10 ppm), Orbitrap (below 5 ppm) and also FT-ICR machines (sub ppm). While the sensitivity of the present mass spectrometers is outstanding, developments in ion guide and focusing devices offer the prospect of even better sensitivity and a simplification of sample preparation. The ultimate sample preparation should be reduced to only sample dilution, if necessary, in order to avoid clean-up artefacts. Most importantly, hybrid mass spectrometers with sophisticated scan functions are increasingly used as these devices combine the advantages of different mass spectrometer types into one instrument (*e.g.* Q-ToF, Q-trap, Paul trap–Orbitrap, ion mobility-triple quadrupole, *etc.*).

Mass spectrometry is one of the most versatile identification and quantification techniques nutritionists have today and, more importantly, it ranges among the very few analytical tools that can deliver information-rich data at high throughput. Its potential as a knowledge provider is constantly growing due to both the push from instrument manufacturers and the pull of researchers seeking to transform nutrition into a mechanism-based discipline.

References

1. S. Ma, S. K. Chowdhury and K. D. Altria, *Curr. Drug Metab.*, 2006, **7**, 503.
2. C. Dass, *Fundamentals of Contemporary Mass Spectrometry*, Wiley & Sons, Hoboken, NJ, 2007.
3. B. M. Ham, *Even Electron Mass Spectrometry with Biomolecule Applications*, Wiley & Sons, Hoboken, NJ, 2008.

4. F. A. Mellon, R. Self and J. R. Startin, *Mass Spectrometry of Natural Substances in Food,* Royal Society of Chemistry, Cambridge, UK, 2000.

5. R. A. W. Johnstone and M. E. Rose, *Mass Spectrometry for Chemists and Biochemists*, Cambridge University Press, Cambridge, UK, 1996.

6. B. Munson and F. H. Field, *J. Am. Chem. Soc.*, 1966, **88**, 2621.

7. S. Abdel-Baky and R. W. Giese, *Anal. Chem.*, 1991, **63**, 2986.

8. R. B. Cole, *Electrospray Ionization Mass Spectrometry. Fundamentals, Intrumentation & Applications*, Wiley & Sons, New York, 1997.

9. P. Kebarle, *J. Mass Spectrom.*, 2000, **35**, 804.

10. J. V. Iribarne and B. A. Thompson, *J. Chem. Phys.*, 1976, **64**, 2287.

11. W. J. Griffiths, A. P. Jonsson, S. Liu, K. R. Dilip and Y. Wang, *Biochem. J.*, 2001, **355**, 545.

12. J. M. Wiseman, D. R. Ifa, Q. Song and R. G. Cooks, *Angew. Chem. Int. Ed Engl.*, 2006, **45**, 7188.

13. R. G. Cooks, Z. Ouyang, Z. akats and J. M. Wiseman, *Science*, 2006, **311**, 1566.

14. J. C. Schwartz, M. W. Senko and J. E. P. Syka, *J. Am. Soc. Mass Spectrom.*, 2002, **13**, 659.

15. R. S. Brown and J. J. Lennon, *Anal. Chem.*, 1995, **67**, 1998.

16. A. N. Verentchikov, W. Ens and K. G. Standing, *Anal. Chem.*, 1994, **66**, 126.

17. A. Makarov, E. Denisov, A. Kholomeev, W. Balschum, O. Lange, K. Strupat and S. Horning, *Anal. Chem.*, 2006, **78**, 2113.

18. K. L. Busch, G. L. Glish and S. A. McLuckey, *Mass Spectrometry/Mass Spectrometry: Techniques and Applications of Tandem mass spectrometry*, VCH, New York, 1988.

19. J. V. Johnson, R. A. Yost, P. E. Kelley and D. C. Bradford, *Anal. Chem.*, 1990, **62**, 2162.

20. K. F. Medzihradszky, J. M. Campbell, M. A. Baldwin, A. M. Falick, P. Juhasz, M. L. Vestal and A. L. Burlingame, *Anal. Chem.*, 2000, **72**, 552.

21. E. Lichtfouse, *Rapid Commun. Mass Spectrom.*, 2000, **14**, 1337.

22. W. A. Brand, *Rapid Commun. Mass Spectrom.*, 1996, **31**, 225.

23. J. T. Brenna, T. N. Corso, H. J. Tobias and R. J. Caimi, *Mass Spectrom. Rev.*, 1997, **16**, 227.

24. R. A. Werner and W. A. Brand, *Rapid Commun. Mass Spectrom.*, 2001, **15**, 501.

25. J. Santrock, S. A. Studley and M. J. Hayes, *Anal. Chem.*, 1985, **57**, 1444.

26. D. E. Matthews and J. M. Haynes, *Anal. Chem.*, 1978, **50**, 1465.

27. M. Sano, Y. Yotsui, H. Abe and S. Sasaki, *Biomed. Mass Spectrom.*, 1976, **3**, 1.

28. A. L. Sessions, *J. Sep. Sci.*, 2006, **29**, 1946.

29. S. Asche, A. L. Michaud and J. T. Brenna, *Curr. Org. Chem.*, 2003, **7**, 1527.

30. R. S. Gohlke and F. W. McLafferty, *J. Am. Soc. Mass Spectrom.*, 1993, **4**, 367.

31. C. Milo and I. Blank, presented at the 214th ACS Meeting, Las Vegas, 1997.

32. L. Maeztu, C. Sanz, S. Andueza, M. P. De Pena, J. Bello and C. Cid, *J. Agr. Food Chem.*, 2001, **49**, 5437.
33. P. Chen and F. P. Abramson, *Anal. Chem.*, 1998, **70**, 1664.
34. J. T. Simpson, D. S. Torok and S. P. Markey, *J. Am. Soc. Mass Spectrom.*, 1995, **6**, 525.
35. W. W. Christie, *Lipids*, 1998, **33**, 343.
36. L. B. Fay and U. Richli, *J. Chromatogr.*, 1991, **541**, 89.
37. J. H. Pazur, J. H. Miskiel and B. Liu, *J. Chromatogr.*, 1987, **396**, 139.
38. D. C. Dejongh, T. Radford, J. D. Hribar, S. Hanessian, M. Bieber, G. Dawson and C. C. Sweeley, *J. Am. Chem. Soc.*, 1969, **91**, 1728.
39. J. B. German and S. M. Watkins, *Trends Food Sci. Technol.*, 2004, **15**, 541.
40. F. K. Yeboah and Y. Konishi, *Anal. Lett.*, 2003, **36**, 3271.
41. K. W. Turteltaub, J. S. Felton, B. L. Gledhill, J. S. Vogel, J. R. Southon, M. W. Caffee, R. C. Finkel, D. E. Nelson, I. D. Proctor and J. C. Davis, *Proc. Nat. Acad. Sci. U. S. A.*, 1990, **87**, 5288.
42. K. H. Dingley, M. L. Roberts, C. A. Velsko and K. W. Turteltaub, *Chem. Res. Toxicol.*, 1998, **11**, 1217.
43. S. A. Ross, P. R. Srinivas, A. J. Clifford, S. C. Lee, M. A. Philbert and R. L. Hettich, *J. Nutr.*, 2004, **134**, 681.
44. J. S. Vogel, K. W. Turteltaub, R. Finkel and D. E. Nelson, *Anal. Chem.*, 1995, **67**, 353A.
45. J. S. Vogel and K. W. Turteltaub, in *Mathematical Modelling in Experimental Nutrition*, ed. A. J. Clifford and H.G-Müller, Plenum Press, New York, 1998, vol. 445, pp. 397–408.
46. G. Lappin and R. C. Garner, *Nat. Rev. Drug Discov.*, 2003, **2**, 233.
47. H. M. Crews, V. Ducros, J. Eagles, F. A. Mellon, P. Kastenmayer, J. B. Luten and B. A. McGraw, *Analyst*, 1994, **119**, 2491.
48. T. Walczyk, *Fresenius' J. Anal. Chem.*, 2001, **370**, 444.
49. P. V. Dael, D. Barclay, K. Longet, S. Metairon and L. B. Fay, *J. Chromatogr. B*, 1998, **715**, 341.
50. R. S. Houk, V. A. Fassel, G. D. Flesch, H. J. Svec, A. L. Gray and C. E. Taylor, *Anal. Chem.*, 1980, **52**, 2283.
51. R. S. Houk, *Anal. Chem.*, 1986, **58**, 97A.
52. S. Sturup, *Anal. Bioanal. Chem.*, 2004, **378**, 273.
53. B. O. Axelsson, M. Jornten-Karlsson, P. Michelsen and F. Abou-Shakra, *Rapid Commun. Mass Spectrom.*, 2001, **15**, 375.
54. W. M. A. Niessen, *Current Practice of Gas Chromatography-Mass Spectrometry,* Marcel Dekker, New York, 2001.
55. L. T. Vuong, B. A. Buchholz, M. W. Lamé and S. R. Dueker, *Nutr. Rev.*, 2004, **62**, 375.

Section 2
Mass Spectrometry Analysis of Food Ingredients

CHAPTER 2

Mass Spectrometry for Food Analysis: The Example of Fat Soluble Vitamins A and K

GREGORY G. DOLNIKOWSKI

Jean Mayer USDA Human Nutrition Research Center on Aging at Tufts University, 711 Washington Street, Boston, MA 01760, USA

The chemical nature of food is extremely complex. Nutritive compounds are classically divided into macro- and micronutrients. The macronutrients are the main fuel for the body and represent the major complements of food ingredients. They are classified into proteins, carbohydrates and lipids. Micronutrients encompass notably vitamins, minerals and phytonutrients.

Vitamins are an extremely diverse set of small molecules. The human body has an absolute requirement for them. A review article by C. J. Bates[1] in 2000 summarizes copious information about vitamins and about how they have been analyzed. Bates wrote that the combination of mass spectrometry and stable isotopes was becoming increasingly attractive as an analytical approach for vitamins. Fat soluble vitamins A and K represent a good example of compounds that are nutritionally extremely important and which require the use of almost all the analytical capabilities of mass spectrometric technology in order to elucidate their structures and metabolic fate. In that regard, vitamins A and K perfectly illustrate how mass spectrometry has become the workhorse of any food analysis or nutrition laboratory.

RSC Food Analysis Monographs No. 9
Mass Spectrometry and Nutrition Research
Edited by Laurent B. Fay and Martin Kussmann
© The Royal Society of Chemistry 2010
Published by the Royal Society of Chemistry, www.rsc.org

Vitamins A and K are actually groups of related compounds that have metabolic activity in the body. These related molecules are often called vitamers. Therefore, those vitamins are defined by their metabolic activity rather than just by their chemical structure. Often the metabolic activity of vitamins is difficult to determine because the levels of these compounds in body fluids are typically very low, and because before the advent of stable isotope techniques, it was impossible to differentiate exogenous vitamins that have been newly ingested from endogenous vitamin from body stores.

The body requires vitamin A (retinol) to prevent blindness, particularly in children.[2,3] Retinol is available in the diet both as the intact molecule and as provitamin A carotenoids. β-carotene, the most important provitamin A carotenoid, is commonly found in plant food sources and can be converted by the body to retinol using a symmetric cleavage enzyme. β-carotene can also be cleaved asymmetrically to produce apo-carotenals.[4,5] Retinol is part of a larger group of compounds called the retinoids, which also have important functions in the body; for example, retinaldehyde represses adipogenesis and diet-induced obesity.[6]

Vitamin K is necessary to allow the body to make blood clots. An entire class of vitamin K-dependent proteins has been discovered that regulate hemostasis, bone metabolism, tissue calcification and cell cycle regulation.[7–9] Phylloquinone is the primary dietary source of vitamin K and is present at highest concentrations in green leafy vegetables and certain plant oils.[10]

Many mass spectrometric technologies have been deployed to analyze these two fat-soluble vitamins. Concerning vitamin A, mass spectrometric analysis of retinol and related compounds (retinoids) was first achieved by gas chromatography-mass spectrometry (GC-MS) after derivatization, typically with silylation reagents to produce retinol trimethylsilyl ethers. This trimethylsilyl group protects the retinol from degradation during its passage through the GC column. Then under electron capture negative chemical ionization (ECNCI) conditions, the group is cleaved off to produce the retinol $(M-H)^-$ ion.[11] Cool on-column injection was often used to reduce decomposition and increase sensitivity, and the highest detection levels (picogram range) were obtained using ECNCI.

With the development of atmospheric pressure chemical ionization (APCI), retinoids were analyzed by liquid chromatography-mass spectrometry (LC-MS)[12] and were quantified in foods,[13] dietary supplements,[14] fish eggs[15] and cell extracts.[16] Recently, vitamin K isomers were analyzed by GC-MS[17] and LC-MS[18,19] after APCI. The best separations were typically accomplished with C-18 or C-30 stationary phases, and typical on-column detection limits for phylloquinone by LC-MS are in the picogram range.

Carotenoids represent an interesting analytical challenge since they degrade at GC-MS temperatures and therefore must be analyzed by milder methods such as LC-MS.[20] Either conventionally packed columns or capillary columns can be used, but the best separations are typically accomplished with C-30 stationary phases that permit *cis* and *trans* (double bond) isomers to be distinguished.[21–23] While it is possible to detect carotenoids with an electrospray ion source (ESI), atmospheric pressure chemical ionization is much more sensitive.[20] Carotenoids have been quantified in a large variety of food samples including dietary supplements,[24,25] spinach,[26] carrots,[27,28] kale,[29] leafy greens[30] (with and without olive oil), fish eggs,[31] tomatoes,[32] golden rice,[33] green and yellow vegetables,[2] thermally processed vegetables,[34] and red palm oil.[35]

Nutrition research has investigated the fate of vitamins in the human body using either stable isotope labelled standards (typically D or ^{13}C) and "conventional" mass spectrometry or isotope ratio mass spectrometry to provide information about vitamin absorption, metabolism and excretion *in vivo* without the drawbacks of radioactivity. Stable isotope-labelled vitamins have been produced chemically for decades for research use. Recently, the advent of intrinsically labelled foods[36] grown with D_2O or $^{13}CO_2$ has provided powerful new sources for delivering labelled vitamins to study subjects.

When fed to humans, labelled β-carotene was used to track the conversion of β-carotene to retinol *in vivo*,[37,38] and thereby to determine the vitamin A equivalence of β-carotene by measuring retinol with GC-MS using a stable isotope reference method.[39,40] The purity of the labelled material was determined by fast atom bombardment mass spectrometry[37] and APCI-MS.[39] This technique has been recruited to demonstrate that:

- green and yellow vegetables can maintain body stores of vitamin A in Chinese children;[3]
- spinach and carrot can supply significant amounts of vitamin A;[41] and
- golden rice is an effective source for vitamin A.[33]

GC-MS, LC-MS and accelerator mass spectrometry have shown that the absorption of β-carotene and its retinol equivalence in humans is influenced by prior dietary vitamin A intake.[42–44] Moreover, it as been shown with GC-MS that, despite the relatively poor conversion factor of β-carotene to retinol, humans can obtain significant amounts of vitamin A by eating vegetables containing β-carotene that are cooked in oil.[41]

Deuterated retinol has also been used as an ingested tracer with GC-MS detection to perform a quantitative assessment of total body stores of vitamin A with the use of a deuterated-retinol-dilution procedure.[45,46] This technique has been used to:

- measure the dietary vitamin A intakes of Filipino elders with adequate or low liver vitamin A concentrations;[47]
- monitor the vitamin A status of Nicaraguan schoolchildren one year after initiation of the Nicaraguan National Program of Sugar Fortification with Vitamin A;[48] and
- determine that carotene-rich plant foods ingested with fat enhance total-body vitamin A stores in Filipino schoolchildren.[49]

Mass spectrometry has served to investigate other carotenoids such as lycopene and lutein. The bioavailability of synthetically and biosynthetically deuterated lycopene in humans,[50] and the chemical structure and bioavailability of lutein in humans from intrinsically labelled vegetables[51,52] were measured using LC-MS.

The kinetics of vitamin K absorption and metabolism, with stable isotope-labelled phylloquinone as an ingested tracer,[29,53–56] has also been determined by GC-MS. Isotope-labelled vegetables allow researchers to discriminate intake of phylloquinone in the form of food from phylloquinone stored in the body and, ultimately, to determine bioavailability of phylloquinone from foods. Changes in deuterated phylloquinone concentrations in plasma and triglyceride-rich lipoprotein parallel changes in triglyceride concentrations, which suggest a close association with fat absorption. These results were obtained using a two-step approach: first phylloquinone concentrations in plasma and lipoprotein sub-fractions were measured by high performance liquid chromatography (HPLC) and, second, the ion abundances of deuterated and endogenous phylloquinone were determined using GC-MS.

The examples described briefly above highlight the challenges of analyzing food components either directly in the food matrix or in biological samples after food consumption. For most classes of macronutrients and micronutrients, the entire range of MS technologies has been recruited to circumvent the analytical hurdles generated by thermal and light instability, low volatility, high reactivity or low abundance of these compounds. In this section of the book, mass spectrometry is described for the analysis of food macronutrients such as carbohydrates (Daniel Kolarich and Nicolle Packer) and proteins (Claudio Corradini, Lisa Elviri and Antonella Cavazza), and for some nutritionally important phyto-micronutrients (Jean-Luc Wolfender, Aude Violettea and Laurent Fay). Because of the biological importance and complexity of *in vivo* reactions from ingested or *in situ* oxidized foods, mass spectrometric analysis of lipid oxidation products is extensively covered in foods and in biological samples by Arnis Kuksis.

The analysis of free oligosaccharides and protein-bound sugars in secretions such as milk still poses analytical challenges, even for such an advanced technology like mass spectrometry. Daniel Kolarich and Nicolle Packer (Chapter 3) discuss the benefits and limitations of different sample preparations for the mass spectrometric analysis of free oligosaccharides and glycoproteins. Different MS techniques and instrument combinations already successfully applied to the analysis of milk oligosaccharides and tandem and multi stage mass spectrometry (MS^n) applications for the differentiation of structural isomers are also described.

Claudio Corradini, Lisa Elviri and Antonella Cavazza (Chapter 4) highlight the fact that proteins, peptides and amino acids in the diet contribute to physical properties, biological activities and sensory characteristics of food. Such functional properties can be related to their ability to form viscoelastic networks, bind water, emulsify fat and oil, entrap flavours and form stable foams. In addition, the denaturation levels of food proteins translate into the sensory and technological properties of food products. Moreover, the breakdown of proteins to smaller peptides and free amino acids during proteolysis influences the quality of fermented or otherwise ripened dairy products, fish, meat, cereals and vegetables. Furthermore, proteins are widely used in formulated food due their high nutritional value and because they are "generally recognized as safe" (GRAS). Because of this technological and biological diversity of food proteins and peptides, Corradini and colleagues argue that mass spectrometry is the tool of choice to study these qualities. They discuss different proteomic workflows and describe how they can be and have been applied to functional food protein characterization.

The nature and biological consequences of *in vivo* reactions of ingested or *in situ* oxidized foods are extensively discussed by Arnis Kuksis (Chapter 5). He states that many oxo-lipids are known to interact with biological materials to cause cellular damage and describes bi-functional secondary products of lipid oxidation (*e.g.* malondialdehyde and hydroxyalkenals) as powerful cross-linking agents that react with amino groups of enzymes, proteins and nucleic acids. Kuksis argues that the full spectrum of interactions between oxo-lipids and cellular components has only become apparent using combinations of chromatography and mass spectrometry. This recent development culminated in lipidomic studies of dietary and tissue lipid oxidation, and a metabolomic assessment of its biological consequences.

Kuskis' article opens with a brief discussion of the mechanism of lipid autoxidation and the main types of autoxidized fatty acids in dietary fats and oils. The review then considers the dietary oxo-lipid transformation following ingestion as well as the absorption of the water and lipid-soluble products. The discussion focuses on lipidomics to identify the secondary autoxidation products of unsaturated fatty acids, acylglycerols, glycerophospholipids and cholesteryl esters, which have emerged as the primary agents of disease and ageing. Kuksis concludes with a discussion of the limits of endogenous and exogenous antioxidant systems.

In the last chapter of this section, Jean-Luc Wolfender, Aude Violettea and Laurent Fay draw the attention to mass spectrometric analysis of a major class of micronutrients, namely phytonutrients—non-essential food components that are found in vegetables, fruits, spices and traditional ingredients. Based on the main type of natural products for which health claims exist, these authors summarize key mass spectrometric techniques used for their detection, quantification and identification in both food products and biological fluids. They then review phytonutrients claimed for a given health benefit and discuss their bioavailability and efficacy on disease risk reduction.

References

1. C. J. Bates, in *Encyclopaedia of Analytical Chemistry*, ed. R. A. Meyers, John Wiley Chichester, UK, 2000, pp. 7390–7425.
2. G. Tang, X. F. Gu, S. M. Hu, Q. M. Xu, J. Qin, G. G. Dolnikowski, C. R. Fjeld, X. Gao, R. M. Russell and S. A. Yin, *Am. J. Clin. Nutr.*, 1999, **70**, 1069–1076.

3. L. Davidsson, P. Adou, C. Zeder, T. Walczyk and R. Hurrell, *Br. J. Nutr.*, 2003, **90**, 337–343.

4. O. Ziouzenkova, G. Orasanu, G. Sukhova, E. Lau, J. P. Berger, G. W. Tang, N. I. Krinsky, G. G. Dolnikowski and J. Plutzky, *Mol. Endocrinol.*, 2007, **21**, 77–88.

5. C. C. Ho, F. F. de Moura, S. H. Kim and A. J. Clifford, *Am. J. Clin. Nutr.*, 2007, **85**, 770–777.

6. O. Ziouzenkova, G. Orasanu, M. Sharlach, T. E. Akiyama, J. P. Berger, J. Viereck, J. A. Hamilton, G. W. Tang, G. G. Dolnikowski, S. Vogel, G. Duester and J. Plutzky, *Nat. Med. (N.Y.)*, 2007, **13**, 695–702.

7. G. Ferland, *Nutr. Rev.*, 1998, **56**, 223–230.

8. B. Furie, B. A. Bouchard and B. C. Furie, *Blood*, 1999, **93**, 1798–1801.

9. C. Vermeer, B. L. Gijsbers, A. M. Craniun, M. M. Groenen-van Dooren and M. H. Knapen, *J. Nutr.*, 1996, **126**, 1187S–1191S.

10. S. L. Booth and J. W. Suttie, *J. Nutr.*, 1998, **128**, 785–788.

11. G. W. Tang, J. Qin, G. G. Dolnikowski and R. M. Russell, *Am. J. Clin. Nutr.*, 2003, **78**, 259–266.

12. R. Andreoli, P. Manini, D. Poli, E. Bergamaschi, A. Mutti and W. M. A. Niessen, *Anal. Bioanal. Chem.*, 2004, **378**, 987–994.

13. A. J. Edwards, C. S. You, J. E. Swanson and R. S. Parker, *Am. J. Clin. Nutr.*, 2001, **74**, 348–355.

14. C. J. Blake, *J. AOAC Int.*, 2007, **90**, 897–910.

15. H. X. Li, S. T. Tyndale, D. D. Heath and R. J. Letcher, *J. Chromatogr. B Analyt. Technol. Biomed. Life Sci.*, 2005, **816**, 49–56.

16. R. Ruhl, *Rapid Commun. Mass Spectrom.*, 2006, **20**, 2497–2504.

17. A. T. Erkkila, A. H. Lichtenstein, G. G. Dolnikowski, M. A. Grusak, S. M. Jalbert, K. A. Aquino, J. W. Peterson and S. L. Booth, *Metab. Clin. Exp.*, 2004, **53**, 215–221.

18. S. Ahmed, N. Kishikawa, K. Nakashima and N. Kuroda, *Anal. Chim. Acta*, 2007, **591**, 148–154.

19. X. Fu, J. W. Peterson, M. Hdeib, S. L. Booth, M. A. Grusak, A. H. Lichtenstein and G. G. Dolnikowski, *Anal. Chem.*, 2009, **81**, 5421–5425.

20. Z. G. Hao, B. Parker, M. Knapp and L. L. Yu, *J. Chromatogr. A*, 2005, **1094**, 83–90.

21. L. Q. Fang, N. Pajkovic, Y. Wang, C. G. Gu and R. B. van Breemen, *Anal. Chem.*, 2003, **75**, 812–817.

22. K. Frohlich, J. Conrad, A. Schmid, D. E. Breithaupt and V. Bohm, *Int. J. Vitamin Nutr. Res.*, 2007, **77**, 369–375.

23. R. B. van Breemen, X. Y. Xu, M. A. Viana, L. W. Chen, M. Stacewicz-Sapuntzakis, C. Duncan, P. E. Bowen and R. Sharifi, *J. Agric. Food Chem.*, 2002, **50**, 2214–2219.

24. D. E. Breithaupt and J. Schlatterer, *Eur. Food Res. Technol.*, 2005, **220**, 648–652.

25. H. Kroll, J. Friedrich, M. Menzel and P. Schreier, *J. Agric. Food Chem.*, 2008, **56**, 4198–4204.

26. M. Dachtler, T. Glaser, K. Kohler and K. Albert, *Anal. Chem.*, 2001, **73**, 667–674.
27. A. J. Edwards, C. H. Nguyen, C. S. You, J. E. Swanson, C. Emenhiser and R. S. Parker, *J. Nutr.*, 2002, **132**, 159–167.
28. O. Livny, R. Reifen, I. Levy, Z. Madar, R. Faulks, S. Southon and B. Schwartz, *Eur. J. Nutr.*, 2003, **42**, 338–345.
29. A. C. Kurilich, S. J. Britz, B. A. Clevidence and J. A. Novotny, *J. Agric. Food Chem.*, 2003, **51**, 4877–4883.
30. R. Lakshminarayana, M. Raju, T. P. Krishnakantha and V. Baskaran, *J. Agric. Food Chem.*, 2007, **55**, 6395–6400.
31. H. X. Li, S. T. Tyndale, D. D. Heath and R. J. Letcher, *J. Chromatogr. B Analyt. Technol. Biomed. Life Sci.*, 2005, **816**, 49–56.
32. D. Naviglio, T. Caruso, P. Iannece, A. Aragon and A. Santini, *J. Agric. Food Chem.*, 2008, **56**, 6227–6231.
33. G. W. Tang, J. Qin, G. G. Dolnikowski, R. M. Russell and M. A. Grusak, *Am. J. Clin. Nutr.*, 2009, **89**, 1776–1783.
34. A. A. Updike and S. J. Schwartz, *J. Agric. Food Chem.*, 2003, **51**, 6184–6190.
35. C. S. You, R. S. Parker and J. E. Swanson, *Asia Pac. J. Clin. Nutr.*, 2002, **11**, S438–S442.
36. M. A. Grusak, *J. Nutr. Biochem.*, 1997, **8**, 164–171.
37. Y. M. Lin, S. R. Dueker, B. J. Burri, T. R. Neidlinger and A. J. Clifford, *Am. J. Clin. Nutr.*, 2000, **71**, 1545–1554.
38. Y. Wang, X. Y. Xu, M. van Lieshout, C. E. West, J. Lugtenburg, M. A. Verhoeven, A. F. L. Creemers and R. B. van Breemen, *Anal. Chem.*, 2000, **72**, 4999–5003.
39. G. Tang, J. Qin, G. G. Dolnikowski and R. M. Russell, *Eur. J. Nutr.*, 2000, **39**, 7–11.
40. Z. X. Wang, S. Yin, X. F. Zhao, R. M. Russell and G. W. Tang, *Br. J. Nutr.*, 2004, **91**, 121–131.
41. G. W. Tang, J. Qin, G. G. Dolnikowski, R. M. Russell and M. A. Grusak, *Am. J. Clin. Nutr.*, 2005, **82**, 821–828.
42. S. L. Lemke, S. R. Dueker, J. R. Follett, Y. M. Lin, C. Carkeet, B. A. Buchholz, J. S. Vogel and A. J. Clifford, *J. Lipid Res.*, 2003, **44**, 1591–1600.
43. R. J. Pawlosky, V. P. Flanagan and J. A. Novotny, *J. Lipid Res.*, 2000, **41**, 1027–1031.
44. S. J. Hickenbottom, S. L. Lemke, S. R. Dueker, Y. M. Lin, J. R. Follett, C. Carkeet, B. A. Buchholz, J. S. Vogel and A. J. Clifford, *Eur. J. Nutr.*, 2002, **41**, 141–147.
45. H. C. Furr, M. H. Green, M. Haskell, N. Mokhtar, P. Nestel, S. Newton, J. D. Ribaya-Mercado, G. W. Tang, S. Tanumihardjo and E. Wasantwisut, *Public Health Nutr.*, 2005, **8**, 596–607.
46. J. D. Ribaya-Mercado, F. S. Solon, G. E. Dallal, N. W. Solomons, L. S. Fermin, M. Mazariegos, G. G. Dolnikowski and R. M. Russell, *Am. J. Clin. Nutr.*, 2003, **77**, 694–699.

47. J. D. Ribaya-Mercado, F. S. Solon, L. S. Fermin, C. S. Perfecto, J. A. A. Solon, G. G. Dolnikowski and R. M. Russell, *Am. J. Clin. Nutr.*, 2004, **79**, 633–641.

48. J. D. Ribaya-Mercado, N. W. Solomons, Y. Medrano, J. Bulux, G. G. Dolnikowski, R. M. Russell and C. B. Wallace, *Am. J. Clin. Nutr.*, 2004, **80**, 1291–1298.

49. J. D. Ribaya-Mercado, C. C. Maramag, L. W. Tengco, G. G. Dolnikowski, J. B. Blumberg and F. S. Solon, *Am. J. Clin. Nutr.*, 2007, **85**, 1041–1049.

50. G. W. Tang, A. L. A. Ferreira, M. A. Grusak, J. Qin, G. G. Dolnikowski, R. M. Russell and N. I. Krinsky, *J. Nutr. Biochem.*, 2005, **16**, 229–235.

51. A. Lienau, T. Glaser, G. W. Tang, G. G. Dolnikowski, M. A. Grusak and K. Albert, *J. Nutr. Biochem.*, 2003, **14**, 663–670.

52. K. Putzbach, M. Krucker, K. Albert, M. A. Grusak, G. W. Tang and G. G. Dolnikowski, *J. Agric. Food Chem.*, 2005, **53**, 671–677.

53. G. G. Dolnikowski, Z. Y. Sun, M. A. Grusak, J. W. Peterson and S. L. Booth, *J. Nutr. Biochem.*, 2002, **13**, 168–174.

54. A. T. Erkkila, A. H. Lichtenstein, G. G. Dolnikowski, M. A. Grusak, S. M. Jalbert, K. A. Aquino, J. W. Peterson and S. L. Booth, *Metab. Clin. Exp.*, 2004, **53**, 215–221.

55. K. E. Hansen, A. N. Jones, M. J. Lindstrom, L. A. Davis, J. A. Engelke and M. M. Shafer, *J. Bone Miner. Res.*, 2008, **23**, 1052–1060.

56. K. S. Jones, L. J. C. Bluck, L. Y. Wang and W. A. Coward, *Eur. J. Clin. Nutr.*, 2008, **62**, 1273–1281.

CHAPTER 3

Mass Spectrometry for the Analysis of Milk Oligosaccharides

DANIEL KOLARICH AND NICOLLE H. PACKER

Department of Chemistry and Biomolecular Sciences, Macquarie University, Sydney, NSW, 2109, Australia

3.1 Mass Spectrometry for the Analysis of Carbohydrates

Mass spectrometry (MS) has become the most popular technique for the analysis of complex biomolecules such as (glyco)proteins and carbohydrates. Standardized methods for the assessment of the protein backbone using proteolytically derived peptides were developed in the last 20 years and are the foundation of high throughput proteomics workflows.[1]

Tandem MS fragmentation allows new approaches to peptide sequencing and protein identification. The ability to identify proteins and the success of these workflows depend on the sequence of the genome being known. In contrast to peptides, oligosaccharides do not follow a template driven biosynthesis and are rarely assembled in a linear way, making structural determination from tandem MS spectra more challenging.[2] In addition, the isomeric monosaccharide building blocks of oligosaccharides are not distinguishable by their mass alone,[3] and monosaccharide analysis by itself does not give any information on the structures of the resulting oligosaccharides.[4,5] Fortunately for

RSC Food Analysis Monographs No. 9
Mass Spectrometry and Nutrition Research
Edited by Laurent B. Fay and Martin Kussmann
© The Royal Society of Chemistry 2010
Published by the Royal Society of Chemistry, www.rsc.org

the analyst, Nature does not use all the theoretically possible repertoire of monosaccharides to synthesize eukaryotic milk oligosaccharides but uses a rather limited collection (Table 3.1).[6] In addition, the pathways of oligosaccharide biosynthesis—especially for protein-bound glycans—are highly conserved, and this knowledge assists the assignment of structures from MS and/or tandem MS data. Though MS technology offers capability only imagined just two decades ago, the crucial key to successful analysis is still in the initial sample preparation. Enrichment of the target compounds and simultaneous depletion of contaminating substances is the essential step that must be optimized for each sample matrix.

3.2 Mankind, Milk and Carbohydrates

Humans and all other mammals have mother's milk as the first, and one of the most important, nutrients. In the first phase of life, milk not only supplies the right amount and composition of nutrients beneficial for infant development, but also provides many factors that support the undeveloped immune system.[7] Proteins, glycoproteins, free glycans and glycolipids are all part of the phalanx of immune active components that together contribute to a reduced risk of gastrointestinal and immunological diseases of the newborn.[8,9]

Cow's milk and a wide range of animal milk products have evolved in many cultures as a major part of the daily diet. A variety of bacteria, yeasts and fungi have been used for the production of cheeses, yoghurts, probiotic drinks, *etc.* and the variety of milk-derived products takes up considerable space in supermarket shelves. This is reflected by the 560 million tonnes of cow's milk production worldwide in 2007 (www.dairyco.org.uk/datum/milk-supply/milk-production/world-milk-production.aspx), not including milk from different animal sources such as, for example, goat, sheep or buffalo. Thus animal milk plays a central role in our diet and knowledge of the different components and their biological functions are of great interest, both scientifically and economically. All milks are a rich source of oligosaccharides and glycoconjugates such as glycoproteins and free oligosaccharides. These components are of special interest both nutritionally and as part of the antipathogenic protection of our offspring.[8] Stomach flora are also affected by milk, as the growth of various strains of bifidobacteria have been shown to be influenced by different carbohydrates.[10]

In recent years, MS has emerged as an indispensable tool for the analysis of biomolecules in general due to its sensitivity, versatility and its ease of applicability to complex samples.[11–14] This review summarizes the state-of-the-art in the analysis of glycans (free or protein bound) from milk as an example of a complex sample matrix.

3.3 Free Oligosaccharides in Milk

Free oligosaccharides are found in milk from all mammals analyzed so far.[15,16] However, compared to human milk ($7–12\,\mathrm{g\,L^{-1}}$), the concentration of these

Table 3.1 Monosaccharide building blocks and oligosaccharide modifications in milk.

	Residue mass			Symbol[a]	
	Monoisotopic	Average	Monosaccharide	Colour	Black/white
Hexose	162.0528	162.1424	Glucose	Blue circle	●
			Galactose	Yellow circle	○
			Mannose	Green circle	◑
N-Acetyl hexosamine	203.0794	203.1950	*N*-Acetyl glucosamine	Blue square	■
			N-Acetyl galactosamine	Yellow square	□
Desoxyhexose	146.0579	146.1430	Fucose	Red triangle	◀
Neuraminic acid	291.0954	291.2579	*N*-Acetyl neuraminic acid	Purple diamond	◆
	307.0903	307.2573	*N*-Glycolyl neuraminic acid[b]	White diamond	◇
Phosphate	79.9663	79.9799		P	P
Sulfate	79.9568	80.0642		S	S

[a]Symbols used are as suggested by the Consortium for Functional Glycomics (CFG) (www.functionalglycomics.org).
[b]N-Glycolyl neuraminic acid is not a building block of human oligosaccharides.

components in milk from most relevant domestic mammals is one to two orders of magnitude less.[6] They are composed of a comparably limited number of different monosaccharide building blocks (Table 3.1):

- hexoses—D-glucose (Glu) and D-galactose (Gal);
- N-acetylhexosamines—N-acetylglucosamine (GlcNAc);
- L-fucose (Fuc); and
- neuraminic acid—N-acetyl neuraminic acid (NeuAc).

N-acetylgalactosamine (GalNAc) has not been found to be part of free oligosaccharides in human milk, but is part of the protein bound oligosaccharides, and N-glycolyl neuraminic acid (NeuGc) is not usually found in humans.[6,17] Though the number of building blocks is relatively small and the number of possible structures theoretically huge, to date approximately 200 different human milk oligosaccharides have been identified.[6]

Free oligosaccharides share the presence of lactose at the reducing end and different glycosyltransferases extend this core disaccharide with variable numbers of LacNAc (GlcNAc-Gal disaccharide) repeats.[6] On this backbone, various extensions of α-glycosidic linked fucose—forming different Lewis antigen structures—are commonly seen on human milk oligosaccharides, but are more an exception in milk from most common dairy animals.[6,14] On the non-reducing end of the free oligosaccharides, NeuAc (and NeuGc in animals) is often found "capping" these structures. Oligosaccharides containing sialic acid residues form the acidic oligosaccharide fraction and this feature has often been used for their specific enrichment. The analysis by MS of the sialylated oligosaccharides is complicated by these charged residues (see Sections 3.8 and 3.9).

3.4   Protein Bound Oligosaccharides

Free and protein bound oligosaccharides use similar biosynthetic enzymes and monosaccharide building blocks (Table 3.1), but the glycans attached to proteins are distinctively different in their structures from their free counterparts.[18,19] They are classified by their attachment to proteins as so-called N-linked and O-linked glycans.[2] The attachment of N-linked glycans requires a particular sequon on the protein, . . . Asn-X-Ser/Thr. . . (X = any amino acid except Pro), where a high mannose structure is attached during protein synthesis to the Asn residue.[2] This high mannose structure is then further subjected to controlled degradation and complex rebuilding in the Golgi apparatus where the complex type oligosaccharide structures are built.[2] The common chitobiose pentasaccharide core is extended with different degrees of GlcNAc-Gal disaccharides, partially decorated with fucoses in different linkages and positions with the addition of sialic acid usually on the non-reducing termini.[2]

O-Linked glycans are attached to the protein via an O-glycosidic bond to either Ser or Thr residues, but no unique amino acid sequon is known so far.[20] The initial protein O-glycosylation step occurs in the Golgi apparatus by the attachment to the protein usually of a GalNAc residue, although other attachments such as fucose have been reported.[21] These residues are then further extended stepwise by a range of glycosyltransferases.[2] In contrast to the *N*-glycans, *O*-glycans do not share a common core structure and eight core structures have been described.[20]

Both the *N*- and *O*- glycosylation profile of human milk fat globule membrane proteins have been shown to differ significantly from cow's milk (Figures 3.1 and 3.2). In particular, human milk contains only core 2 type *O*-linked structures. In bovine milk, however, predominantly core 1 type structures are found on the same subset of proteins in peak lactation.[21] The formation of core 2 or higher type structures is considered a prerequisite for the further complex extension of the protein-bound *O*-glycans, leading to different numbers of LacNAc repeats and further attachment of fucose and sialic acid residues. These types of structures are the basis of the Lewis type antigens that are recognized by bacterial lectins. Free glycans of considerable size (M > 3000 Da) with a very high level of fucosylation have been identified in human milk and are an example of the great diversity of structures that can be present in milk of human origin.[22]

3.5 It's All About Sample Preparation—General Considerations

In any kind of mass spectrometric analysis, sample preparation is the first critical step. Depending on the sample matrix and the concentration of the compounds of interest, some kind of isolation or enrichment and/or depletion of interfering substances is required to enable MS analysis. These steps also reduce sample complexity by the separation of the sample into different fractions. However, reducing sample complexity and enrichment of specific compounds runs the risk of losing sample or introducing a bias towards particular components. The desired speed of analysis presents an additional compromise between in-depth analysis and assay needs. The experiment planning phase helps determine the analytical techniques to be used, for example:

- What information is required?
- Is a qualitative profile required or relative quantitative distribution information?
- How much sample is needed?
- How quickly can the sample be processed?
- How many samples can be processed in what time frame?
- What analyzes can be performed on the available instrumentation?
- Cost of analysis?

3.6 Sample Preparation Options for Free Milk Oligosaccharides

Free oligosaccharides in milk have been the subject of numerous studies and to date around 200 distinct structures have been identified in human milk.[6] Due to

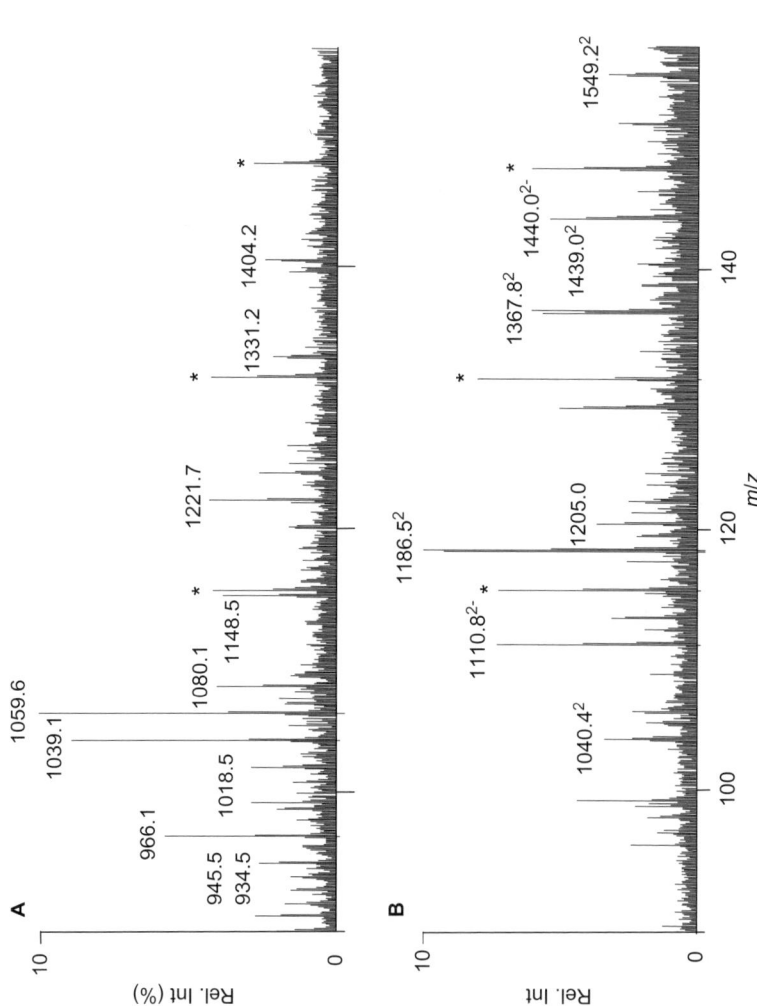

Figure 3.1 Combined LC-ESI-MS spectra (RT 10–30 min) of globally released *N*-linked oligosaccharides from (A) bovine milk fat globule membranes and (B) human milk fat globule membranes. * Indicates hexose polymer contamination. Reproduced from ref. 21 with permission of the American Chemical Society (license no: 2316831176509).

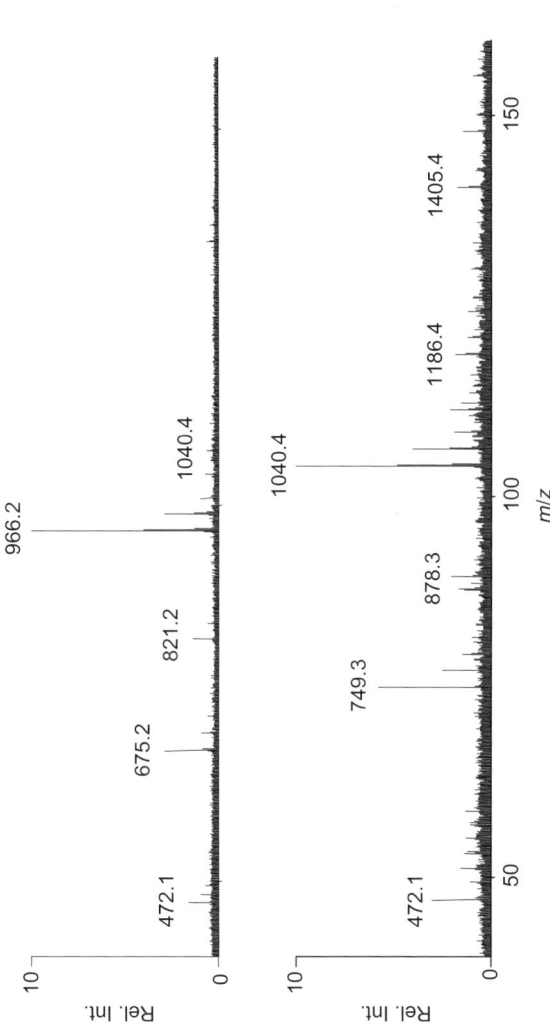

Figure 3.2 Combined LC-ESI-MS spectra (RT 10–30 min) of globally released *O*-linked oligosaccharides from (**A**) bovine milk fat globule membranes and (**B**) human milk fat globule membranes. Reproduced from ref. 21 with permission of the American Chemical Society (license no: 2316831176509).

the hydrophilic nature of free oligosaccharides, most published protocols remove the fat component first.[19,23–31] Centrifugation of the milk sample at 4 °C, for usually 30–60 min, is followed by careful removal of the fat layer from the top of the milk sample. Besides the free glycans, water-soluble components such as proteins, glycoproteins and salts are still highly abundant in the remaining aqueous fraction.

(Glyco)proteins are often depleted from this skimmed milk by overnight cold ethanol precipitation. Before free glycans can be analyzed, further pre-fractionation steps or the application of other specific sample preparation strategies are required.[19,23,24,26,29] For example, Niñonuevo *et al.* separated the protein fraction from any remaining lipids in skimmed milk by performing a chloroform : methanol extraction.[25] Proteins were subsequently precipitated with cold ethanol from the aqueous fraction and the free oligosaccharides were recovered after drying the supernatant. The free oligosaccharides were reduced before carbon solid phase extraction (SPE) and analysis by graphitized carbon liquid chromatography-mass spectrometry (LC-MS).[25]

The presence of neuraminic acid enables further fractionation of the free oligosaccharide pool using ion exchange chromatography, and results in neutral and acidic oligosaccharide fractions.[16,30] Essentially pure single oligosaccharide structures have been obtained using TSK gel Amido-80 high performance liquid chromatography (HPLC) separation for detailed structural elucidation by NMR.[16] The amount of sample, cost of instrumentation and expertise required limits this approach as the first method of choice for most characterization studies.

3.7 Sample Preparation for Protein Bound Glycans

In contrast to free oligosaccharides that are found in the aqueous layer, protein bound oligosaccharides—due to their attachment to a protein carrier—are present in both the aqueous and the fat component of milk (Kolarich *et al.* manuscript in preparation). Depending on the study objective, cold centrifugation delipidation of milk allows a quick and easy sub-fractionation of many milk glycoproteins based on their aqueous solubility. Several proteomic studies have used skim milk as their starting material for protein separation by electrophoretic (one-dimensional or two-dimensional polyacrylamide gel electrophoresis, PAGE) or chromatographic methods.[32–37] The proteins of the milk fat globule membranes are contained in the fat layer of non-homogenized milk and can be separated by a series of centrifugation and buffering steps.[21,38–42]

The detailed MS structural analysis of protein bound oligosaccharides is more easily accomplished after their release from the protein backbone. Comprehensive site-specific glycosylation analysis of a glycoprotein is usually only successful on purified, known protein sequences. The enzyme peptide-*N*-glycosidase F (PNGase F) has become the gold standard for releasing *N*-glycans from animal protein backbone since it allows analysis of both the released *N*-glycan as well as the deglycosylated peptide/protein.[2] Enriched or purified milk glycoproteins can be de-*N*-glycosylated either in solution[36,43–45] or in gel after electrophoretic separation[38,46] before the released *N*-glycans and degly-cosylated proteins/peptides are analyzed separately. Chemical de-glycosylation using hydrazine has also been used;[32,47] Wilson *et al.* used PNGase F to release *N*-glycans[21] after immobilizing milk proteins on polyvinylidene fluoride (PVDF) membrane. This approach allows a global glycomic profile of the *N*-glycans and the *O*-glycans on the same protein sample to be obtained.[21,48]

There is no commercially available enzyme analogous to PNGase F for the release of *O*-linked glycans.[2] The available *O*-glycanases are all strictly glycan structure specific and thus not applicable for a global release protocol.[49] Chemical methods such as β-elimination or hydrazinolysis are able to release *O*-linked glycans from their protein backbone,[20,50–52] but the protein is usually destroyed and cannot be further analyzed. Reductive β-elimination, performed using sodium borohydride in potassium or sodium hydroxide, releases the *O*-glycans and simultaneously reduces them to protect the oligosaccharides from chemical peeling reactions (degradation of the released glycans). Due to its simplicity and reproducibility, reductive β-elimination is the most commonly used method of release of *O*-glycans from glycoproteins.[53] However, since reductive β-elimination does not allow for subsequent fluorescent labelling (*e.g.* with 2-aminobenzamide, see Section 3.11) of the reducing terminus of *O*-gly-cans, β-elimination using hydrazine has been widely explored to release *O*-glycans whilst retaining the reducing end.[54,55]

3.8 Analysis of Underivatized Glycans using ESI-MS

In our opinion, MS is the best method of direct, sensitive (fmol to pmol range) analysis of underivatized glycans.[11,12,56–58] Analysis of released glycans without further derivatization reduces sample preparation steps and electrospray ioni-zation (ESI) MS has successfully been used for their characterization from milk protein samples.[23,25–28,30,31,59–62]

Certain considerations need to be taken into account if both neutral and acidic glycans are to be analyzed simultaneously by mass spectrometry. In ESI positive mode, neutral and acidic glycans are ionized well in on-line LC-MS conditions, with the signal intensities of each mass corresponding to the relative structure distribution within the sample.[63,64] On-line LC-ESI-MS analysis offers the additional advantage of reducing sample complexity by chromato-graphic separation of the individual oligosaccharides allowing for separate tandem mass spectrometry (MS/MS) fragmentation analyzes.

Porous graphitized carbon (PGC) LC-MS is now the most widely used method for analysing underivatized glycans by LC-ESI-MS; its applications, benefits and limitations for the analysis of glycans have been recently reviewed by Ruhaak *et al.*,[65] with specific approaches described by numerous groups.[21,23,25,27,28,48,65–69] Reduced oligosaccharides are not hydrophobic and are not well retained by reversed phase chromatographic matrices, but they are well retained by PGC, and isobaric structures differing only in linkage can be separated and analyzed independently. For PGC chromatography, it is desirable to reduce both the *N*- and *O*-linked oligosaccharides prior to analysis, since the α- and β-anomers of the same structural isomer are separated on the carbon stationary phase and the resulting chromatogram contains two peaks for each structure.[65] Released *O*-glycans are already in the reduced form if reductive β-elimination release is used, and enzymatically released *N*-glycans or free glycans can be simply reduced with sodium borohydride before analysis.[23,25,70] This step comes with an additional benefit: reduction adds a + 2 Da mass tag on the reducing sugar allowing unambiguous identification of this glycan terminus in MS/MS spectra, thus facilitating their interpretation.

Negative ionization MS is often considered to be less sensitive compared to its positive ion counterpart. However, with negatively charged acidic glycans, negative ionization produces increased low fmol sensitivity and offers an attractive alternative for their detection and analysis.[13,67,69,71–74] Signal intensities of negatively charged glycans under these conditions are usually higher compared with similar amounts of neutral oligosaccharide structures. Nevertheless, both neutral and acidic oligosaccharides ionize well in negative ion mode and relative quantitation can be obtained using normalization factors.[75] Negative ionization PGC LC-ESI-MS analysis also facilitates the detection of sulfated and phosphorylated glycan structures.[76,77] This approach was successfully applied to study the *N*- and *O*-glycans released from milk fat globule membranes of human and bovine milk.[21] Both human and bovine milk fat globule membrane protein-bound oligosaccharides were identified to be highly sialylated, but with distinct differences in core *O*-glycan structures between bovine and human milk. For example, bovine MUC1 at peak lactation presented mono- and disialylated core 1 type *O*-linked structures whereas human MUC1 mainly contained core 2 type structures with different extensions on the C6 branch and sialylation on the C3 branch.[21]

Positive carbon LC-ESI-MS was used by Niñonuevo *et al.* to analyze free oligosaccharides in human milk.[23,25] Tao *et al.* undertook more extensive work on free oligosaccharides in bovine milk using the same chip time-of-flight (chip-ToF) MS system as Niñonuevo *et al.*, but using off-line nano ESI Fourier transform ion cyclotron resonance (FTICR) MS for high mass accuracy profiling in both positive and negative mode.[27,28] They again confirmed cow's milk as being far less complex in terms of oligosaccharide composition and abundance compared with human milk, with 40 identified compositions in bovine milk compared to about 200 found in the human sample.[27] Significant differences in sialylation and fucosylation were found between bovine and human milk free oligosaccharides. Sialylation of the free glycans was more abundant in cows whereas fucosylated oligosaccharides, which make up about 70% of the human structures, was essentially absent in bovine milk.[27]

Wuhrer *et al.* have presented an alternative approach for LC-ESI-MS analysis of released, underivatized, protein-bound glycans using homemade Amide-80 nano columns that allow normal-phase nanoscale separation of oligosaccharides based on polar interactions of the oligosaccharide hydroxyl groups with the stationary phase.[78] The same chromatographic material has often been used for larger scale separations of free milk oligosaccharides.[16,79]

Liquid chromatographic separation is often coupled with online ESI-MS analysis, but significant work on the elucidation of oligosaccharide structures has also been done by off-line, direct injection nano ESI-MS analysis. Nano ESI-MS offers high sensitivity combined with a minimal amount of sample consumption, whilst still providing sufficient MS analysis time.[80,81] Various off-line purification steps precede this type of MS analysis of milk oligosaccharides and this approach has provided useful data on the fragmentation patterns of milk oligosaccharide structures under various ionization conditions.[59,60,82–86] More cross-ring fragmentation cleavages have been found to be produced from negatively charged ions under low collision induced dissociation (CID) conditions compared to positive mode fragmentation.[82,87] Pfenninger *et al.* have evaluated the off-line analysis of milk oligosaccharides by nano electrospray ionization multi stage mass spectrometry (ESI-MS[n]) in negative mode on an ion trap instrument.[59,60] Specific fragmentation patterns were identified and used for the determination of isomeric mixtures of milk oligosaccharides. Chai *et al.* performed their analyzes on oligosaccharides using a Q-ToF instrument, also showing that specific MS/MS fragmentation masses can assist in determining the branching pattern of milk oligosaccharides.[82–85,88] Definition of general fragmentation rules for oligosaccharides is thus seen to be very useful for structural assignment, but varies with ionization conditions and between instruments.

3.9   Analysis of Underivatized Glycans by MALDI-MS

Matrix-assisted laser desorption ionization (MALDI) MS is an often used, contamination-tolerant and robust method for analysis of oligosaccharides without chemical modification.[30,31,56,57,61,89] In contrast to ESI, where singly and multiple protonated $[M + xH]^{x+}$ and deprotonated $[M - yH]^{y-}$ ions are the major ions produced during the positive and negative ionization process respectively, singly charged metal adduct ions are primarily formed in MALDI-MS analysis of oligosaccharides.[12] Sodiated ions $[M + Na]^+$ are generally most abundant, but this can be shifted towards different cationic ions by doping the matrix with the appropriate salt.[12] Sialylated ions usually produce a mixture of ions $\{[M - H]^-, [M + Na]^+, [M - nH + (n + 1Na)]^+\}$ due to salt formation and are also more susceptible to in-source and post-source fragmentation with the concomitant loss of sialic acid.[12,13] In the use of reflectron ToF instruments, post-source fragmentation is reported to produce metastable peaks.[12] More detail about the theory of MALDI ionization with regard to oligosaccharides can be read in the excellent reviews by Harvey and Zaia.[12,13,90,91]

Though a large number of different MALDI matrices have been published for oligosaccharide analysis, 2,5-dihydroxybenzoic acid (DHB) remains the most commonly used.[11] Matrices more suitable for sialylated oligosaccharides are 2,4,6-trihydroxyacetophenone (THAP), which has been shown to give best results when mixed with ammonium citrate.[92] The matrix 2,6-azathiothymine has also successfully been used for both sialylated and neutral oligosaccharide analysis by MALDI-MS. In addition, as 2,6-azathiothymine is a non-acidic matrix, different exoglycosidase treatments used for determining specific linkages can be applied directly to the oligosaccharides on the MALDI plate.[93] Pfenninger *et al.* undertook a thorough study of different matrices and conditions to optimize conditions for milk oligosaccharide analysis, determining that DHB, 3-aminoquinoline, 2,6-azathiothymine and 5-chloromercapto-benzothiazole are the most suitable ones for neutral oligosaccharides.[61] DHB was the only matrix that was improved by additives such as NaCl. In their hands, 2,6-azathiothymine with diammonium hydrogencitrate (DAHC) as an additive, was their preferred matrix for acidic oligosaccharides.[61] Tzeng *et al.* applied alkali-hydroxide doped matrices for the structural characterization of neutral underivatized oligosaccharides using MALDI-ToF-MS.[94] The partial alkaline degradation that occurred upon laser desorption/ionization facilitated their structural characterization by PSD-ToF-MS.[94]

A slightly different approach in terms of pre-MALDI sample preparation was taken by Dreisewerd *et al.*, in which the native glycans were separated by high performance thin layer chromatography (TLC) before MALDI analysis, with the TLC plate used directly as the MALDI target.[58] Using glycerol as a liquid matrix, a homogenous wetting of the silica gel was achieved before the sample was ionized using an infrared laser and a Q-ToF as a mass analyzer. They also succeeded in using a 337 nm UV-laser by applying a composite matrix of α-cyano-4-hydroxycinnamic acid and glycerol to the plate. Using this novel TLC separation step before MS analysis, Dreisewerd *et al.* were able to establish "mobility profiles" by scanning the laser beam across the separated analyte bands.[58]

3.10   Derivatization of Glycans for MS Analysis

Though both MALDI and ESI ionization techniques can be used in the analysis of underivatized glycans, chemical modifications of the glycans can provide the oligosaccharides with different MS properties. Permethylation of glycans is one widely used process that methylates all hydroxyl groups on the monosaccharide components of the glycans.[95] Besides making the hydrophilic molecule more hydrophobic, this step also neutralizes the negative charge of sialic acids and thus affects their ionization behaviour.[11,13] Sialic acids are stabilized after permethylation and do not show the in-source and post-source fragmentation often detected for unmodified glycans in MALDI-MS.[13] Permethylation allows direct comparison of signal intensities for relative quantitation of different structures, since neutral oligosaccharide signal intensities by MALDI-ToF-MS can be correlated with their relative amount in a sample.[11] For data interpretation, it needs to be considered that permethylation results in an overall mass increase for every glycan structure.[3]

Pfenninger *et al.* used MALDI-FTICR-MS to take a closer look at large, permethylated, free milk oligosaccharides; the high resolution and mass accuracy of this instrument was key to being able to distinguish between multiple *N*-acetyl lactosamine repeats and highly fucosylated structures of oligosaccharides with up to ten fucose residues and seven *N*-acetyllactosamine repeats.[22] Parry *et al.* used this approach to investigate the *N*-glycans of MUC1 from epithelial cells and milk, and found a marked difference in glycosylation of the membrane-bound and the secreted form of MUC1.[45] Using purified human milk oligosaccharides as standards, Mechref *et al.* showed that per-methylation of oligosaccharides was beneficial for their MALDI-ToF/ToF-MS/MS approach to distinguish between different sialic acid linkages by distinct diagnostic ions.[96] Permethylation of glycans is known to facilitate interpretation and structural determination of glycans by CID MS/MS analysis as internal fragment ions produce unique mass values.[11,13,97,98] Reinhold's group [97,99–101] uses off-line injection of permethylated glycans in combination with MS[n] analysis to accomplish detailed structural assignment of oligo-saccharides. Compared with LC-ESI-MS approaches that use on-line separa-tion before analysis, the total glycan pool is analyzed simultaneously during off-line injection, making it more challenging to distinguish between isobaric structures—especially if there is a considerable concentration difference between structures present in the mixture.

Sulfated sugars so far have only been reported in dog milk,[102] but per-methylation conditions usually used for MS analysis of oligosaccharides do not "neutralize" the negative charge of sulfated glycans and these structures are often overlooked by positive mode MALDI-MS.[103] Recently published approaches from Lei *et al.* and Yu *et al.* address this issue.[104,105] Lei *et al.* applied a sequential process of permethylation using methyl iodide, de-sulfation and subsequent deuteromethyl iodide methylation to identify the degree and position of sulfation on oligosaccharides by MALDI-ToF-MS.[104] Yu *et al.* permethy-lated the sulfate on the glycans using the sodium hydroxide slurry method, which also allowed them to distinguish between sulfated and phosphorylated glycans; they used high energy CID on a MALDI-ToF/ToF instrument to obtain sequence and linkage information of their glycans by tandem MS.[105]

3.11 MS Analysis of Reductively Aminated Glycans

Before modern day mass spectrometric instruments became widely available, reductive amination was a much used method to label glycans for analysis by liquid chromatography and fluorescent detection.[106,107] Numerous compounds have been described in the literature for labelling and analysing reductively aminated oligosaccharides and most derivatives are compatible with mass spectrometric detection.[108–113] Labelling with a chemical compound changes the properties of the glycans in many ways and can, in some cases, have ben-eficial effects that improve analysis:[114–116]

- glycans can be made more hydrophobic;
- a charge can be introduced;

- sensitive detection by fluorescence can be used; and
- ionization can be enhanced.

However, reductive amination requires an large excess of the labelling reagent for the reaction to proceed which needs to be removed before further analysis.[108] Unwanted modifications to the glycan and sample losses may also occur under the derivatization conditions.

Pabst *et al.* performed a comprehensive comparative study of 15 compounds frequently used for oligosaccharide labelling, comparing the fluorescence properties as well as the MS and LC performance. In their hands, the differences between native and labelled oligosaccharides using ESI (positive and negative mode) were minor, with the most sensitive label resulting in a ion intensity signal roughly only twice that of the underivatized glycans.[108] A similar result was obtained for MALDI-ToF positive ion mode analysis, with the spot preparation and selection of "hot spots" having more influence on the signal intensity than the derivatization labels. However, in MALDI-ToF analysis using negative ionization and THAP as a matrix, derivatization of the oligosaccharides with labels containing one acidic group performed significantly better than labels containing three acidic groups.[108]

MALDI-MS of milk oligosaccharides labelled with 2-aminopyridine has been performed by Suzuki *et al.* using MALDI-QIT-ToF-MS, allowing the differentiation of linkage isomers of isobaric structures.[117] The MS^n capabilities of this instrument were of considerable advantage to this analysis. Kinoshita *et al.* analyzed the high molecular fraction of seal milk by a combination of offline HPLC, MALDI-ToF MS and sequential exoglycosidase digestion of the 2-aminobenzamide labelled glycans.[118] High sensitivity detection of *N,N*-dimethylated milk glycans down to 30 fmol was reported by Broberg *et al.* using PGC on-line separation.[119]

3.12   Conclusions

The introduction of the soft ionization techniques of ESI and MALDI initiated an avalanche in biological MS technology and application development. The tremendous progress made in ion detection and fragmentation options accompanying this expansion has allowed a new dimension, not just of sensitivity and selectivity, but also of the range of biomolecules that can be analyzed by MS. The ultimate capability of this technology has yet to be fully realized.

Despite these developments in instrumentation, the first crucial step for a successful MS analysis remains in the preparation of the sample. Nevertheless, many of the MS analytical approaches described for milk in this review can equally be applied to the analysis of free and protein-bound oligosaccharides from different nutritional sources.

References

1. M. Mann, R. C. Hendrickson and A. Pandey, *Annu. Rev. Biochem.*, 2001, 437.

2. A. Varki, R. D. Cummings and G. Hart, *Essentials of Glycobiology*, Cold Spring Harbor Laboratory Press, New York, 2nd Edition, 2002, p. 101–127.
3. A. Dell, *Methods Enzymol.*, 1990, **193**, 647.
4. B. Lindberg and J. Lonngren, *Methods Enzymol.*, 1978, **50**, 3.
5. R. Townsend, in *Carbohydrate Analysis: High Performance Liquid Chromatography and Capillary Electrophoresis*, ed. Z. E. Rassi, Elsevier, Amsterdam, 1995, pp. 181–209.
6. G. Boehm and B. Stahl, *J. Nutr.*, 2007, **137**(3 Suppl 2), 847S.
7. D. S. Newburg, *J. Anim. Sci.*, 2009, **87**(13 Suppl), 26.
8. D. S. Newburg, *J. Nutr.*, 2005, **135**, 1308.
9. S. Salvatore, B. Hauser, T. Devreker, S. Arrigo and Y. Vandenplas, *Nutrition*, 2008, **24**, 1205.
10. R. E. Ward, M. Ninonuevo, D. A. Mills, C. B. Lebrilla and J. B. German, *Mol. Nutr. Food Res.*, 2007, **51**, 1398.
11. D. J. Harvey, *Expert Rev. Proteomics*, 2005, **2**, 87.
12. D. J. Harvey, *Proteomics*, 2005, **5**, 1774.
13. J. Zaia, *Mass Spectrom. Rev.*, 2004, **23**, 161.
14. B. Casado, M. Affolter and M. Kussmann, *J. Proteomics*, 2009, **73**, 196.
15. T. Urashima, T. Saito, T. Nakamura and M. Messer, *Glycoconj. J.*, 2001, **18**, 357.
16. T. Urashima, in *Experimental Glycoscience: Glycochemistry*, ed. N. Taniguchi, A. Suzuki, Y. Ito, H. Narimatsu, T. Kawasaki and S. Hase, Springer, New York, 2008, pp. 82–86.
17. M. Rivero-Urgell and A. Santamaria-Orleans, *Early Hum. Dev.*, 2001, **65**(Suppl), S43.
18. C. Kunz and S. Rudloff, *Acta Paediatr.*, 1993, **82**, 903.
19. P. Chaturvedi, C. D. Warren, M. Altaye, A. L. Morrow, G. Ruiz-Palacios, L. K. Pickering and D. S. Newburg, *Glycobiology*, 2001, **11**, 365.
20. P. H. Jensen, D. Kolarich and N. H. Packer, *FEBS J.*, 2010, **277**, 81.
21. N. L. Wilson, L. J. Robinson, A. Donnet, L. Bovetto, N. H. Packer and N. G. Karlsson, *J. Proteome Res.*, 2008, **7**, 3687.
22. A. Pfenninger, S.-Y. Chan, M. Karas, B. Finke, B. Stahl and C. E. Costello, *Int. J. Mass Spectrom.*, 2008, **278**, 129.
23. M. Ninonuevo, H. An, H. Yin, K. Killeen, R. Grimm, R. Ward, B. German and C. Lebrilla, *Electrophoresis*, 2005, **26**, 3641.
24. M. R. Ninonuevo, Y. Park, H. Yin, J. Zhang, R. F. Ward, B. H. Clowers, J. B. German, S. L. Freeman, K. Killeen, R. Grimm and C. B. Lebrilla, *J. Agric. Food Chem.*, 2006, **54**, 7471.
25. M. R. Ninonuevo, P. D. Perkins, J. Francis, L. M. Lamotte, R. G. LoCascio, S. L. Freeman, D. A. Mills, J. B. German, R. Grimm and C. B. Lebrilla, *J. Agric. Food Chem.*, 2008, **56**, 618.
26. M. R. Ninonuevo, R. E. Ward, R. G. LoCascio, J. B. German, S. L. Freeman, M. Barboza, D. A. Mills and C. B. Lebrilla, *Anal. Biochem.*, 2007, **361**, 15.
27. N. Tao, E. J. DePeters, S. Freeman, J. B. German, R. Grimm and C. B. Lebrilla, *J. Dairy Sci.*, 2008, **91**, 3768.

28. N. Tao, E. J. DePeters, J. B. German, R. Grimm and C. B. Lebrilla, *J. Dairy Sci.*, 2009, **92**, 2991.

29. P. Chaturvedi, C. D. Warren, G. M. Ruiz-Palacios, L. K. Pickering and D. S. Newburg, *Anal. Biochem.*, 1997, **251**, 89.

30. B. Finke, B. Stahl, A. Pfenninger, M. Karas, H. Daniel and G. Sawatzki, *Anal. Chem.*, 1999, **71**, 3755.

31. B. Stahl, S. Thurl, J. Zeng, M. Karas, F. Hillenkamp, M. Steup and G. Sawatzki, *Anal. Biochem.*, 1994, **223**, 218.

32. G. Spik, G. Strecker, B. Fournet, S. Bouquelet, J. Montreuil, L. Dorland, H. van Halbeek and J. F. Vliegenthart, *Eur. J. Biochem.*, 1982, **121**, 413.

33. C. J. Hogarth, J. L. Fitzpatrick, A. M. Nolan, F. J. Young, A. Pitt and P. D. Eckersall, *Proteomics*, 2004, **4**, 2094.

34. D. J. Palmer, V. C. Kelly, A. M. Smit, S. Kuy, C. G. Knight and G. J. Cooper, *Proteomics*, 2006, **6**, 2208.

35. L. Blackberg and O. Hernell, *Eur. J. Biochem.*, 1981, **116**, 221.

36. E. Landberg, P. Pahlsson, H. Krotkiewski, M. Stromqvist, L. Hansson and A. Lundblad, *Arch. Biochem. Biophys.*, 1997, **344**, 94.

37. R. M. Jones, F. Schweikart, S. Frutiger, J. C. Jaton and G. J. Hughes, *Biochim. Biophys. Acta*, 1998, **1429**, 265.

38. J. Charlwood, S. Hanrahan, R. Tyldesley, J. Langridge, M. Dwek and P. Camilleri, *Anal. Biochem.*, 2002, **301**, 314.

39. T. A. Reinhardt, A. G. Filoteo, J. T. Penniston and R. L. Horst, *Am. J. Physiol., Cell. Physiol.*, 2000, **279**, C1595.

40. T. A. Reinhardt and J. D. Lippolis, *J. Dairy Sci.*, 2008, **91**, 2307.

41. S. Quaranta, M. G. Giuffrida, M. Cavaletto, C. Giunta, J. Godovac-Zimmermann, B. Canas, C. Fabris, E. Bertino, M. Mombro and A. Conti, *Electrophoresis*, 2001, **22**, 1810.

42. D. Fortunato, M. G. Giuffrida, M. Cavaletto, L. P. Garoffo, G. Della-valle, L. Napolitano, C. Giunta, C. Fabris, E. Bertino, A. Coscia and A. Conti, *Proteomics*, 2003, **3**, 897.

43. G. J. Hughes, A. J. Reason, L. Savoy, J. Jaton and S. Frutiger-Hughes, *Biochim. Biophys. Acta*, 1999, **1434**, 86.

44. Y. Mechref, P. Chen and M. V. Novotny, *Glycobiology*, 1999, **9**, 227.

45. S. Parry, F. G. Hanisch, S. H. Leir, M. Sutton-Smith, H. R. Morris, A. Dell and A. Harris, *Glycobiology*, 2006, **16**, 623.

46. L. Royle, A. Roos, D. J. Harvey, M. R. Wormald, D. van Gijlswijk-Janssen, E. -R. M. Redwan, I. A. Wilson, M. R. Daha, R. A. Dwek and P. M. Rudd, *J. Biol. Chem.*, 2003, **278**, 20140.

47. A. Mizoguchi, T. Mizuochi and A. Kobata, *J. Biol. Chem.*, 1982, **257**, 9612.

48. N. L. Wilson, B. L. Schulz, N. G. Karlsson and N. H. Packer, *J. Proteome Res.*, 2002, **1**, 521.

49. Y. Endo and A. Kobata, *J. Biochem.*, 1976, **80**, 1.

50. A. Pierce-Cretel, J. P. Decottignies, J. M. Wieruszeski, G. Strecker, J. Montreuil and G. Spik, *Eur. J. Biochem.*, 1989, **18**, 457.

51. A. Pierce-Cretel, M. Pamblanco, G. Strecker, J. Montreuil and G. Spik, *Eur. J. Biochem.*, 1981, **114**, 169.

52. H. van Halbeek, J. F. Vliegenthart, A. M. Fiat and P. Jolles, *FEBS Lett.*, 1985, **187**, 81.

53. Y. C. Lee, and K. G. Rice, in *Glycobiology: A Practical Approach*, ed. M. Fukuda and A. Kobata, Oxford University Press, New York, 1993, pp. 127–163.

54. Y. Huang, Y. Mechref and M. V. Novotny, *Anal. Chem.*, 2001, **73**, 6063.

55. N. Kuraya and S. Hase, *J. Biochem.*, 1992, **112**, 122.

56. D. Kolarich and F. Altmann, *Anal. Biochem.*, 2000, **285**, 64.

57. B. Kuster, A. P. Hunter, S. F. Wheeler, R. A. Dwek and D. J. Harvey, *Electrophoresis*, 1998, **19**, 1950.

58. K. Dreisewerd, S. Kolbl, J. Peter-Katalinic, S. Berkenkamp and G. Pohlentz, *J. Am. Soc. Mass Spectrom.*, 2006, **17**, 139.

59. A. Pfenninger, M. Karas, B. Finke and B. Stahl, *J. Am. Soc. Mass Spectrom.*, 2002, **13**, 1341.

60. A. Pfenninger, M. Karas, B. Finke and B. Stahl, *J. Am. Soc. Mass Spectrom.*, 2002, **13**, 1331.

61. A. Pfenninger, M. Karas, B. Finke, B. Stahl and G. Sawatzki, *J. Mass Spectrom.*, 1999, **34**, 98.

62. Y. Xie and C. B. Lebrilla, *Anal. Chem.*, 2003, **75**, 1590.

63. M. Pabst and F. Altmann, *Anal. Chem.*, 2008, **80**, 7534.

64. L. Huang and R. M. Riggin, *Anal. Chem.*, 2000, **72**, 3539.

65. L. R. Ruhaak, A. M. Deelder and M. Wuhrer, *Anal. Bioanal. Chem.*, 2009, **394**, 163.

66. N. G. Karlsson and N. H. Packer, *Anal. Biochem.*, 2002, **305**, 173.

67. N. G. Karlsson, B. L. Schulz and N. H. Packer, *J. Am. Soc. Mass Spectrom.*, 2004, **15**, 659.

68. N. G. Karlsson, B. L. Schulz, N. H. Packer and J. M. Whitelock, *J. Chromatogr. B Analyt. Technol. Biomed. Life Sci.*, 2005, **824**, 139.

69. N. G. Karlsson, N. L. Wilson, H. J. Wirth, P. Dawes, H. Joshi and N. H. Packer, *Rapid Commun. Mass Spectrom.*, 2004, **18**, 2282.

70. P. H. Jensen, N. G. Karlsson, D. Kolarich and N. H. Packer, in *Methods in Molecular Biology, Therapeutic Glycoproteins*, ed. T. Stadheim and H. Li, Humana Press, in press.

71. D. J. Harvey, *J. Am. Soc. Mass Spectrom.*, 2005, **16**, 647.

72. D. J. Harvey, *J. Am. Soc. Mass Spectrom.*, 2005, **16**, 631.

73. D. J. Harvey, *J. Am. Soc. Mass Spectrom.*, 2005, **16**, 622.

74. J. Peter-Katalinic, *Mass Spectrom. Rev.*, 1994, **13**, 77.

75. F. J. Olson, M. Backstrom, H. Karlsson, J. Burchell and G. C. Hansson, *Glycobiology*, 2005, **15**, 177.

76. N. G. Karlsson and K. A. Thomsson, *Glycobiology*, 2009, **19**, 288.

77. K. A. Thomsson, B. L. Schulz, N. H. Packer and N. G. Karlsson, *Glycobiology*, 2005, **15**, 791.

78. M. Wuhrer, C. A. Koeleman, A. M. Deelder and C. H. Hokke, *Anal. Chem.*, 2004, **76**, 833.

79. T. Urashima, T. Saito, K. Ohmisya and K. Shimazaki, *Biochim. Biophys. Acta*, 1991, **1073**, 225.
80. D. Sagi, J. Peter-Katalinic, H. S. Conradt and M. Nimtz, *J. Am. Soc. Mass Spectrom.*, 2002, **13**, 1138.
81. S. Y. Vakhrushev, M. Mormann and J. Peter-Katalinic, *Proteomics*, 2006, **6**, 983.
82. W. Chai, V. Piskarev and A. M. Lawson, *Anal. Chem.*, 2001, **73**, 651.
83. W. Chai, V. E. Piskarev, B. Mulloy, Y. Liu, P. G. Evans, H. M. Osborn and A. M. Lawson, *Anal. Chem.*, 2006, **78**, 1581.
84. W. Chai, V. Piskarev and A. M. Lawson, *J. Am. Soc. Mass Spectrom.*, 2002, **13**, 670.
85. H. Kogelberg, V. E. Piskarev, Y. Zhang, A. M. Lawson and W. Chai, *Eur. J. Biochem.*, 2004, **271**, 1172.
86. J. L. Seymour, C. E. Costello and J. Zaia, *J. Am. Soc. Mass Spectrom.*, 2006, **17**, 844.
87. D. J. Harvey, L. Royle, C. M. Radcliffe, P. M. Rudd and R. A. Dwek, *Anal. Biochem.*, 2008, **376**, 44.
88. W. Chai, V. E. Piskarev, Y. Zhang, A. M. Lawson and H. Kogelberg, *Arch. Biochem. Biophys.*, 2005, **434**, 116.
89. D. Rendic, I. B. Wilson, G. Lubec, M. Gutternigg, F. Altmann and R. Leonard, *Electrophoresis*, 2007, **28**, 4484.
90. D. J. Harvey, *Mass Spectrom. Rev.*, 1999, **18**, 349.
91. D. J. Harvey, *Int. J. Mass Spectrom.*, 2003, **226**, 1.
92. D. I. Papac, A. Wong and A. J. Jones, *Anal. Chem.*, 1996, **68**, 3215.
93. H. Geyer, S. Schmitt, M. Wuhrer and R. Geyer, *Anal. Chem.*, 1999, **71**, 476.
94. Y. K. Tzeng, Z. Zhu and H. C. Chang, *J. Mass Spectrom.*, 2009, **44**, 375.
95. A. K. Powell and D. J. Harvey, *Rapid Commun. Mass Spectrom.*, 1996, **10**, 1027.
96. Y. Mechref, P. Kang and M. V. Novotny, *Rapid Commun. Mass Spectrom.*, 2006, **20**, 1381.
97. D. J. Ashline, A. J. Lapadula, Y. H. Liu, M. Lin, M. Grace, B. Pramanik and V. N. Reinhold, *Anal. Chem.*, 2007, **79**, 3830.
98. D. M. Sheeley and V. N. Reinhold, *Anal. Chem.*, 1998, **70**, 3053.
99. D. Ashline, S. Singh, A. Hanneman and V. Reinhold, *Anal. Chem.*, 2005, **77**, 6250.
100. A. J. Lapadula, P. J. Hatcher, A. J. Hanneman, D. J. Ashline, H. Zhang and V. N. Reinhold, *Anal. Chem.*, 2005, **77**, 6271.
101. H. Zhang, S. Singh and V. N. Reinhold, *Anal. Chem.*, 2005, **77**, 6263.
102. W. A. Bubb, T. Urashima, K. Kohso, T. Nakamura, I. Arai and T. Saito, *Carbohydr. Res.*, 1999, **318**, 123.
103. J. Mitoma, X. Bao, B. Petryanik, P. Schaerli, J. M. Gauguet, S. Y. Yu, H. Kawashima, H. Saito, K. Ohtsubo, J. D. Marth, K. H. Khoo, U. H. von Andrian, J. B. Lowe and M. Fukuda, *Nat. Immunol.*, 2007, **8**, 409.
104. M. Lei, Y. Mechref and M. V. Novotny, *J. Am. Soc. Mass Spectrom.*, 2009, **20**, 1660.

105. S. Y. Yu, S. W. Wu, H. H. Hsiao and K. H. Khoo, *Glycobiology*, 2009, **19**(10), 1136.
106. S. Hase, T. Ikenaka and Y. Matsushima, *J. Biochem.*, 1979, **85**, 989.
107. S. Hase, T. Ikenaka and Y. Matsushima, *J. Biochem.*, 1979, **85**, 995.
108. M. Pabst, D. Kolarich, G. Poltl, T. Dalik, G. Lubec, A. Hofinger and F. Altmann, *Anal. Biochem.*, 2009, **384**, 263.
109. J. C. Bigge, T. P. Patel, J. A. Bruce, P. N. Goulding, S. M. Charles and R. B. Parekh, *Anal. Biochem.*, 1995, **230**, 229.
110. D. J. Harvey, *Rapid Commun. Mass Spectrom.*, 2000, **14**, 862.
111. K. R. Anumula, *Anal. Biochem.*, 1994, **220**, 275.
112. J. Charlwood, J. M. Skehel and P. Camilleri, *Anal. Biochem.*, 2000, **284**, 49.
113. W. Morelle, M. C. Slomianny, H. Diemer, C. Schaeffer, A. van Dorsselaer and J. C. Michalski, *Rapid Commun. Mass Spectrom.*, 2005, **19**, 2075.
114. M. Seveno, G. Cabrera, A. Triguero, C. Burel, J. Leprince, C. Rihouey, L. P. Vezina, M. A. D'Aoust, P. M. Rudd, L. Royle, R. A. Dwek, D. J. Harvey, P. Lerouge, J. A. Cremata and M. Bardor, *Anal. Biochem.*, 2008, **379**, 66.
115. A. Klein, A. Lebreton, J. Lemoine, J. M. Perini, P. Roussel and J. C. Michalski, *Clin. Chem.*, 1998, **44**, 2422.
116. D. J. Harvey, *J. Am. Soc. Mass Spectrom.*, 2000, **11**, 900.
117. Y. Suzuki, M. Suzuki, E. Ito, H. Ishii, K. Miseki and A. Suzuki, *Glycoconj. J.*, 2005, **22**, 427.
118. M. Kinoshita, H. Ohta, K. Higaki, Y. Kojima, R. Urashima, K. Nakajima, M. Suzuki, K. M. Kovacs, C. Lydersen, T. Hayakawa and K. Kakehi, *Anal. Biochem.*, 2009, **388**, 242.
119. A. Broberg, *Carbohydr. Res.*, 2007, **342**, 1462.

CHAPTER 4

Mass Spectrometry in Protein, Peptide and Amino Acid Analysis

CLAUDIO CORRADINI, LISA ELVIRI AND
ANTONELLA CAVAZZA

Dipartimento di Chimica Generale e Inorganica, Chimica Analitica, Chimica
Fisica, Università di Parma, Parma, Italy

4.1 Proteins in Food

The importance of proteins, peptides and amino acids in the diet has been increasingly acknowledged as a result of new scientific findings in the field of nutrition over the last two decades. These compounds contribute to the physical properties, biological activities and sensory characteristics of food. Such functional properties can be related to their ability to form viscoelastic networks, bind water, emulsify fat and oil, entrap flavours and form stable foams. In addition, the denaturation levels of food proteins are closely related to the sensory and technological properties of food products. Moreover, the breakdown of proteins to smaller peptides and free amino acids during proteolysis influences the quality of fermented or otherwise ripened dairy products, fish, meat, cereals and vegetables. Texture and physical properties change, amino acids and peptides are released, and volatile aroma components can be formed by catabolism of proteolysis products. Numerous studies on the isolation and characterization of peptides and proteins are conducted to understand relationships between peptide structures and their activities in foods.[1]

RSC Food Analysis Monographs No. 9
Mass Spectrometry and Nutrition Research
Edited by Laurent B. Fay and Martin Kussmann
© The Royal Society of Chemistry 2010
Published by the Royal Society of Chemistry, www.rsc.org

Furthermore, proteins are widely used in formulated food due their high nutritional value and because they are "generally recognized as safe" (GRAS). The functional properties of food proteins such as emulsification, gelation, foaming and water binding, make them convenient and widely used ingredients in the food industry.[2–4]

The labels of many foods now list the amount but not the quality of protein in the formulation. Because of the current emphasis on the need for labelling foods for both the quantity and nutritional quality of protein and because many food proteins are adversely affected by food processing, further development of methods for evaluation of protein quality is desirable. From the nutritional and functional points of view, proteins are the most essential components of food.

Recently, nutrigenomics has emerged as a new field that focuses on the study of the interaction between nutrition and the human genome; this includes the tools and applications of genomic technologies for the investigation of nutritional issues.[5] The generic name of "genomic technologies" includes the investigation of transcriptomes, proteomes and metabolomes—all of them subsequent steps of gene expression which define the biochemical status of the cell, tissue or organism investigated. The study of these complex interactions requires the development of advanced analytical approaches combined with bioinformatics.[6] Accordingly, the application of proteomics to the study of the nutritional components of food is closely linked to the analysis of previous steps in gene expression (transcriptomics). It is also necessary to understand the repercussions that, at the nutritional level, may derive from either changes in protein concentrations or post-translational protein modifications.

4.2   Mass Spectrometry in Food Protein Analysis

Improvements in instrumentation, advances in online separation techniques and in data processing have contributed to the great expansion of MS in food-related analysis and have promoted MS as an indispensable tool in peptide and protein characterization.

The growing availability on the market of novel instrumentations with high sensitivity and resolution, together with database generation, allows expansion of the concept of protein analysis. Proteomics is now a concept involving several aspects of protein analysis—the study of the protein structure, localization, modification, interactions and functions. Whereas MS-based proteomics techniques are well demonstrated to be fundamental tools for cancer and other medical and biological research, their powerful role in food analysis is just emerging.

Several applications of MS techniques in food analysis have recently been successful. An important example indicating the potential of MS-based proteomics is the capability to identify and quantify food allergens. Food allergy is an increasingly important issue for the consumers and, therefore, for the food industry. MS-based methods allowed the identification of several allergenic proteins [NB enzyme-linked immunosorbent assays (ELISAs) often do not work in food matrices] in different food samples and can be used as techniques complementary to commonly used immunoassays to confirm the presence of hidden allergens in complex matrices (Figure 4.1).[7,8]

The analysis of the complete proteome of foods can characterize foods with unique quality, food produced by new technologies or food originating from specific geographical areas (authenticity). Proteomics has also been used in market surveillance of genetically modified foods. Another new challenge for proteomics has recently been recognized in the study of molecular interactions and differences in food proteomes relevant for nutrition.

As in the case of food analysis, proteomic techniques present several advantages and can provide useful information to identify proteins in food matrix, to study protein–protein interactions in raw and processed foods, as well as interactions between proteins and other food components, or to identify covalently bound constituents that may be produced during processing. Additionally, the presence and the relative ratios of bioactive compounds (ingredients for functional foods) can be screened by proteomic techniques.

4.3 Gel-based Proteomics

To address proteomics effectively, a full range of analytical biochemical methodologies has to be employed, namely:

- differential centrifugation;
- immunoprecipitation;
- one- and two-dimensional gel electrophoresis; and
- various types of chromatography.

The traditional approach used in proteomics to separate several thousands of proteins in complex mixtures is based on two-dimensional (2D) gel electrophoresis. In one dimension, the proteins are separated by charge up to their isoelectric point with immobilized pH gradients that provide high resolution, reproducibility and loading capacity for preparative purposes; in the second dimension, the proteins are separated by mass. The protein-containing gel can then be stained to reveal the presence of the different proteins for subsequent spot excision and identification by MS.[9] However, any given proteome exhibits the problem of a wide dynamic range (from $g\,mL^{-1}$ to $pg\,mL^{-1}$ or less). Thus, effective pre-analysis fractionation is required in order to observe the low abundant proteins. Isoelectric focusing (IEF) separation enables post-translationally modified forms of the same protein (*i.e.* phosphorylation of serine, threonine and tyrosine) to be separated from each other. Differential gel electrophoresis (DIGE) technology can reveal directly the effect of a treatment *versus* a control procedure. In this methodology, instead of running the samples on two different gels, the proteins labelled separately with the different dyes are pooled and run on the same gel.

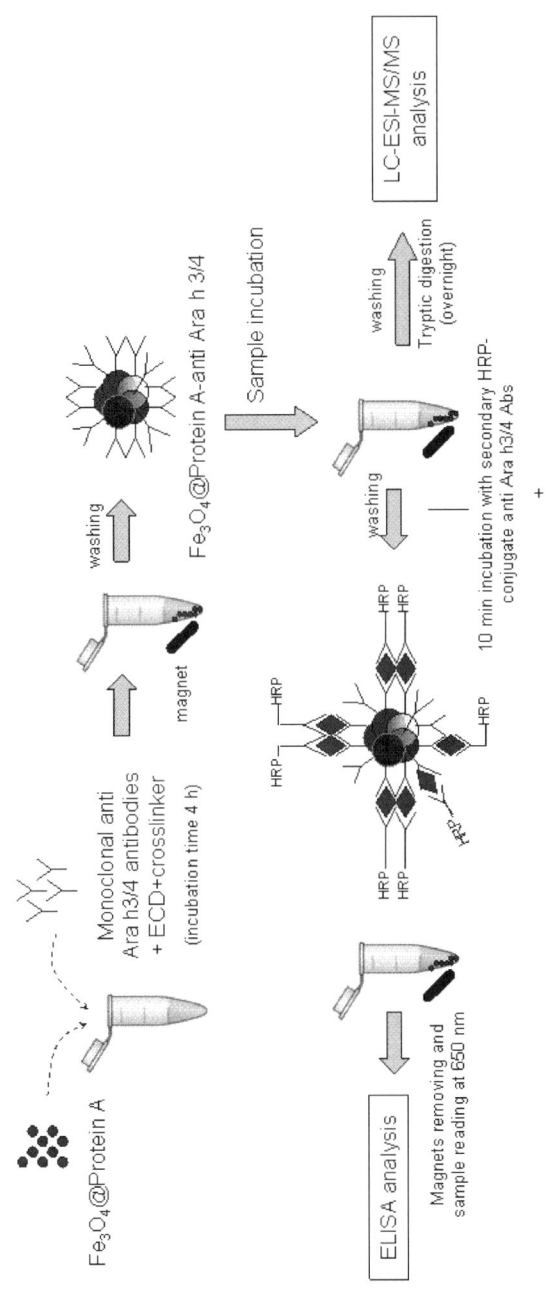

Figure 4.1 MS-based methods can identify allergenic proteins in food samples and are complementary to commonly used immunoassays. Scheme of an analytical protocol: Fe$_3$O$_4$@protein A-anti Ara h3/4 IgG as affinity probes for selective enrichment and determination of Ara h3/4 protein from a food sample using LC-ESI-MS/MS and ELISA methods (from ref. 8).

After separation, the protein spot is excised, the protein is digested with a protease (usually trypsin) and the digested peptides are then analyzed by MS (bottom–up approach). New possibilities under development include the use of trypsin digestion of proteins *in situ* on the gel, followed by blotting and MS analysis of whole gels, which may lead to an increase in reproducibility.[10]

4.4   Liquid Chromatography and Mass Spectrometry in Proteomics

A standard approach used in proteomics to separate proteins in a complex mixture is based on various liquid chromatography (LC) separation approaches, including:

- size exclusion chromatography (SEC);
- reversed phase LC (RPLC);
- strong cation exchange (SCX); and
- affinity chromatography (AC).

More recently, ultra-high performance LC (UPLC) was introduced for efficient and rapid peptide separation. In particular, a Multidimensional Protein Identification Technology (MudPIT approach), based on the 2D-LC separation of a proteolytic digest, has been successfully introduced in proteomic analysis. In this "shotgun" approach peptides are eluted from a strong cation exchange phase and immediately captured on an analytical reverse-phase column before electrospray ionization tandem MS analysis.[11] MudPIT has been combined with automated software (*e.g.* Mascot, SEQUEST) for the identification of peptide sequences. Each peptide is first examined to determine its molecular weight. A statistical scoring scheme is then used to identify which peptide MS/MS spectrum most closely matches that of the unknown peptide. Apart from protein identification, MudPIT methodology has been widely applied to identify post-translational protein modifications.[12]

Proteomics has been enhanced by the development of two ionization sources—matrix-assisted laser desorption ionization (MALDI) and electrospray ionization (ESI)—because they cause little or no fragmentation of the molecule during the ionization and desorption processes. Proteolytic digestion of gel-separated proteins into peptides followed by mass analysis of the peptides provides a peptide mass fingerprint. Modern MALDI time-of-flight (ToF) instruments can measure the molecular weight of each observable peptide with an accuracy in the low parts per million (ppm). However, the complexity of proteomes and today's size of proteomic and genomic databases require confirmation of the peptide identity by sequence information derived from tandem mass spectrometric (MS/MS) analysis.

The development of innovative mass spectrometric techniques such as FTICR (Fourier transform ion cyclotron resonance) or linear ion trap mass analyzers enables the so-called "top–down" proteomics approach,[13] *i.e.* direct analysis of intact proteins. This method circumvents some of the steps described above, in particular 2D electrophoresis and 2D-LC tandem MS, but to be effective a primary separation of proteins is still required.[14] In the top–down approach, multiply charged, intact protein molecular ions are fragmented in the gas phase by using electron capture dissociation (ECD) or electron transfer dissociation (ETD). Upon capture of a low-energy ($<1\,eV$) electron by an ESI-generated polycation, cleavage of the N–C_α backbone bonds occurs in ECD within 10^{-12} seconds. The key to success of this approach is the selection of protein molecular ions and the identification of the ion states of the resulting fragment ions.

While obtaining intact gas phase protein ions even with a high degree of modification is possible with modern soft ionization techniques, MALDI and ESI mass spectrometric post-translational modification (PTM) analysis remains a challenge. This is mainly due to the lability in the gas phase of many modifications, which are easily lost under vibrational (collisional or infrared) excitation without revealing their positions. At the same time, the alternative to vibrational excitation techniques, electron capture dissociation has shown its high potential for PTM characterization on several occasions. With ECD, modifications that are rapidly lost upon vibrational excitation such as phosphorylation,[15,16] N- and O-glycosylation,[17,18] sulfation[19] and γ-carboxylation[20] can be easily preserved during peptide backbone fragmentation. Besides PTM determination in several mostly synthetic peptides, ECD has been used for determination of phosphorylation sites in the 24 kDa protein, bovine β-casein.[15,16] While direct ECD of the intact protein (top–down approach) identified the position of one phosphorylation and restricted three modifications to four possible sites and one more to four sites,[16] ECD of HPLC-separated enzymatic peptides of the same molecule (bottom–up) gave the exact positions of all five phosphorylations.

4.5 Quantitative Proteomics

The quantification of proteins in response to gene- or nutrient-induced changes represents information of fundamental interest. A number of methods to perform relative or absolute protein quantification, most of which based on differential stable isotope labelling, have been proposed (Table 4.1).[21] The isotope-coded affinity technology (ICAT) is widely used and based on the labelling of Cys thiol groups with a tri-functional reagent: one part reacts with the SH group, an alkyl bridge can be labelled with deuterium or ^{13}C, and a biotin group for affinity capture.[26] In the experiment, one sample (*i.e.* the control) is reacted with a "light" ICAT that has natural isotope distribution, whereas the treated sample is reacted with "heavy" $^{13}C4$- or $^2H8$-labelled ICAT. The relative peak areas accounting for the light and heavy labelled reporter peptides are used for quantification.

Another approach is based on the use of known amounts of synthetic ^{13}C-labelled peptides corresponding to the peptides of interest. In this way, such as

Table 4.1 Overview of commonly used methods of quantitative proteomics (from ref. 21)

Category and subcategory	Method	Brief description	Comment
2D gels	Compare densities of stained proteins	Standard and sample are run on separate gels. After staining with Coomassie blue, silver or SYPRO® Ruby, protein spots from different gels are matched and quantitated. Selected spots are analyzed by MS for protein identification.	Easy to set up in a biology laboratory. Low-throughput protein identification.
Label-free			Suffers from gel-to-gel variation. Interference from co-migrating proteins. Hydrophobic proteins and proteins of extreme pH separate poorly.[22]
Chemical labelling (*in vitro* labelling)	DIGE (quantitate in gel by fluorescence labelling)	Standard and sample are labelled with different fluorescent dyes and run on the same gel. Select spots are analyzed by MS for protein identification.	Gel-to-gel variations are eliminated. Fluorescent dyes are expensive and may interfere with protein identification.[22]
Stable isotope labelling Metabolic labelling (*in vivo* labelling)	^{14}N or ^{15}N	Standard and sample are cultured (or fed) with ^{14}N- or ^{15}N-labelled medium (or food), mixed in an early step of sample preparation. Digested mixtures are analyzed by LC-MS/MS. Peptide pairs are identified and relative ratios are calculated, often from intensities of precursor ions.	Consistent and accurate; hundreds or up to a thousand proteins can be quantified in one experiment. Suitable for cultured bacteria, yeast, *Caenorhabditis elegans*, *Drosophila*, *Arabidopsis*, mice and rats.[23–25]
	Stable-isotope labelling of amino acids in cell culture (SILAC)	Cultured cells are labelled with light or heavy isotope-labelled amino acids (usually arginine and lysine).	Consistent and accurate. Suitable for cultured cells.

Chemical labelling (*in vitro* labelling)	ICAT	Standard and sample are labelled with light or heavy ICAT reagents before they are combined and digested. Only cysteine-containing peptides are retained for quantitation.[26]	Reduced sample complexity. Quantitation of proteins may rely on a single peptide.
	Isobaric tag for relative and absolute quantitation (iTRAQ) or tandem mass tag (TMT)	Standard and sample are digested. Exposed *N*-termini of peptides are labelled with reagents and combined for LC-MS/MS analysis. Quantitation and identification are based on the second MS scan.	Better sequence coverage than with ICAT. More proteins are quantified on the basis of multiple peptides. Up to eight samples can be analyzed simultaneously.[27,28]
	^{16}O or ^{18}O	Standard and sample are digested by trypsin in the presence of ^{16}O- or ^{18}O-H_2O, allowing incorporation of light and heavy isotopes to the C-termini of peptides.[29]	As with iTRAQ, isotopically labelled peptides are mixed after digestion, which leaves room for errors during sample preparation.
Label-free MS-based quantitation	Peak intensity or peak area	Samples are digested and analyzed separately by LC-MS/MS. Ion chromatographs are reconstructed for each peptide. Peak intensities or peak areas are compared across samples.	Subject to performance variations of HPLC and mass spectrometer.
	Spectral counting	A simple count of the number of times a peptide or protein is detected, regardless of signal intensity.[30]	Simple. Not reliable for small changes.

for quantitative analysis of small molecules, the amount of the biologically derived peptide can be calculated (AQUA, absolute quantitative analysis).[31]

To quantify proteins in a complex mixture by LC-MS/MS, it is important to choose a fast-scanning mass spectrometer to identify as many peptides as possible for quantitation during peak elution. Peptide quantification is typically performed in a triple quadrupole mass analyzer with the signals acquired in multiple-reaction ion-monitoring (MRM) mode in order to produce highly sensitive and selective ion chromatograms. The validation of the peptide and MRM transition selection is crucial. Hybrid instruments combining a fast scanning, low-resolution ion trap and a slow-scanning, high resolution mass analyzer such as the LTQ-FT or the LTQ-Orbitrap (Thermo Electron, San Jose, CA) are also suitable for this purpose.

4.6   Protein–Peptide Related Food and Nutritional Applications

The application of MS-based techniques to the determination of the complex protein and peptide mixtures present in food matrices and products can address a wide range of analytical questions which include the assessment of food quality and safety, food chain traceability (authenticity) and functionality. The role of MS in nutritional proteomics is pivotal because of the inherent complexity of typical samples collected in nutritional studies, including blood, cells and tissues. Quantitative proteomic data in combination with unambiguous identification provide a more complete understanding of the effects of nutrition on cells, tissues and organisms.

4.6.1   Plant Proteins

Recently, the 2D gel approach was used together with MALDI-ToF-MS analysis to obtain a comprehensive protein profile of wheat grains (*Triticum aestivum* L., cv. Titlis) from organic or conventional agriculture.[32] Abundances of 1049 proteins were recorded in wheat grains of two growing seasons and 25 proteins (storage proteins, enzymes of carbohydrate metabolism, one peroxidase, and proteins of unknown function) were found differentially expressed between organic and conventional wheat in both years. Total protein content was demonstrated to be lower in organic wheat, but these differences are negligible with regard to total protein in an average diet. The authors concluded that food quality of conventional and organic wheat grown in the field trial was equal. Among the 25 proteins identified, 16 were retained with different levels in organic and conventional wheat and have the potential to afford a signature to prove authenticity of organic wheat.

A recently published review discusses the strengths and weaknesses of proteomic technologies in soybean biology.[33] Information from soybean proteomics useful in predicting the function of plant proteins and for molecular cloning of the corresponding genes is presented.

4.6.2 Dairy Proteins and Bacteria

Important proteomic development has been achieved in the field of bacteria used in dairy products. Proteomics has resulted in the establishment of reference maps to detect strain-to-strain variations and to elucidate the mechanisms of *in vitro* and *in vivo* adaptation to environmental conditions. Proteomic analysis of bacteria entrapped in cheese has been accomplished and has revealed which predominant metabolic pathways are active depending on the strain.

Many new low-abundant proteins have been identified in milk over the last five years, in particular in the bovine milk fat globule membrane (MFGM) and in human colostrum. In addition, peptidomics on dairy products has provided useful information about product authenticity, origin, and history, and the presence of bioactive peptides.

As far as dairy bacteria are concerned, *in vitro* proteomic studies have confirmed—in a larger number of species including probiotics—the general mechanisms of stress adaptation. The exploration of the proteomes of bacteria entrapped in cheese also resulted in interesting findings, highlighting the overexpression of different metabolic pathways depending on the strain and the nature of the stresses encountered in the cheese.[34]

Various bacteria—lactic acid bacteria, bifidobacteria, and propionic dairy bacteria—are used in dairy processing for their technological or probiotic properties. Reference proteomic maps have been established from strains grown either in synthetic medium or in milk. Expression patterns differ depending on the medium, strain and the stress applied, and hypotheses have been created regarding the metabolic pathways and proteins involved. However, some limitations have to be underlined: so far, 2D proteomic approaches have allowed the simultaneous detection of a maximum of about 700 proteins, which is about one-third of the open reading frames deduced from the genomes of most dairy bacteria. In particular, envelope proteins are mostly excluded from the analysis because of their hydrophobicity.

4.6.3 Technology Complementarities

The use of liquid chromatography (LC), gas chromatography (GC), capillary electrophoresis (CE) and electrochromatography (CEC) as analytical separation techniques coupled to mass spectrometry as the detection method in the separation and characterization of proteins, peptides and amino acids of food interest has been extensively reviewed and discussed.[35–39] Determination of molecular masses, protein patterns and protein modifications during seed maturation, qualitative and quantitative determination of gliadins in cereal-based food, as well as direct verification of gene-derived sequences in cereal have been investigated by MALDI and ESI mass spectrometry.[40–42] MALDI is less sensitive to impurities and therefore more compatible with direct food analysis without complex sample preparation. Moreover, MALDI yields mainly singly charged peptide ions and therefore simpler spectra than ESI, which in turn yields higher charge states and can therefore be more powerful for MS/MS (doubly charged peptide ions typically fragment more readily than singly charged). ESI interfaces easily with the HPLC instrument to the mass spectrometer and is routinely employed to sequence peptides. Overall, ESI and MALDI for peptide analysis can be considered complementary to each other.

4.6.4 Protein Processing

Employing both approaches, numerous studies highlight the influence of milk protein variety on the composition and technological properties of milk, its biological activities and the sensory characteristics of milk-based food. Industrial-scale methods have been developed for native whey proteins such as immunoglobulins, lactoferrin, lactoperoxidase, α-lactalbumin and β-lactoglobulin. Thus, MS-based strategies focused on either LC/ESI-MS/MS or MALDI-ToF have been utilized to evaluate the impact of different processing technologies on whey protein, verifying that it contains a multitude of different modifications including glycosylation, phosphorylation and oxidation.

The site specificity of modification of β-casein (β-CN) by glucose and methylglyoxal (MGO) has been determined by analyzing tryptic digests with UPLC-ESI-MS. This enabled the sites of formation of the Amadori product, N^{ε}-(fructosyl)lysine (FL), and the advanced glycation end products, N^{ε}-(carboxymethyl)lysine (CML), MGO-derived dihydroxyimidazolidine (MG-DH) and MGO-derived hydroimidazolone (MG-HI), to be located.[43]

The formation of Amadori compounds between whey proteins and lactose upon heat treatment has also been studied by mass spectrometry coupled to electrospray ionization, generating information on changes in the physicochemical properties of the whey proteins upon lactosylation and providing new insights into the lactosylation reaction. In order to identify bovine β-lactoglobulin (β-Lg) peptides glycated with glucose (an aldohexose) or tagatose (a ketohexose), a model system reproducing gastrointestinal digestion has recently been developed, with the "digested" proteins being characterized by LC coupled to positive electrospray ion trap MS.[44] Bovine β-Lg is widely used as a functional ingredient in whey-containing food formulations; its thermal behaviour has been extensively investigated because heating affects its functional properties. The glycation extent of undigested bovine β-Lg incubated with galactose or tagatose at 40 °C for one day has been assessed by MALDI-MS analyses. The spectra showed that there were 14 carbohydrate molecules bound to β-Lg when incubation was performed with galactose and three when it was performed with tagatose (Figure 4.2),[44] *i.e.* galactose exhibited a much higher reactivity than tagatose. Furthermore, unglycated peptides were identified by LC-ESI tandem mass spectrometry, whereas the fragmentation behaviour of galactosylated (through Amadori rearrangement) and tagatosylated (Heyns rearrangement) has been investigated by LC-ESI-MS[n] using a three-dimensional (3D) ion trap MS. MS[3] spectra from collisionally generated imminium ions were proposed for sequencing peptides glycated with aldohexoses and

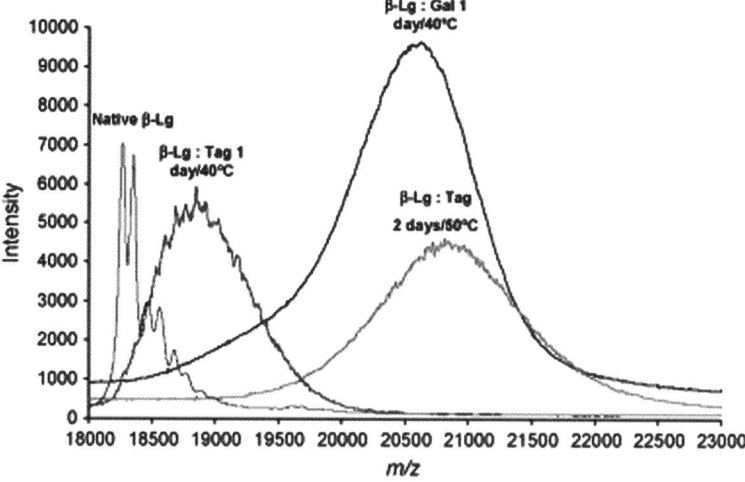

Figure 4.2 Superimposed MALDI-MS spectra of native γ-lactoglobulin and â-lacto-globulin incubated with Gal or Tag at 40 °C for one day or at 50 °C for two days (from ref. 44). The obtained spectra highlighted the presence of 14 carbohydrate molecules bound to â-lactoglobulin when incubation was performed with galactose and three when it was performed with tagatose.

ketohexoses. In addition, identification and characterization of doubly glycated peptides may be achieved by sequential product-ion spectra (MS[4]).[44]

4.6.5 Proteolysis and Ripening

The breakdown of proteins to smaller peptides and free amino acids during proteolysis influences the quality of fermented or otherwise ripened dairy products. Proteolysis is regarded as one of the most important biochemical events during cheese ripening. The biochemical changes that occur during the ripening process of cheese other than proteolysis include glycolysis and lipolysis. However, proteolysis is the most complex and, in most varieties, the most important of the three primary processes that occur during cheese ripening.

The proteolytic digestion of proteins during cheese ripening is a multi-step reaction involving the formation of rather large well-defined peptides, and their subsequent digestion into smaller peptides and free amino acids before final transformation into various volatile aroma compounds. Primary proteolysis includes hydrolysis of caseins to specific degradation products. The degree of the primary proteolysis is influenced by:

- the concentration of rennet used in cheese manufacturing;
- the retention and the activity of the rennet in the curd, which are pH dependent; and

- the concentration and the activity of the indigenous milk proteinase plasmin, which are influenced by many factors, of which stage of lactation and somatic cell counts are the most important.

Secondary proteolysis includes the degradation of proteins and large peptides to smaller peptides and amino acids by the activities of intracellular and extracellular enzymes from the starter and non-starter lactic acid bacteria which occur in cheese. Urea-polyacrylamide gel electrophoresis (urea-PAGE) of the casein fractions, and the extraction and quantification of the nitrogen fractions released during cheese ripening are widely used for the determination of proteolysis. However, the use of such techniques involves considerable analysis time.

ESI and MALDI-MS have been widely used for the structural characterization of peptides released during cheese ripening. MS-based milk protein investigation includes identification of milk protein variants and glycoforms, the degree of glycoforms, as well as the detection of milk adulteration.[45] ESI-MS analyses have been successfully applied to all major bovine milk proteins, including caseins (α_{s1}-, α_{s2}-, β- and κ-caseins) and whey proteins (α-La, β-Lg). ESI-MS of these proteins generally gives strong signals in both negative and positive ion mode, based on their high contents in acid (aspartic and glutamic acids and phosphoric groups) and basic (arginine, lysine, histidine) amino acid residues, respectively. More recently, MALDI-ToF-MS, which is usually employed for the determination of the purity and molecular masses of proteins, has also been proposed for the identification and quantification of proteins of milk from different species such as cow, goat and ewe.[46] This approach has focused on the quantitative and multivariate use of MALDI-ToF-MS spectra of skimmed milk samples.

4.6.6 Proteins in Food Technology

As mentioned above, the functional properties of food proteins such as emulsification, gelation, foaming and water binding capacity render them widely used ingredients in the food industry. Recently, there is a growing nutritional interest in proteins obtained by concentrating whey, which originates as a by-product of the cheese and casein manufacturing process. Normally, whey proteins are isolated after removal of lactose and other low molecular weight constituents. The interest regarding the principal whey proteins β-lactoglobulin (β-Lg), α-lactalbumin (α-La), bovine serum albumin (BSA) and immunoglobulin (Ig) is focused on understanding their basic chemical and structural properties as well as their possible use as functional ingredients in foods and, more recently, neutraceuticals releasing bioactive peptides. Knowledge building has impacted:

- procedures for the isolation, purification and characterization of the individual whey proteins; and

- whey fractionation technologies used for manufacturing whey protein concentrates with improved chemical and functional properties in food systems.

For instance, high hydrostatic pressure treatments can improve the foaming properties and stability of whey protein concentrate (WPC), which are correlated to changes in their tertiary structure. ESI-MS has been employed to measure changes in the tertiary structure of WPC exposed to high-pressure processing. Results reported by Alvarez *et al.*[47] showed that exposure of β-La to high pressure resulted in changes in the tertiary structure, rendering some amino acid side chains normally buried in the hydrophobic core of the native protein (and therefore inaccessible to the solvent) to be exposed and, consequently, to acquire charges upon electrospray ionization. The charged-state distribution of proteins under ESI-MS measurement provides an effective means of measuring changes in tertiary structure that take place in response to physicochemical changes as resulting effects of different high pressure treatments. Thus, ESI-MS can reveal changes in the tertiary structure of α-La, even when no change in rheological properties has been observed after high pressure treatment. On the other hand, changes in the viscoelastic properties of β-Lg solutions reflect changes in the tertiary structure after different high pressure treatments.

4.6.7 Low-abundant Milk Proteins

Besides the main milk proteins (caseins and whey proteins), there is a complex low-abundant proteome that is far from being analytically fully elucidated and exploited for nutritional benefits. Peptides, which are inactive within the sequence of the parent protein, can be released by enzymatic proteolysis, for example, during gastrointestinal digestion or during food processing. Once they are liberated in the body, bioactive peptides may act as regulatory compounds with hormone-like activity. Thus, these peptides represent potential health-enhancing nutraceuticals for food and pharmaceutical applications.

4.7 Food-borne Microorganisms and their Fermenting Effects on the Food Proteome

Bioactive peptides are defined as peptides that are able to modulate physiological functions through binding to specific receptors or target cells leading to induction of physiological responses. Most of the bioactive peptides of nutritional interest are hidden in the original parent protein structure and may be released through one of the following conditions:

- hydrolysis by digestive enzymes such as trypsin and pepsin;
- food processing; or
- hydrolysis by proteolytic microorganisms or through the action of proteolytic enzymes.

Bioactive peptides of nutritional interest are typically leveraged in the probiotic food sector and generally segment into two main lines:

- yoghurt (spoonable and drinkable); and
- single serve delivery drinks such as fruit juice drinks (non-dairy products).

Such probiotic foods contain probiotic bacteria that are defined as "living microorganisms which, upon ingestion in certain numbers, exert health benefits beyond inherent basic nutrition".[48] Foods containing probiotic bacteria are categorized as functional foods and such products are gaining widespread popularity and acceptance throughout the developed world. Probiotics became one of the most important health promoting food fortifiers in recent years, especially, as previously reported, for dairy foods.

Consumption of fermented milks containing probiotic microorganisms has the potential to improve lactose digestion, as well as to modulate immune function and the gastrointestinal flora. Ensuring the viability and functionality of the probiotics until they reach their destination in the human gut is one of the key requirements. In general, it is proposed that dairy products should contain at least 10^7 viable probiotics cell per mL. However, their levels detected in yoghurt and fermented milk products are often much lower due to the adverse conditions during product manufacturing and storage. Furthermore, the extreme acidic conditions in the human stomach can seriously reduce the number of living cells reaching the intestine.

Last but not least, probiotic foods are not restricted to fermented milk drinks and yoghurt, which have limited shelf life. A number of studies have addressed the development of probiotic cheese such as Cheddar cheese, goat cheese, Crescenza cheese, cottage cheese and fresh cheese.

The beneficial health effects of these functional foods may be a result of the production of bacterial metabolites such as cell wall components and the hydrolytic action of cell-free extracts exerting proteinase/peptidase activities on milk protein substrates.[49] In addition, casein/whey protein-derived peptides can be released from milk proteins during dairy fermentations by various bacterial strains. Numerous peptides and peptide fractions with bioactive properties have been isolated from fermented dairy products. These bioactivities include immunomodulatory, anticancer, hypocholesteremic and hypercholesteremic, antimicrobial (including bacteriocins), peptidase inhibitory and bone formation activities.[50–55]

4.8   Food Peptidomics

These findings indicate that bioactive peptides and proteins can affect both the functional and biological properties of food products, and a new "omic" technique named "peptidomics" has emerged: the food peptidome can be defined as the entire peptide pool present in food products or raw materials, or obtained during processing and storage. Food peptidomics provides information about product authenticity, origin and history, biological activities of peptides, functional properties, allergenicity and sensory properties. In dairy products, many studies have been dedicated to identify peptides displaying various biological activities.[56,57] Peptides result from the activity of various proteolytic enzymes. For example, in addition to the action of plasmin in milk, other proteinases can be present in milk after leakage of somatic cells—which comprise neutrophils, macrophages, lymphocytes and some epithelial cells— during mastitis.

Apart from the mass spectrometric peptide analysis itself, peptidomics also covers the development of computational tools for the identification of (native and not necessarily tryptic, as in bottom–up proteomics) peptides and their protein precursors, including genetic variants and chemical and enzymatic modifications by MS.

As for proteomics, the main identification tool in peptidomics is mass spectrometry. Peptide mixtures can be analyzed directly by MS (both ESI and MALDI), requiring very little or no manipulation of the sample. However, in order to identify the whole peptidome, a separation step prior to the mass spectrometric investigation is usually required. For instance, a nanoscale HPLC system can be coupled directly to a quadruple/time-of-flight (QqToF) mass spectrometer. Putative peptides are then selected for tandem mass analysis.

In a related approach, the main peptides produced by hydrolysis of water buffalo (WB) β-casein with plasmin were characterized by capillary electro-phoresis and mass spectrometry, and compared with their bovine homo-logues.[58] The analysis of the main products of plasmin activity, *i.e.* γ-caseins (γ-CNs), is an important tool for the detection of the adulteration of goat's, ewe's and water buffalo's milk by means of adding cow milk, which is not allowed in the manufacture of a number of typical dairy products. The proposed method has been carried out by capillary isoelectric focusing (cIEF) and MS, with the aim of verifying the fraudulent addition of cow milk to buffalo milk in the production of buffalo Mozzarella cheese.[58] cIEF is similar to classical isoelectric focusing where peptides and proteins are sepa-rated on the basis of isoelectric point (pI) by generating a pH gradient using ampholytes. The difference is that, instead of creating a pH gradient in a gel, it is created inside a capillary as in capillary electrophoresis, which has an intrinsic high resolving power particularly suitable for the separation of peptides. An updated overview of capillary electrophoresis and mass spec-trometry in food analysis has recently been published.[39] As depicted in Figure 4.3,[58] the cIEF method can separate and identify all the plasmin γ-CN primary products from bovine and water buffalo β-CN produced by hydro-lysis of the two species, which have been identified by MALDI-ToF analysis. A comparison of the results obtained by cIEF with those obtained by ultra-thin-layer isoelectric focusing (the reference method for routine analysis of plasmin digests of casein) suggests that cIEF may constitute a successful alternative to the traditional slab gel electrophoresis analysis of plasmin digests of casein.

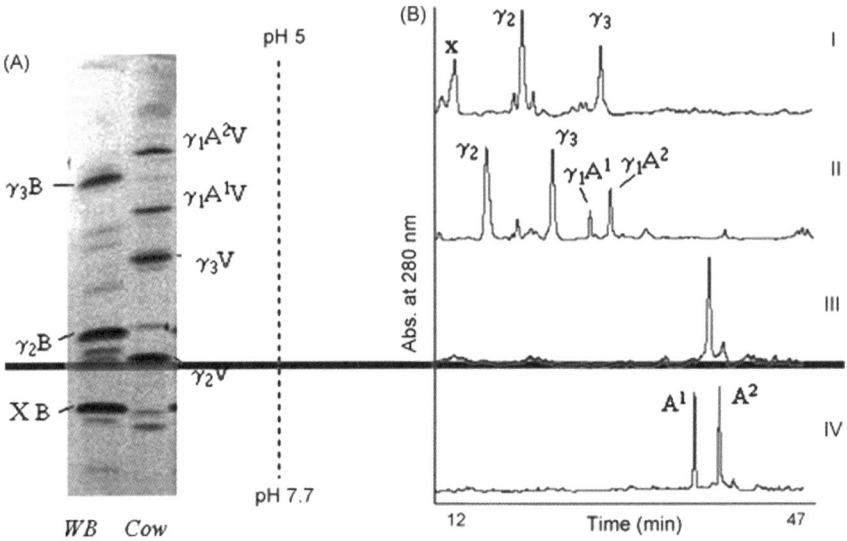

Figure 4.3   Isoelectric focusing (IEF) can separate and identify all plasmin γ-casein primary products from bovine and water buffalo β-CN: (A) ultra-thin layer IEF of plasmin-digested whole cow and water buffalo β-caseins obtained by fast protein liquid chromatography (FPLC); (B) cIEF of the plasmin digests of FPLC purified whole water buffalo (I) and cow (II) casein, intact water buffalo β-casein (III) and intact cow β-casein (IV) (from ref. 58).

4.9   ACE Inhibitors

Peptides generated by protein hydrolysis or fermentation may represent only minor constituents in a highly complex matrix and, therefore, identification of *in situ* generated biologically active peptides in food matrices is a challenging task in food technology. Identification of bioactive peptides in fermented dairy products or milk protein hydrolysates generated by the action of several and possibly unspecific enzymes is a labour-intensive and difficult task. In this context, mass spectrometry has developed into a necessary tool to determine the presence and behaviour of functional components such as these bioactive peptides.

Various peptides derived from proteolytic digestion of whey proteins have been shown to have inhibitory activity against angiotensin-converting enzyme (ACE), which plays a major role in the regulation of blood pressure and thereby hypertension.[59,60] Currently, enzymatically hydrolyzed milk proteins as well as milk fermentation by selected lactic bacteria are the main sources for ACE-inhibitory peptides in functional foods. ACE-inhibitory peptides have also been identified from various plant food sources including soybean, fermented soybean products, sunflower, rice, corn, wheat, buckwheat, broccoli, mushroom, garlic, spinach and wine, and after the hydrolysis of gliadins with an acidic protease.

A recent paper reported a study of the release of ACE-inhibitory peptides in Cheddar cheeses produced with different probiotic microorganisms and ripened at two different temperatures (4 and 8 °C) for 24 weeks.[61] ACE-inhibitory peptides in water-soluble cheese extracts were isolated by RP-HPLC and the fractions with the highest ACE-inhibitory activity, after further separation and purification, were identified using chemically assisted fragmentation (CAF) and MALDI-ToF-MS. CAF-MALDI is based on the introduction of a negatively charged group (sulfonation) to the *N*-terminus of peptides generated by tryptic digestion. The presence of a strong acidic group greatly enhances the fragmentation ability of tryptic peptides which, as a consequence, produces almost exclusively b and y fragments. Moreover, the negative charge at the *N*-terminus counterbalances the positive charge of the captured proton in the b fragments, rendering them neutral and thus undetectable. In this way, CAF peptides yield post-source decay (PSD) spectra containing exclusively y ions. A mass spectral identification of an ACE-inhibitor peptide is depicted in Figure 4.4:[61] the spectrum reported in panel (a) corresponds to the purified peptide exhibiting a molecular mass of 854.674 Dalton; panel (b) shows the peptide sequencing by CAF-MALDI (Ala-Arg-His-Pro-His). Following the sequence interpretation and molecular weight determination, this peptide was identified as Ala-Arg-His-Pro-His-Pro-His, originating from k-casein (CN).

4.10   Antimicrobial and Other Bioactive Peptides[62–66]

Other peptides related to nutrition are antimicrobial peptides (AMPs) which are synthesized by microorganisms and multicellular organisms. These peptides can have linear, cyclic or open-ended cyclic structures with one or more di-sulfide bridges. They exhibit α-helical conformations, β-sheets, amphipathic β-hairpin like β-sheets, and α-helix/β-sheet mixed folds. Some of them are rich in proline, histidine, arginine or lysine, in addition to containing hydrophobic amino acid residues. AMPs exhibit two main modes of action, one involving an intracellular target, and another the interaction with the cytoplasmic membrane from microorganisms. Membrane-active peptides that include hormones, signal sequences and lytic agents interact electrostatically with the cellular external membrane and eventually partition into the hydrophobic lipid bilayer where they express their activity.

Multifunctional peptides are widely recognized in food sources. Owing to the wide reactivity of endogenous peptides generated in cells, appropriate oligo-peptide-restricted peptidases either give rise to shorter bioactive peptides or contribute to the complete degradation of oligopeptides.[67,68] Finally, based on their original structures, several of these peptidases have been engineered to produce peptide derivatives characterized by pronounced aggregation resistance and/or increased biological activity. For example, bacteriocins, which have bactericidal or bacteriostatic effects, have been fully characterized by amino acid analysis, Edman sequencing and mass spectrometry.[69]

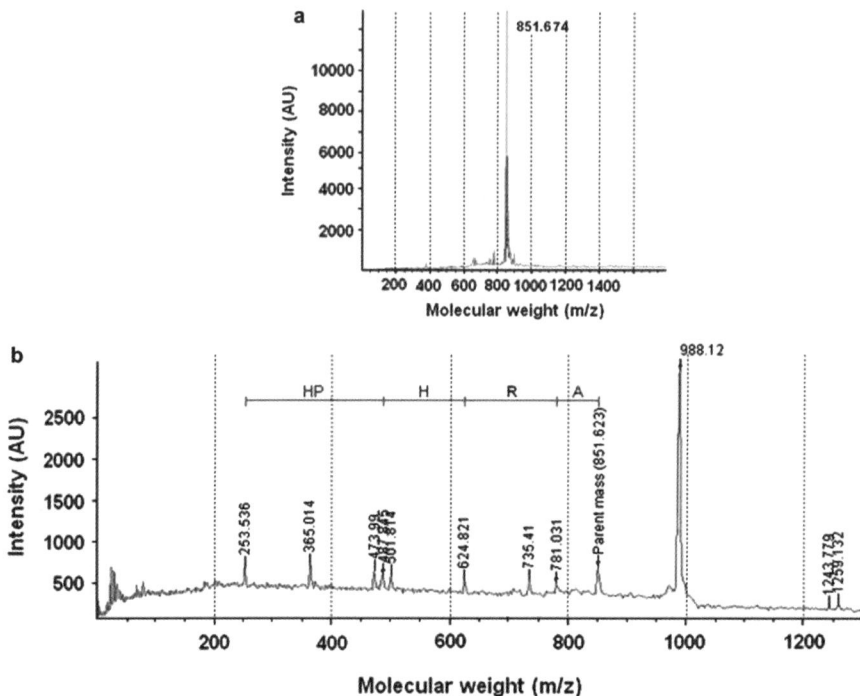

Figure 4.4   Mass spectral identification of an ACE-inhibitor peptide. (a) Molecular weight of the purified peptide obtained using MALDI-TOF-MS (condition as described in the text). (b) Chemically assisted fragmentation (CAF) of the purified peptide; the first five amino acids of the *N*-terminal were identified as Ala-Arg-His-Pro-His (CAF condition as described in the text). Based on sequence readout and molecular weight determination, the peptide was identified as Ala-Arg-His-Pro-His-Pro-His (from ref. 61).

4.11   Mass Spectrometry in Food Protein Status and Quality

The traditional determination of primary structure of proteins and peptides has been performed by Edman degradation, which can be carried out by a fully automated protein sequencer at pmol level. However, amino acid sequence analysis by tandem mass spectrometry is much more sensitive and provides additional information concerning amino acid identity and the location of post-translational modifications. Recently, several methods for amino acids analysis in food products employing mass spectrometric techniques have been performed by coupling LC-MS[70] or CE-MS.[71] On the other hand, amino acid analysis enables food fermentations to be followed during technological processes or storage. Amino acid analysis has been also used to identify genetically modified varieties of maize and soybean.[72]

A metabolomic study of fermented food samples has been recently performed by LC-MS on Japanese fermented food (miso) at different stages of ripeness; amino acids, but also citric acid and Amadori compounds, have been found responsible for characteristic tastes at different stages of ripeness.[73] The presence of the amino acid derivative, 3-methyhistidine, is used as an indicator of possible addition of non-meat proteins or hydrolysates to meat products since this compound is present only in myosin and actin but not in collagen.[74] Similarly, addition of milk powder to cheese was identified by checking for the presence of lysinoalanine.[75] A similar approach has been used to establish the presence of cow's milk in fresh cheeses derived from sheep milk.

Determination of amino acid enantiomers is also an important topic in food analysis, since the presence of D-isomers may indicate adulteration, microbiological contamination, or uncontrolled fermentation processes. Therefore, the D- and L-enantiomer content can be a useful marker for quality control, contamination detection, and process monitoring.[76] Amino acids are moreover involved in several reactions occurring during food processing, particularly cooking and high temperature treatments. Variations in pH values and heating processes are responsible for reduced digestibility of some essential amino acids because of cross-link formation (lysinoalanine or histidylalanine). Additional negative health effects are due to reduced bioavailability of minerals such as copper, iron and zinc.

A method able to reveal the progress of protein glycation in dairy products has been developed based on liquid chromatography-tandem mass spectrometry; the method enables simultaneous determination of lysine, N^{ε}-(fructosyl)lysine, N^{ε}-(carboxymethyl)lysine and pyrraline (Pyr) in dairy products. It has been applied to the analysis of raw and processed cow milk, evaluating differences between raw milk after mild heat treatment, *i.e.* pasteurization ultra-high-temperature (UHT) treatment, and after more severe heat treatment as done with condensed milk.[77]

A study on the effect of dietary non-essential amino acid composition on the delta ^{13}C values of individual amino acids in rainbow trout using liquid chromatography coupled to isotope ratio mass spectrometry (LC-IRMS) demonstrates that measuring isotopic amino acid signatures by LC-IRMS is a promising new technique for nutritional physiologists.[78] The aspect of incorporation of amino acids and peptides into the body's protein pool and related questions of protein turnover are discussed in a separate, dedicated chapter on protein and proteome turnover by Michael Affolter (Chapter 10) in the nutritional health aspect section of this book.

4.12 Conclusions

Mass spectrometry has become the core technology for the characterization of food proteins and peptides. The development of soft ionization techniques such as ESI and MALDI, as well as the introduction of multistage and "hybrid" analyzers able to generate sequence information has revolutionized the field of protein and peptide analysis. In particular, qualitative and quantitative MS analysis of the complex protein mixtures contained in food play a key role in understanding their nature, structure, functional and nutritional properties, and impact on human health. Protein databases interrogated by high-resolution and accuracy mass spectrometric data have resulted in new possibilities of protein characterization including post-translational modifications, protein conformations, and protein–protein and protein–ligand interactions. Finally, MS-based strategies adapted to food and nutrition proteomics are now capable of addressing a wide range of analytical questions which include issues related to food quality and safety, certification and traceability of (typical) products, and to the definition of the structure–function relationship of food proteins and peptides. All these aspects can be effectively understood only by using integrated and up-to-date analytical approaches.

References

1. A. H. Pripp, T. Isaksson, L. Stepaniak, T. Sorhaug and Y. Ardö, *Trends Food Sci. Tech.*, 2005, **16**, 484.
2. C. M. Bryant and D. J. McClement, *Trends Food Sci. Tech.*, 1998, **9**, 143.
3. E. A. Foegeding, J. P. Davis, D. Doucet and M. K. McGuffey, *Trends Food Sci. Tech.*, 2002, **13**, 151.
4. E. Dickinson, in *Food colloids, biopolymers and materials*, ed. E. Dickinson and T. van Vlie, Royal Society of Chemistry, Cambridge, 2003, p. 68.
5. H. M. Roche, *J. Sci. Food Agric.*, 2006, **88**, 1156.
6. V. Garcìa-Cañas, C. Simò, C. Leòn and A. Cifuentes, *J. Pharmaceut. Biomed. Anal.*, 2010, **51**, 290.
7. M. Careri, A. Costa, L. Elviri, J. B. Lagos, A. Mangia, M. Terenghi, A. Cereti and L. P. Garoffo, *Anal. Bioanal. Chem.*, 2007, **389**, 1901.
8. M. Careri, L. Elviri, J. B. Lagos, A. Mangia, F. Speroni and M. Terenghi, *J. Chromatogr. A*, 2008, **1206**, 89.
9. A. C. Gorg, G. Obermaier, A. Boguth, B. Harder, R. Scheibe, R. Wildgruber and W. Weiss, *Electrophoresis*, 2000, **21**, 1037.
10. W. V. Bienvenut, J. C. Sanchez, A. Karmime, V. Rouge, K. Rose, P. A. Binz and D. F. Hochstrasser, *Anal. Chem.*, 1999, **71**, 4800.
11. A. J. Link, J. Eng, D. M. Schieltz, E. Carmack, G. J. Mize, D. R. Morris, B. M. Garvik and J. R. Yates III, *Nat. Biotechnol.*, 1999, **17**, 676.
12. M. J. MacCoss, W. H. McDonald, A. Saraf, R. Sadygov, J. M. Clark, J. J. Tasto, K. L. Gould, D. Wolters, M. Washburn, A. Weiss, J. I. Clark and J. R. Yates III, *Proc. Natl. Acad. Sci. U. S. A.*, 2002, **99**, 7900.
13. D. M. Horn, R. A. Zubarev and F. W. McLafferty, *Proc. Natl. Acad. Sci. U. S. A*, 2000, **97**, 10313.
14. F. Meng, B. J. Cargile, S. M. Patrie, J. R. Johnson, S. M. McLoughlin and N. L. Kelleher, *Anal. Chem.*, 2002, **74**, 2923.
15. A. Stensballe, O. N. Jensen, J. V. Olsen, K. F. Haselmann and R. A. Zubarev, *Rapid Commun. Mass Spectrom.*, 2000, **14**, 1793.

16. S. D.-H. Shi, M. E. Hemling, S. A. Carr, D. M. Horn, I. Lindh and F. W. McLafferty, *Anal. Chem.*, 2001, **73**, 19.

17. K. Håkansson, H. J. Cooper, M. R. Emmett, C. E. Costello, A. G. Marshall and C. L. Nilsson, *Anal. Chem.*, 2001, **18**, 4530.

18. K. F. Haselmann, B. A. Budnik, J. V. Olsen, M. L. Nielsen, C. A. Reis, H. Clausen, A. H. Johnsen and R. A. Zubarev, *Anal. Chem.*, 2001, **73**, 2998.

19. N. L. Kelleher, R. A. Zubarev, K. Bush, B. Furie, B. C. Furie, F. W. McLafferty and C. T. Walsh, *Anal. Chem.*, 1999, **71**, 4250.

20. H. Niiranen, B. A. Budnik, R. A. Zubarev, S. Auriola and S. Lapinjoki, *J. Chromatogr. A*, 2002, **962**, 95.

21. J. J. Moresco, M. Q. Dong and J. R. Yates III, *Am. J. Clin. Nutr.*, 2008, **88**, 597.

22. P. G. Righetti, A. Castagna, F. Antonucci, C. Piubelli, D. Cecconi, N. Campostrini, P. Antonioli, H. Astner and M. Hamdan, *J. Chromatogr. A*, 2004, **1051**, 3.

23. W. W. Wu, G. Wang, S. J. Baek and R. F. Shen, *J. Proteome Res.*, 2006, **5**, 651.

24. J. Krijgsveld, R. F. Ketting, T. Mahmoudi, J. Johansen, M. Artal-Sanz, C. P. Verrijzer, R. H. A. Plasterk and A. J. R. Heck, *Nat. Biotechnol.*, 2003, **21**, 927.

25. C. J. Nelson, E. L. Huttlin, A. D. Hegeman, A. C. Harms and M. R. Sussman, *Proteomics*, 2007, **7**, 1279.

26. S. P. Gygi, B. Rist, S. A. Gerber, F. Turecek, M. H. Gelb and R. Aebersold, *Nat. Biotech.*, 1999, **17**, 994.

27. L. Choe, M. D'Ascenzo, N. R. Relkin, D. Pappin, P. Ross, B. Williamson, S. Guertin, P. Pribil and K. H. Lee, *Proteomics*, 2007, **7**, 3651.

28. L. Dayon, A. Hainard, V. Licker, N. Turck, K. Kuhn, D. F. Hochstrasser, P. R. Burkhard and J. C. Sanchez, *Anal. Chem.*, 2008, **80**, 2921.

29. I. I. Stewart, T. Thomson and D. Figeys, *Rapid Commun. Mass Spectrom.*, 2001, **15**, 2456.

30. H. Liu, R. G. Sadygov and J. R. Yates III, *Anal. Chem.*, 2004, **76**, 4193.

31. S. A. Gerber, J. Rush, O. Stemman, M. W. Kirschner and S. P. Gygi, *Proc. Natl. Acad. Sci. U. S. A.*, 2003, **100**, 6940.

32. C. Zorb, T. Betsche and G. Langenkamper, *J. Agric. Food Chem.*, 2009, **57**, 2932.

33. S. Komatsu and N. Ahsan, *J. Proteomics*, 2009, **72**, 325.

34. V. Gagnaire, J. Jardin, G. Jan and S. Lortal, *J. Dairy Sci.*, 2009, **92**, 811.

35. G. Mamone, G. Picariello, S. Caira, F. Addeo and P. Ferranti. *J. Chromatogr. A*, 2009, in press.

36. A. Bendixen, *Meat Sci.*, 2005, **71**, 138.

37. A. D. Zamfir, *J. Chromatogr. A*, 2007, **1159**, 2.

38. R. Haselberg, G. J. de Jong and G. M. Somsen, *J. Chromatogr.*, 2007, **1159**, 81.

39. C. Simò, C. Barbas and A. Cifuentes, *Electrophoresis*, 2005, **26**, 1306.

40. A. Hermando, I. Valdes and E. Méndez, *J. Mass Spectrom.*, 2003, **38**, 862.

41. Y. Qian, K. Preston, O. Krokhin, J. Mellish and W. Ens, *J. Am. Soc. Mass Spectrom.*, 2008, **19**, 1542.
42. D. J. Skylas, D. Van Dyk and C. W. Wrigley, *J. Cereal Sci.*, 2005, **41**, 165.
43. M. Lima, C. Moloney and J. M. Ames, *Amino Acids*, 2009, **36**, 475.
44. M. Corzo-Martinez, R. Lebron-Aguilar, M. Villamiel, J. E. Quintanilla-Lopez and F. J. Moreno, *J. Chromatogr. A*, 2009, **1216**, 7205.
45. H. F. Alomirah, I. Alii and Y. Konishi, *J. Chromatogr. A*, 2000, **893**, 1.
46. K. H. Liland, B.-H. Mevik, E. O. Rukke, T. Almøy and T. Isaksson, *Chemometr. Intell. Lab. Syst.*, 2009, **99**, 39.
47. P. A. Alvarez, H. S. Ramaswamy and A. A. Ismail, *Int. Dairy J.*, 2007, **17**, 881.
48. F. Guarner and G. J. Schaafsma, *Int. J. Food Microbiol.*, 1998, **39**, 237.
49. M. Gobbetti, L. Stepaniak, M. De Angelis, A. Corsetti and R. Di Cagno, *Crit. Rev. Food Sci. Nutr.*, 2002, **42**, 223.
50. M. Yoshikawa, H. Fujita, N. Matoba, Y. Takenaka, T. Yamamoto, R. Yamauchi, H. Tsuruki and K. Takahata, *BioFactors*, 2000, **12**, 143.
51. F. Bäckhrd, R. E. Ley, J. L. Sonnenburg, D. A Peterson and J. I. Gordon, *Science*, 2005, **307**, 1915.
52. S. Severin and X. Wenshui, *Crit. Rev. Food Sci. Nutr.*, 2005, **45**, 645.
53. S. Peng, J.-Y. Lin and M.-Y. Lin, *J. Agric. Food Chem.*, 2007, **55**, 5092.
54. J. Meltretter, A. Schmidt, A. Humeny, C.-M. Becker and M. Pischetsrieder, *J. Agric. Food Chem.*, 2008, **56**, 2899.
55. F. C. Prado, J. L. Parada, A. Pandey and C. R. Soccol, *Food Res. Int.*, 2008, **41**, 111.
56. B. A. Murray and R. J. Fitzgerald, *Curr. Pharm. Des.*, 2007, **13**, 773.
57. H. Korhonen and A. Pihlanto, *Int. Dairy J.*, 2006, **16**, 945.
58. A. Somma, P. Ferranti, F. Addeo, R. Mauriello and L. Chianese, *J. Chromatogr. A*, 2008, **1192**, 294.
59. M. Gobbetti, P. Ferranti, E. Smacchi, F. Goffredi and F. Addeo, *Appl. Environ. Microbiol.*, 2000, **66**, 3898.
60. O. N. Donkor, A. Henriksson, T. K. Singh, T. Vasiljevic and N. P. Shah, *Int. Dairy J.*, 2007, **17**, 1321.
61. L. Ong and N. P. Shah, *Food Sci. Technol.*, 2008, **41**, 1555.
62. R. J. Xu, *Food Rev. Int.*, 1998, **14**, 1.
63. J. C. K. Chan and E. C. Y. Li-Chan, in *Nutraceutical Proteins and Peptides in Health and Diseases,* ed. Y. Mine and F. Shahidi, Taylor and Francis, New York, 2006, pp. 99–136.
64. L. Losito, T. Carbonara, M. Domenica De Bari, M. Gobbetti, F. Palmesano, C. G. Rizzello and P. G. Zambonin, *Rapid Commun. Mass Spectrom.*, 2006, **20**, 447.
65. R. Hartmann and H. Meisel, *Curr. Opin. Biotech.*, 2007, **18**, 163.
66. I. de Nom and R. Floris, *Int. Dairy J.*, 2007, **17**, 504.
67. H. Meisel, *Livest. Prod. Sci.*, 1997, **50**, 125.
68. L. de Vuyst and F. Leroy, *J. Mol. Microbiol. Biotech.*, 2007, **13**, 19.
69. K. B. McCann, B. J. Shiell, W. P. Michalski, A. Lee, J. Wan, H. Roginski and M. J. Coventry, *Int. Dairy J.*, 2004, **15**, 133.

70. B. Thiele, K. Füllner, N. Stein, M. Oldiges, A. J. Kuhn and D. Hofmann, *Anal. Bioanal. Chem.*, 2008, **391**, 2663.
71. A. V. Colnaghi Simionato, E. P. Moraes, E. Carrilho, M. F. Maggi Tavares and E. Kenndler, *Electrophoresis*, 2008, **29**, 2051.
72. J. L. Bernal, M. J. Nozal, L. Toribio, C. Diego, R. Mayo and R. Maestre, *J. Chromatogr. A.*, 2008, **1192**, 266.
73. H. Yoshida, J. Yamazaki, S. Ozawa, T. Mizukoshi and H. Miyano, *J. Agric. Food Chem.*, 2009, **57**, 1119.
74. F. Kvasnička, *J. Sep. Sci.*, 2005, **28**, 813.
75. M. G. Calabrese, G. Mamone, S. Caira, P. Ferranti and F. Addeo, *Food Chem.*, 2009, **116**, 799.
76. C. Simó, A. Rizzi, C. Barbas and A. Cifuentes, *Electrophoresis*, 2005, **26**, 1432.
77. J. Hegele, V. Parisod, J. Richoz, A. Förster, S. Maurer, R. Krause, T. Henle and T. Bütler, *Ann. N. Y. Acad. Sci.*, 2008, **1126**, 300.
78. J. McCullagh, J. Gaye-Siessegger and U. Focken, *Rapid Commun. Mass Spectrom.*, 2008, **22**, 1817.

CHAPTER 5

Lipidomics and Metabolomics of Dietary Lipid Peroxidation

ARNIS KUKSIS

Banting and Best Department of Medical Research, University of Toronto, Toronto, Ontario, Canada

5.1 Introduction

The biological consequences of *in vivo* reactions of ingested or *in situ* oxidized foods have long attracted the attention of biochemists and food scientists. Many oxo-lipids are known to interact with biological materials to cause cellular damage. Bi-functional secondary products of lipid oxidation such as malondialdehyde and hydroxyalkenals are powerful cross-linking agents and react with the amino groups of enzymes, proteins and nucleic acids. The biological aspects of lipid oxidation have thus become the subject of very active research and many reviews have appeared on the mechanism and potential consequences of lipid peroxidation.

The full spectrum of interactions between oxo-lipids and cellular components, however, have only become apparent using combinations of chromatographic and mass spectrometric methods. The new methodology has permitted a lipidomic approach to primary and secondary oxidation of dietary and tissue lipids, and a metabolomic assessment of its biological consequences.

This review opens with a brief discussion of the mechanism of lipid autoxidation, followed by the identification of the main types of autoxidized fatty acids found in dietary fats and oils. The review continues with a consideration of the dietary oxo-lipid transformation following ingestion and the absorption

RSC Food Analysis Monographs No. 9
Mass Spectrometry and Nutrition Research
Edited by Laurent B. Fay and Martin Kussmann
© The Royal Society of Chemistry 2010
Published by the Royal Society of Chemistry, www.rsc.org

of the water as well as lipid-soluble products and evidence for their presence in tissues. The discussion focuses on the use of lipidomics in the identification of the secondary products of autoxidation of unsaturated fatty acids, acylglycerols, glycerophospholipids and cholesteryl esters—which of all the oxo-lipids have emerged as the primary agents of disease and ageing. The review concludes with a consideration of the limits of endogenous and exogenous antioxidant systems, which may become overwhelmed by overdosing with dietary polyunsaturates, exposing the cells and tissues to an uninhibited attack by oxolipids.

5.2 Mechanism of Lipid Peroxidation

The exact mechanism of oxo-lipid formation is not known with certainty, but plausible hypotheses, although not without challenge,[1] have been advanced for both enzymatic[2] and chemical[3,4] oxidation to account for many of the final structures that have been isolated and identified beyond doubt by chromatographic, nuclear magnetic resonance (NMR) and tandem mass spectrometry (MS/MS) methods.

5.2.1 Free Radical Oxidation

The reaction of oxygen with unsaturated lipids involves free radical initiation, propagation and termination processes.[5] The initiation takes place in the presence of trace metals, light or heat, and involves abstraction of a hydrogen atom carried by a *bis*-allylic carbon belonging to a 1,4-*cis,cis*-pentadiene moiety of a polyunsaturated fatty acid (PUFA). This reaction leads to a fatty acid radical formation, which is stabilized by a molecular rearrangement to form a conjugated diene detectable at $\lambda = 235$ nm. This initial step is followed by a propagation step in which the unstable fatty acid radical reacts with molecular oxygen to give a peroxy fatty acid radical. The hydroperoxide may be reduced to the corresponding hydroxy fatty acid as a termination reaction (Figure 5.1). However, the hydroperoxide may also propagate the oxidative stress reaction by taking a hydrogen from another PUFA, which generates another alkyl radical and a hydroperoxide.[6,7] Lipid hydroperoxides can react again with oxygen to form such secondary products as epoxyhydroperoxides, ketohydroperoxides, dihydroperoxides, cyclic peroxides and bicyclic endoperoxides. All peroxides are subject to decomposition by homolytic cleavage to form alkoxy radicals and their reduction and breakdown products, including aldehydes, ketones, alcohols, hydrocarbons, esters, furans and lactones.

The basic assumption that the main product, hydroperoxide, is responsible for initiation and autocatalytic free radical generation has been questioned on the basis of studies completed on methyl linoleate autoxidation.[1] It was demonstrated that the autoxidation accelerating activity under mild conditions is not found in the chromatographically separated main product hydroperoxide fraction, but is found in other fractions. These fractions were shown to contain

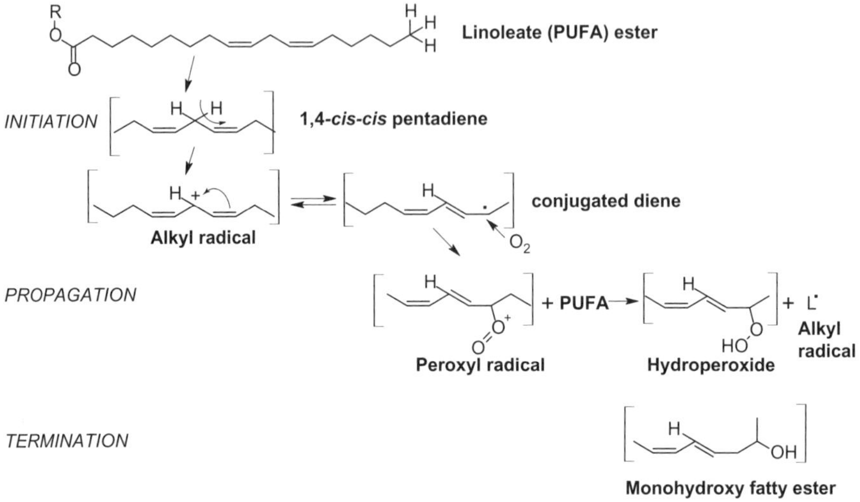

Figure 5.1 Overall mechanism of peroxidation of polyunsaturated fatty acids (PUFA) and their glyceryl and cholesteryl esters. Modified from ref. 6.

a peroxide-linked dimer with two hydroperoxy groups. The formation of a peroxide-linked dimer with two hydroperoxy groups is thought to begin with the addition of a peroxy radical to a conjugated diene. A similar intramolecular addition of a linolenate peroxy radical to its conjugated diene had been earlier shown to result in peroxidic ring compounds.[8,9] The fundamental issues of lipid peroxidation have been recently reviewed with special emphasis on routes to 4-hydroxynonenal.[10]

Initiation takes place by loss of a hydrogen radical in the presence of trace metals, light or heat. The free radicals can react with each other or antioxidants to break this chain reaction to form non-radical products. The non-radical products may include cellular proteins that had interacted with lipid peroxides without leaving a tell-tale trace of covalent binding of oxo-lipids. Decomposition of lipid hydroperoxides produces a multitude of materials that have biological effects and cause flavour deterioration in fat-containing foods.

5.2.2 Singlet State Oxygen Oxidation

Another important way in which unsaturated lipids can be oxidized involves exposure to light and a sensitizer such as chlorophyll. By this non free radical process, oxygen becomes activated to the singlet state by transfer of energy from the photosensitizer. The resulting singlet oxygen ($^1O_2$) produced by this process is extremely reactive. Linoleate is reported to react at least 1500 times faster with $^1O_2$ than with normal oxygen in the triplet ground state ($^3O_2$).[6] This hydroperoxidation reaction is so rapid that it has been postulated to initiate free radical autoxidation. In direct photooxidation, free radicals formed by ultraviolet (UV) radiation decompose oxygen-containing lipids following the free radical chain reaction mechanism. Direct photooxidation of food fats and oils is mostly avoided by appropriately covering samples with glass, plastic or some other material, which prevents light absorption of less than 230 nm.[4] In photosensitized oxidation, photosensitizers absorb visible or near UV light to become electronically excited.

Two forms of excitation have been recognized—the singlet and the triplet state.[6,11] The triplet state has a longer lifetime and is believed to initiate all photosensitized oxidation reactions. It can proceed along two different pathways, depending on the photosensitizer.[4] Pathway 1 leads to a hydrogen or electron transfer to an unsaturated lipid to yield a conjugated lipid radical, which reacts further with oxygen. Hydroperoxides formed by Pathway 1 and by normal free radical oxidation are identical, and differ from hydroperoxides produced by the alternate Pathway 2 reaction (Figure 5.2).[12] According to Pathway 2, unsaturated lipids are attacked on either side of a double bond by electrophilic single oxygen according to a concerted "ene" addition mechanism. The photosensitized oxidations arising from Pathways 1 and 2 cannot be prevented by chain-breaking antioxidants.[4] Typical photosensitizers found in food and biological materials, in addition to chlorophyll, are hemoproteins and riboflavin. Methylene Blue (Pathway 2) dissolved in $CHCl_3/MeOH$ is a popular experimental photosensitizer.[13] The fatty acid hydroperoxides generated by

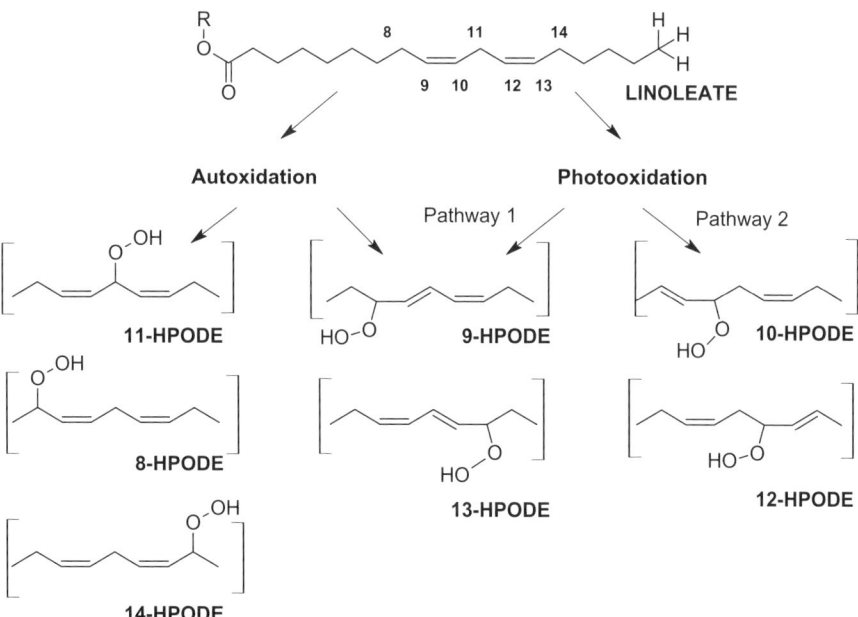

Figure 5.2 Summary of hydroperoxy-octadecadienoic acids (HPODEs) generated by autoxidation and photooxidation. Modified from ref. 12.

chemical oxidation with hydroperoxides (*e.g. tert*-butylhydroperoxide) are believed to be similar to those formed by free radical oxidation.[3,14]

5.2.3 Biological Lipid Oxidation

5.2.3.1 Lipoxygenases

Enzymatic oxidation of unsaturated lipids is catalyzed by different lipoxygenases (LOXs), which are widely distributed in human, animal and plant tissues. These enzymes are non-heme iron-containing proteins, which among other activities, catalyze dioxygenation of the 1,4-*cis-cis*-pentadiene moiety of unsaturated fatty acids such as linoleic acid, linolenic acid, arachidonic acid (ARA) and eicosapentaenoic acid (EPA) to yield hydroperoxides with one pair of conjugated double bonds.[15,16] The primary site of oxidation in ARA is commonly used in naming these enzymes, *e.g.* 15-lipoxygenase (15-LOX) catalyzes the production of 15-hydroperoxy-ARA from ARA. The oxidation process consists of removal of a hydrogen atom to form a free radical, conjugation of the double bonds, rearrangement of the radical electron, and insertion of di-oxygen. Enzymatically catalyzed processes are regio- and stereospecific, producing a variety of positional, geometric and optical isomers.[15–18] Exactly how these enzymes achieve this positional and stereo control is not clear. Animal LOXs generally catalyze oxidation of free fatty acids (FFAs), although an LOX from reticulocytes can also oxygenate their esters.[16]

Hydroperoxidation by soybean LOX can take place following incorporation of PUFAs into saturated diacylglycerols (DAGs).[19,20] Increasing levels of FFA in low density lipoprotein (LDL) enhance the accumulation of hydroperoxides and hydroxides of cholesteryl linoleate (Ch 18 : 2) in LDL exposed to 15-LOX.[21] The oxidation products of Ch 18 : 2, formed over prolonged periods of time, display an entirely or more predominantly non-enzymatic profile, whereas those that are formed within minutes, are entirely enzymatic. These results demonstrate that two mechanisms contribute to 15-LOX induced LDL oxidation. The non-enzymatic oxidation of cholesteryl ester (ChE) is both initiated and promoted by α-tocopherol.[21]

The reaction of docosahexaenoic acid (DHA), docosapentaenoic acid-ω3 (DPAω3) and docosapentaenoic acid-ω6 (DPAω6) with 5-, 12- and 15-LOXs produces oxylipins, which are identified and characterized by liquid chromatography-electrospray ionization-tandem mass spectrometry (LC/ESI-MS/MS).[22] Enzymatically oxygenated derivatives of ω-3 fatty acids DHA and EPA, known as resolvins, have potent inflammation resolution activity.[23]

5.2.3.2 Cyclooxygenases

Cyclooxygenases (COX) catalyze the reaction of achiral PUFA with oxygen to form a chiral peroxide product of high regio- and stereochemical purity. These enzymes also employ free radical chemistry, but execute efficient control during catalysis to form a specific product over a multitude of isomers found in the non-enzymatic reaction.[18] Four mechanistic models have been presented that could account for the specific reactions of molecular oxygen with a fatty acid in the LOX as well as in COX active site. The major puzzling issue in understanding how the LOX and COX enzymes control the regio- and stereochemistry of their catalytic reactions is the uncontrolled access of oxygen to the reactive lipid intermediate.[24]

5.3 Analysis of Lipid Oxidation Products in Food

5.3.1 Methods of Analysis

The general LC-MS and MS/MS methods used in lipidomic analyses of peroxidation of glycerophospholipids and ChEs have been recently reviewed[25] and presented in a protocol form.[26,27] The original detection, isolation and identification of the lipid peroxides is performed using thin layer (TLC) and high performance liquid (HPLC) chromatography techniques in combination with spectrophotometric monitoring. The oxo-fatty acid moieties are identified by gas chromatography-mass spectrometry (GC-MS) following reduction and derivatization of the analytes. Intact peroxides of synthetic and natural glycerolipids and ChE are resolved and identified by HPLC in combination with MS/MS. Novel hybrid instruments are now available, including hybrid linear ion-trap-orbitrap MS (LTQ Orbitrap TM, ThermoElectron, Bremen, Germany), which has been used in both positive and negative ion modes to determine the distribution of oxidized glycerophospholipid species in copper-oxidized (20 h) LDL and in lipid extracts of carotid artery plaques from patients with artherosclerosis.[28]

There have been several recent reviews published on the application of MS in food-related analyses and nutrition research, which have included lipid analyses. These have covered the determination and identification of natural lipids in food,[29] proteonomics,[30] proteonomic and metabolic responses to dietary factors,[31] toxic oxygenated α,β-aldehydes in foods,[32] MS in nutrition,[33] and α,β-unsaturated aldehyde-protein adducts.[34] More recently, a method for the analysis of the homologous series of alkanals, (*E*)-2-alkenals and (*E*,*E*)-2,4-alkadienals has been described[35] utilizing a headspace solid phase microextraction step and on-fibre derivatization with *O*-(2,3,4,5,6-pentafluorobenzyl)-hydroxylamine hydrochloride. Oxime derivatives formed on a fibre were desorbed in the GC injector and analyzed by comprehensive two-dimensional (2D) GC coupled to quadrupole MS. Quantification of the aldehydes utilized a stable isotope dilution analysis assay with octan-d_{16}-al as the isotopomeric internal standard. Structural characterization of α,β-unsaturated aldehydes by GC-MS has been shown to be dependent upon ionization method.[36] A new method for the determination of carbonyls in air using 2,4-dinitrophenylhydrazine (DNPH) has been developed.[37] The improvement involves a transformation of the C=N double bond into a C–N single bond using reductive amination of the hydrazone derivatives.

Two-dimensional sodium dodecyl sulfate polyacrylamide gel electrophoresis (SDS-PAGE) combined with matrix-assisted laser desorption ionization time-of-flight mass spectrometry (MALDI-ToF/ToF-MS) has been used for proteonomic profiling of rat lung epithelial cells exposed to acrolein (ACR).[38] One μl of matrix (α-cyano-4-hydroxycinnamic acid) was applied to the 2D-PAGE sample spot and allowed to dry. The MALDI-ToF/ToF-MS was performed using peptide mass fingerprinting and MS/MS analysis. Subsequent analyses were performed by means of appropriate software systems.

Protein targets of oxidized phospholipids in endothelial cells have been examined in detail by SDS-PAGE and reversed phase LC-ESI-MS/MS following treatment of human aortic endothelial cells (HAECs) with oxidized 1-palmitoyl-2-arachidonoyl-glycerophosphoethanolamine (GroPEtn)-*N*-biotin (Ox-PAPE-*N*-biotin)[39] (see Section 5.4.4.2). The fractions collected were enriched in 1-palmitoyl-2-(5-oxo]valeroyl)/glutaroyl-GroPEtn-*N*-biotin   (POV/GPE-*N*-biotin), 1-palmitoyl-2-([5,6-epoxy]isoprostane E$_2$) GroPEtn-*N*-biotin (PEIPE-*N*-biotin) and in unoxidized PAPE-*N*-biotin.

5.3.2   Oxidation of Dietary Lipids

It was observed in the 1950s that some fats and oils contain relatively high concentrations of non-volatile (compared to volatile) carbonyl compounds. The major non-volatile carbonyl class in milk fat was identified as an oxo-fatty acid (OFA).[40] Few other oxo-lipids were reported to occur naturally until a relatively simple method was described[41] for the detailed routine isolation and estimation of OFA in lipids as their DNPH derivatives. Specifically, milk fat, olive and safflower oil stored at room temperature in their original containers under ordinary laboratory light for 2–3 years yielded OFA values that were 4–25 times higher than those of the original fats and oils (1–3 μmoles g^{-1}). Interestingly, the identification included the semialdehyde DNPH, methyl azelaldehyde, which separated cleanly from model OFA DNPHs on an alumina bed.

An early analysis of fast-food fat (mostly tallow), olive oil and safflower oil heated in air for four days at the cooking temperature (180 °C) yielded OFA, and hydroxy and polyhydroxy fatty acids after transmethylation and preparation of DNPH derivatives. The sum of the concentrations of different OFAs at the time of maximum formation in the oils was approximately 260 μmoles g^{-1} at 48–72 h for the safflower oil, 200 μmoles g^{-1} at 48–72 h for olive oil and 170 μmoles g^{-1} at 72 h for the fast-food fat.[42] Specific identification of the oxidation products, however, could not be made until appropriate analytical methods were developed and model studies completed. It is now known that dehydrated foods, developed for both human and animal consumption, also become oxidized upon removal of water from the finished product.[43,44]

5.3.2.1 Primary Oxidation Products

For the present purposes, primary oxidation products of unsaturated dietary lipids include all peroxidation products except those involving chain cleavage;

these are grouped under secondary oxidation products (see Section 5.3.2.2). The products of free radical oxidation of free fatty acids (FFA) or their methyl esters have been investigated[6,45,46] most extensively and provide reference structures for their more complex glycerol (Gro) and cholesterol (Ch) esters.

5.3.2.1.1 Oxo Fatty Acids. Oleic acid, linoleic acid and linolenic acid, which are major components of vegetable oils, were the first and simplest dietary lipids to be investigated for lipid peroxidation. This work led to the development of appropriate analytical methodology and plausible hypotheses about the mechanisms involved in lipid peroxidation. It has been described in detail and extensive reviews have been prepared.[3,4,6,8,9,12] Only the major oxidation products are referred to here (see Figure 5.2).

Thus, autoxidation (air, O_2) of oleic acid occurs symmetrically about the double bond and yields four regioisomeric hydroperoxides, which are well resolved and identified by GC-MS after reduction and trimethylsilylation.[8] The 8- and 11-hydroperoxy isomers are present in slightly higher abundance than the 9- and 10-hydroperoxy isomers; ^{13}C NMR was used later to show the *cis,trans* geometry of all eight *cis* and *trans* allylic 8-, 9-, 10- and 11-oleate hydroperoxides.[4]

The autoxidation products of linoleic acid are extensively resolved by normal phase HPLC and are eluted in order of increasing polarity:[6]

- 13-hydroperoxy-*cis*-9, *trans*-11-octadecadienoate;
- methyl 13-hydroperoxy-*trans*-9, *trans*-11-octadecadienoate;
- methyl 9-hydroperoxy-*trans*-10, *cis*-12-octadecadienoate; and
- 9-hydroperoxy-*trans*-10, *trans*-12-octadecadienoate.

A complete resolution of the hydroperoxy diene products is obtained following reduction to the corresponding alcohols. A mechanism for the oxidation process has been proposed.[3] Four conjugated diene hydroperoxides result from oxygen addition at either the 9- or the 13- position of the eighteen carbon chain. Two of the products have *cis,trans* geometry in the conjugated diene unit, while the other two display *trans,trans* geometry.

Linolenic acid has two *bis*-allylic methylene groups and reacts twice as fast with oxygen as linoleic acid.[6] Reaction with oxygen at the end carbon positions of each pentadienyl radical produces a mixture of four positional peroxyl radicals leading to the corresponding conjugated diene 9-, 12-, 13- and 16-hydroperoxides containing a third isolated *cis*-double bond. Following reduction to linoleic acid hydroxides, the mixture is resolved by normal phase HPLC into eight major geometric components.[6] The formation of minor amounts of hydroperoxy-cyclic peroxides and other intact chain products of linoleic acid is also known.[9] The cyclic peroxides are resolved either by reversed phase HPLC after reduction to allylic alcohols, or directly on a micro silica column and the isolated fractions are characterized by TLC and GC-MS. The cyclic peroxides and dihydroperoxides were suggested to be important flavour precursors in oxidized fats.[9]

Conjugated linoleic acid is easily oxidized *in vitro* yielding the furan fatty acids: 8,11-epoxy-8,10-, 9,12-epoxy-9,11-, 10,13-epoxy-10,12- and 11,14-epoxy-11,13-octadecadienoic acids, which were isolated and identified by GC-MS.[47]

Arachidonic acid, a major component of glycerophospholipids, reacts with oxygen about twice as fast as linolenic acid because it has three active *bis*-allylic methylene groups and three 1,4-diene systems. ARA produces three penta-dienyl radicals. Oxygen attack at either end of these pentadienyl radicals produces a mixture of six positional isomers with hydroperoxide substitutions on carbon 5, 8, 9, 11, 12 and 15.[6,48] These compounds all have *trans,cis* geometry. The hydroperoxides are eluted from a normal phase HPLC column in order of increasing polarity as follows: *cis-trans*-12-; *cis-trans*-15-; *cis-trans*-11-; *cis-trans*-9-; *cis-trans*-8; and *cis-trans*-5-isomer. A chiral phase resolution of the racemic mixtures of *R*- and *S*-isomers of hydroxy eicosatetraenoic acids produced by photoperoxidation has also been reported.[49] The *S*-isomers emerge ahead of the *R*-isomers. Along with hydroperoxides, ARA yields iso-prostanes (IsoPs), which are prostaglandin-like compounds, by non-enzymatic reactions catalyzed by free radicals.[50] IsoPs include isomers of the D, E, and F prostaglandins. The F_2-IsoPs are isomers of PGF_2. Evidence has been obtained for free radical induced generation of D_2/E_2-isoprostanes *in vivo*, and for their presence in non-esterified and esterified form (as glycerophospholipids).[51]

Other important polyhydroperoxides are produced by autoxidation of eicosapentaenoic acid, a major component of fish oil, which yields the *cis*-5,8,11,14,17- and docosahexaenoic acid, a major component of brain pho-spholipids, which yields the *cis*-4,7,10,13,16,19-isomers found in fish and marine oils and represent the ω3-isomers. The eight hydroperoxides produced from EPA have been identified as 5-, 8-, 9-, 11-, 12-, 14-, 15- and 18-isomers, while the ten hydroperoxides from DHA have been identified as 4-, 7-, 8-, 10-, 11-, 13-, 14-, 16-, 17- and 20-isomers.[52,53] An improved LC-MS method has recently been reported for the analysis of mono- and polyhydroperoxy DHA.[54,55] Similarly to ARA, autoxidation of EPA and DHA generates an array of isoP-like compounds. EPA oxidation *in vitro* yielded F_3-isoPs as established by chemical and MS methods.[56] DHA oxidation *in vitro* and *in vivo* yields a number of different F_4-ring isomers, termed neuroprostanes (NPs).[57] A specific LC-ESI-MS/MS method has been described for a lipidomic analysis of 27 prostanoids and isoprostanes in a variety of biological fluids and extracts of tissues.[58]

5.3.2.1.2 Oxo Glycerolipids. Recent work has shown[59] that heating of trili-noleoyl glycerol (LLL) at 250 °C in an inert atmosphere (N_2) isomerizes the 9c12c fatty acids to the 9c12t, 9t12c and 9t12t acids. Under same conditions, the 9t12t fatty acids in trilinolelaidoyl glycerol isomerized to 9c12t, 9t12c and 9c12c acids. Both linoleic and linolelaidic acids yielded identical conjugated linoleic acids (CLAs) as shown by GC and infrared (IR) spectrometry. The isomerization was accompanied by degradation, which yielded chain-shor-tened aldehydes.

Autoxidation (air, O_2) of trioleoylglycerol (OOO) yielded three peaks corresponding to *tris-*, *bis-* and *mono*-hydroperoxides as shown by reversed phase HPLC and fast atom bombardment (FAB) MS (see below under photoperoxidation). The *sn*-2-isomers were eluted ahead of the corresponding *sn*-1(3)-isomers.

The autoxidation products of trilinoleoyl glycerol (LLL) yielded 28.4% *mono-*, 10.9% *bis-*, and 1.7% *tris*-hydroperoxides by preparative reversed and normal phase HPLC.[60] Subsequent analytical normal phase HPLC and ^{1}H NMR of the monohydroperoxide mixture yielded the eight positional and geometric monohydroperoxide isomers expected from previous studies of the Me ester oxidation. The *sn*-2-isomers were eluted ahead of the corresponding *sn*-1(3)-isomers.

Likewise, reversed phase HPLC effectively resolved the hydroperoxides obtained by autoxidation of trilinolenoyl glycerol (LnLnLn).[61] The intact hydroperoxides were identified spectrophotometrically, and after lipolysis and hydrogenation as the fatty acid trimethylsilyl (TMS) derivatives by gas–liquid chromatography (GLC) and GC-MS. The products included 9-, 12-, 13- and 16-*mono-*, *bis-* and *tris*-hydroperoxides as primary products. Analytical normal phase HPLC and ^{1}H NMR of the monohydroperoxide mixture from LnLnLn yielded the positional and geometric isomers expected from previous studies of the Me ester oxidation. The *sn*-2-isomers were eluted ahead of the corresponding *sn*-1(3)-isomers. In addition, the 9- and 16-hydroperoxy epidioxy, 9- and 16-hydroperoxy bicycloendoperoxy and 9,12-, 13,16- and 9,16-dihydroperoxy linolenoyl glycerols were also isolated following LnLnLn autoxidation.[61] Analytical normal phase HPLC separation of the monohydroperoxy epidioxides of LnLnLn showed eight peaks, as expected from the previous oxidation of the Me ester. The *sn*-2-glycerol isomers of monohydroperoxy epidioxides eluted ahead of the corresponding *sn*-1(3)-glycerol isomers. The products include 9-, 12-, 13- and 16-*mono-*, *bis-*, *tris*-hydroperoxy, and 9- and 16-hydroxyperoxy bicycloendoperoxy linolenoyl glycerols, which again were identified by GC-MS as the TMS derivatives. Reversed phase HPLC was similarly used to identify the autoxidation products of synthetic mixed acid triacylglycerols (TAGs) containing linoleic and linolenic acids, which gave monohydroperoxides and hydroperoxy epidioxides as the main products.[62]

Subsequently, combinations of TLC and HPLC systems with thermospray ionization (TSI) MS and electrospray ionization (ESI) MS were employed for a determination of elution factors of synthetic oxo-TAGs as an aid in identification of peroxidized natural TAGs.[63,64] Synthetic TAGs of known structure were converted to hydroperoxides, hydroxides, epoxides and core aldehydes, and their DNPH derivatives. The oxo-TAGs were identified by LC-TSI-MS and LC-ESI-MS following TLC and reversed phase HPLC. The elution factors were determined in relation to a homologous series of saturated TAGs, ranging from 24 to 54 acyl carbons. The TAGs were oxidized using *tert*-butyl hydroperoxide (*tert*-BHP) and ferrous ions under conditions claimed to mimic lipid peroxidation *in vivo*.[14] Peroxidation of corn oil TAGs yielded monohydroperoxides and dihydroperoxides of different chain length, along with monohydroperoxides and core aldehydes of different chain length, all of which were resolved by TLC. Combined LC-TSI-MS of individual oxo-TAG bands permitted the identification of the major components in each TLC band based on single ion mass (SIM) chromatograms. Due to shielding, the PUFA in the *sn*-2-position of the TAG has been observed to be more stable to oxidation compared with the *sn*-1(3) position.[65]

The photoperoxidation products of soybean, safflower and olive oils also have been identified by reversed phase HPLC and characterized by FAB-MS following reduction.[66] Under these conditions, OOO gave three peaks corresponding to *tris-, bis-* and *mono*-hydroperoxides (in order of elution). *Mono*-hydroperoxides were the major oxidation products in all oils, with the *tris-* and *bis*-peroxides accumulating with prolonged photoperoxidation.

More recently, reversed phase HPLC coupled to MS/MS *via* an atmospheric pressure chemical ionization (APCI) source was used to identify the intact oxidation products of OOO, LLL and LnLnLn.[67] APCI-MS was also used to identify the epidioxides and hydroperoxyepidioxides from peroxidation of LnLnLn, which were not observed for the other TAGs (OOO and LLL).[67] Among these were stable epoxides formed by at least two distinct processes resulting in two types of epoxides. The first process was characterized by formation of the epoxide across the double bond in the TAG molecule, while the second involved formation of the epoxide next to a double bond.

The oxidation products from OOO obtained under frying conditions have also been analyzed by HPLC with APCI-MS detection.[68] The TAG was heated at 190 °C with 2% water added every hour until polar components reached about 30%. The oxidation products included hydroperoxides, epoxides and acetone. Other products included chain addition products formed by addition of acyl chain subunits to intact OOO to form higher molecular weight products.

Extensive analyses of TAG hydroperoxides in vegetable oils have been reported[19,20] using reversed phase HPLC with UV detection (see also below). The *mono-* and *bis*-hydroperoxides of LLL, LLO and LOO were identified by HPLC using a linear gradient of CH_3CN in methylbutyl ether.

The autoxidation products of TAGs from normal and genetically modified canola oil varieties have been reported[69] using HPLC with tandem MS detection. Normal, high stearic acid and high lauric acid canola oils were heated in the presence of air to allow autoxidation to occur. The major autoxidation products that remained intact were epoxides and hydroperoxides. Intact oxidation products resulted mostly from oxidation of oleic acid, while oxidation products of linoleic and linolenic acid chains decomposed to yield chain-shortened species. A dual parallel ESI-APCI-MS, MS/MS and MS/MS/MS was recommended for analysis of TAGs and TAG oxidation products.[70]

LC-ESI-MS and LC-CID-ESI-MS based on a single quadrupole instrument[71,72] and an extensive collection of standards[64] were employed to characterize the hydroperoxides, diepoxides and hydroxides obtained as major components from peroxidation of synthetic TAGs containing one and two double bonds per molecule. Previously unidentified peroxide bridged *tert*-BHP adducts were present in significant amounts in all preparations. Later, mixed hydroperoxides and core aldehydes were quantified as the DNPH derivatives in autoxidized sunflower seed oil by TLC-LC-MS.[73] These separations and peak identification has been reviewed elsewhere.[25-27]

More recently, LC-ESI-MS/MS has been used to characterize the perox-idation of synthetic 1,2-dipalmitoyl-3-oleoylglycerol (PPO) and 1,3-dipalmi-toyl-2-oleoylglycerol (POP) thermo-oxidized under $^{18}O_2$ atmosphere.[74] The hydroperoxide of PPO gave an intense NH_4 adduct at m/z 883 $[M(NH_4)]^+$. MS/MS fragmentation of the precursor ion at m/z 883.0 gave product ions at m/z 866.0, 609.5, 591.8, 575.7 and 551.8, which were attributed to $[M + H]^+$, $[M - P]^+$, $[M - P - H_2O]^+$, $[M - P - H_2O_2]^+$ and $[M - HpoO]^+$, respectively. The LC-ESI-MS/MS data together with Fourier transform IR (FTIR) and NMR analyses confirmed the formation of epoxy (Es)-TAGs containing oleic acid.[75] The $[M - P]^+/[M - Es]^+$ ratios (0.75 and 1.24 for PPEs and PEsP, respectively) differed for individual regio-isomerics, and could be used to dis-tinguish and quantify the two positional isomers.

The polyunsaturated marine oil TAGs containing 20 : 4, 20 : 5, 22 : 5 and 22 : 6 fatty acids yield more complex hydroperoxides and epoxides than the seed oils. Furthermore, they would be anticipated to form the isoprostane and neuroprostane esters, as reported for the corresponding simple methyl esters and glycerophospholipids.[50,51] However, no specific reports of isoprostane or neuroprostane-containing TAGs have been published, although prostaglandin esters of arachidonoyl glycerol are known.[77,78]

Like unesterified PUFA, PUFA glycerophospholipids yield isoprostane and neuroprostane derivatives, which are released by phospholipase A₂.[50,51,76-78] The glycerophospholipids containing the oligo-unsaturated fatty acids yield hydroperoxides and hydroxides similar to those of TAGs with similar fatty acid composition.

5.3.2.1.3 Oxo Cholesteryl Esters. Ch palmitate (Ch-16 : 0) becomes oxi-dized exclusively in the sterol ring, while Ch-18 : 1 becomes oxidized in the sterol ring and in the oleic acid moiety.[79] GC-MS analyses of the TMS deri-vatives of the reduced oxo-sterol moiety released by ChE hydrolase yielded the 7α- and 7β-hydroxy cholesterols and 7-ketocholesterol.

Since the bond dissociation energy of a *bis*-allylic carbon–hydrogen in the PUFA is substantially lower than that of an allylic carbon–hydrogen bond in Ch, the peroxidation of Ch-18 : 2 and Ch-20 : 4 yields predominantly Ch-LOOH and Ch-ARAOOH.[80,81] Ch-18 : 2 hydroperoxide and Ch-20 : 4 hydroperoxide (and their isomers) of oxidized LDL were characterized by GC-MS of the oxo-fatty acid moieties.[80,81] Intact ChE oxidation products have also been resolved by HPLC without reducing the hydroperoxy groups to the more stable hydroxides.[82]

Detailed identification and quantification have been reported[83] for the regioisomeric hydroperoxides of Ch-18 : 2 exposed to aqueous radicals gen-erated by the thermolabile azo compound 2,3'-azobis(2-amidinopropane) dihydrochloride, or a free radical initiator, di-*tert*-butyl hyponitrite (DTBN).

HPLC provided separation of the regioisomeric hydroperoxides and their hydroxide reduction products.

A comparison has been reported of the products obtained by copper and soybean 15-LOX oxidation of LDL Ch-18 : 2 based on HPLC with UV monitoring.[84] It was found that, in LDL oxidized by activated monocytes, CuSO$_4$ or soybean 15-LOX, the most abundant oxidation product was the same Ch-HPODE.

It has been shown that peroxidation of PUFA esters of Ch yields a complex mixture of hydroperoxides and cyclic peroxides, including dozens of diastereomers and regioisomers, which can be successfully analyzed by co-ordination ion-spray mass spectrometry (CIS-MS).[85] Silver ion use results in readily detected Ag$^+$ adducts of peroxides and hydroperoxides. The ions at [M + 107] and [M + 109] undergo fragmentation typical of hydroperoxides and cyclic peroxides. This coupling permitted, for the first time, the assignment of defined structures after separation by conversion of the ChE to the corresponding Me esters, and a comparison with samples previously characterized. Online characterization was performed following semipreparative HPLC separation and collection of peaks, followed by LC-CIS-MS. The Ag$^+$ ion methodology[85] has been used to examine the prostaglandin (isoprostanoid) bicyclic endoperoxides from the peroxidation of Ch-20 : 4 by LC-MS and GC-MS techniques.[86] On the basis of the free radical mechanisms of the transformation, it was concluded that only 12- and 8-peroxyl radicals (those leading to 12-HPTE and 8-HPTE) of Ch-ARA can form these new peroxides. In addition, *in vitro* autoxidation of Ch-ARA produced a novel class of peroxides (dioxolane–isoprostanes),[87] which possessed a bicyclic endoperoxide moiety characteristic of the isoprostanes and a dioxolane peroxide functionality in the same molecule.

5.3.2.2 Secondary Oxidation Products

Chain cleavage of the fatty hydroperoxides results in the formation of volatile compounds of low molecular weight and a TAG, glycerophospholipid or ChE remnant containing a short acyl chain.[88,89] Whereas the former are largely removed from the oil by volatilization during heating, the latter remain in the frying oil and are absorbed by the food. With autoxidized linoleate, 2,4-decadienal and methyl octanoate are produced by a homolytic cleavage on the ester side of 9-hydroperoxide, and 3-nonenal and [9-oxo]-nonanoate by a cleavage on the methyl terminal side; hexanal and [12-oxo]-9-dodecenoate are obtained by cleavage on the ester side and pentane, 1-pentanol, pentanal and [13-oxo]-9,11-tridecadienoate are produced by cleavage on the methyl terminal side of the 13-hydroperoxide.[6] A 4,5-epoxy-*trans*-2-decenal is formed during heating of fats containing linoleic acid. This aldehyde is a key component of flavours of different foods, but also causes dose- and time-dependent apoptosis in endothelial cells[32] (Figure 5.3).

While the volatile aldehydes have been extensively investigated, there has been relatively little work on the high molecular weight core aldehydes. The carboxy-terminated aldehydes[90] are nearly totally esterified to glycerolipids and

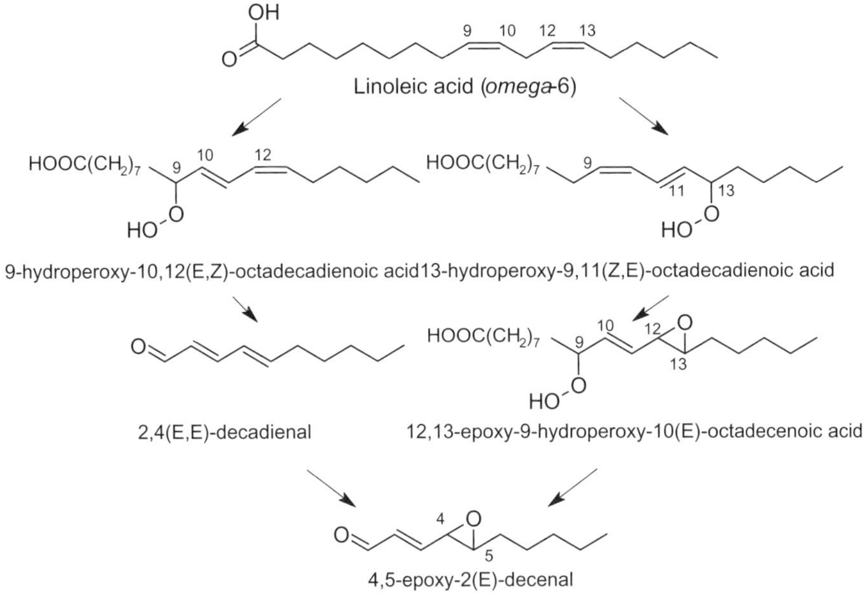

Figure 5.3 Formation of 4,5-epoxy-*trans*-2-decenal during heating of fats containing linoleic acid. This aldehyde is a key component of flavours of different foods, but also causes dose- and time-dependent apoptosis in endothelial cells. Modified from ref. 32.

Ch (core aldehydes); they nevertheless react with membrane proteins or are subject to release as the free carboxylates by lipases (see below).

5.3.2.2.1 HHE and HNE. HHE (4-hydroxy-2-hexenal) and HNE (4-hydroxy-2-nonenal) are produced from ω-3- and ω-6-PUFA, respectively, in foods during processing or storage, as well as in cells and tissues of living organisms.[34] Figure 5.4 shows the formation of the two volatile aldehydes from the ω-3 and ω-6 fatty acids, and their glyceryl and cholesteryl esters. The intermediate steps involving the initial peroxidation, the metal catalyzed hemolytic cleavage of the hydroperoxides and reduction are not included in the figure.

The formation of HNE and HHE in separately thermally oxidized methyl esters of linoleic and linolenic acids has been demonstrated at cooking temperatures (185 °C/0–6 h).[91,92] The unsaturated aldehydes become incorporated into fried foods and can be absorbed from diet (see below). The aldehydes were determined by HPLC as the DNPH derivatives. The formation of HNE and ONE in foods[30] and *in vivo* has been recently reviewed in great detail.[34] The 4-keto analogue of HNE, ONE, has now been shown to be a direct product of lipid oxidation,[89,93] arising independently and not from oxidation of HNE (HNE and ONE appear not to interconvert metabolically).

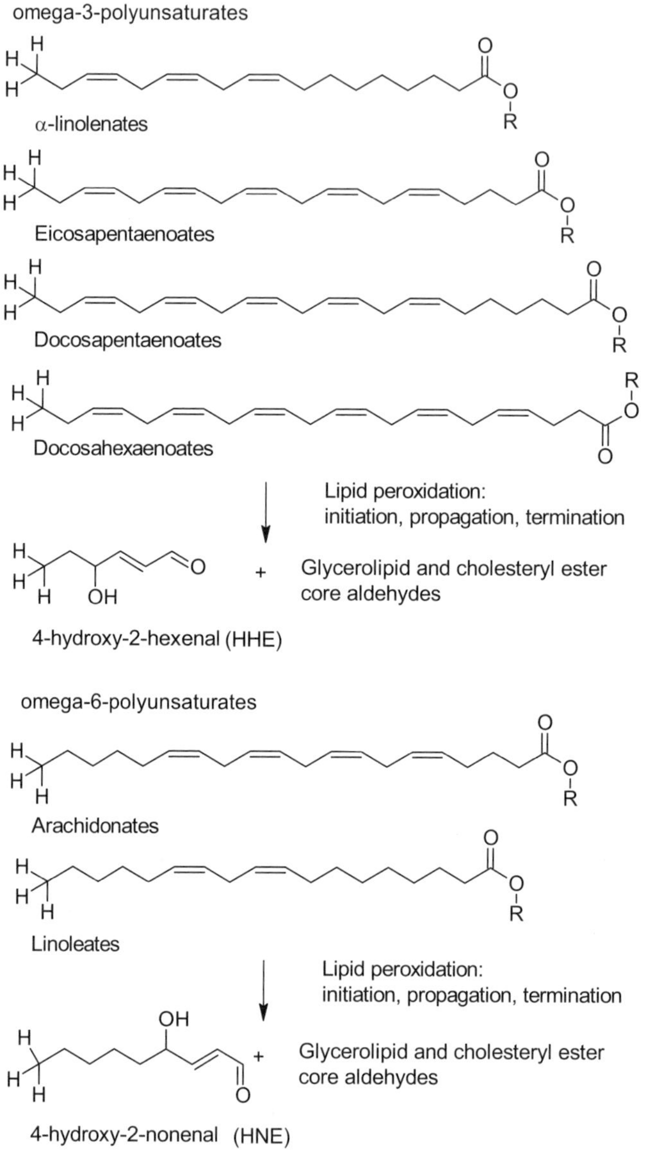

Figure 5.4 Generation of reactive hydroxy alkenals during peroxidation of ω-3 and ω-6 fatty acids and their esters. Structures of mirror image core aldehydes are not shown.

Several mechanisms of HNE formation involving a 4,5-epoxy derivative as an intermediate have been presented.[94] Recently, it was suggested[10] that its formation may occur by several different pathways including a direct non-enzymatic breakdown of the hydroperoxide, an epoxidation mechanism starting with either hydroperoxy- or hydroxy-fatty acids, and a complex dimerization mechanism involving the reaction of two hydroperoxy fatty acids. All these pathways lead to 4-hydroperoxynonenal as the intermediate, which is finally reduced to 4-HNE. The hydroxyl alkenals can be measured in biological fluids or in cell culture media by GC-MS. The aldehydes are derivatized into pentafluorobenzyl oxime directly in the sample.[95,96] Two *syn* and *anti* isomers generated during the derivatization step are extracted with hexane, and the hydroxyl group converted into TMS ether with *N,O-bis*(TMS)-trifluoroacetamide and the derivatives analyzed by negative chemical ionization (NCI) GC-MS. Hydroxy-alkenals are easily detectable and they reflect, separately, the peroxidation of all ω6 (4-HNE as marker) and all ω3 (4-HHE as marker) fatty acids. Quantification of HNE in different fats and oils has met with difficulty because of extensive polymerization of the aldehydes in some specimens and not in others.[91] The estimates have ranged from 10 to 1000 µg kg^{-1} food. However, it must be noted that hydroxyl-alkenals are reactive molecules able to make covalent adducts with amines[97,98] and thiols,[99,100] which may lead to a serious underestimation of their true tissue content.

The use of short-chain aldehydes as biomarkers of lipid peroxidation in comparison with other markers of oxidative stress has been recently reviewed.[43] The authors propose 4-hydroxy-alkenals and their oxidized derivatives as the best choice. The main advantages of these markers are that 4-hydroxy-hexenal and 4-hydroxy-nonenal (as well as their oxidized derivatives) can be determined in one run, providing information on the overall peroxidation of n-3 and n-6 polyunsaturated fatty acids, respectively. The 4-HHE and 4-HNE, measureable in urine, are non-invasive biomarkers that gave an indication of the oxidative stress.

Nucleophilic targets and adduct formation have been recently suggested as universal molecular mechanisms of 4-hydroxy-2-nonenal and acrolein, as well as that of other oxo-lipids generated from polyunsaturated fatty acids.[101] The α,β-unsaturated carbonyl structure provides an electron polarizability, which turns them into soft electrophiles that preferentially form 1,4-Michael type adducts with soft nucleophiles. Cysteine sulfhydryl groups are the primary soft nucleophilic targets of ACR and HNE. Studies on the side-chain modifying chemistry have concluded that ONE is a more reactive protein modifier and cross-linking agent than HNE.[102,103]

5.3.2.2.2 Malondialdehyde and Acrolein. Malondialdehyde (MDA) is produced by the radical breakdown of hydroperoxides resulting from PUFA peroxidation containing at least two double bonds (18 : 2, 18 : 3, 20 : 3, 20 : 4, 22 : 3, 22 : 4, 22 : 5 and 22 : 6).[104] The decomposition of such a *bis*-allylic radical peroxide can be catalyzed by transition metals such as iron (Fe^{2+}), but can also be stimulated by cyclooxygenase and thromboxane synthase in appropriate cells.[105] MDA is the most popular marker of lipid peroxidation, although not the most informative.

The biologically active carbonyl compounds derived from lipid peroxidation include acrolein (ACR).[104,106] The major precursor of ACR, however, is glycerol.[107] The emission of volatile C_3 to C_9 aldehydes from DAG-rich and TAG-rich oils containing fatty acids with 1–3 double bonds during deep-frying has been measured and found not to be affected by the acylglycerol structure.[108]

An effective method for simultaneous determination of ACR, MAD and HNE from lipids peroxidized with Fenton's reagent (H_2O_2/Fe^{2+}) has been developed.[109] The aldehydes were derivatized to *N*-methylpyrazoline, *N*-methylpyrazole and 5-(1′-hydroxyhexyl)-1-methyl-2-pyrazolin with *N*-methylhydrazine, respectively, and determined by capillary GLC with nitrogen-phosphorus detector: maximum amounts of ACR ($9.7 \pm 2.1 \, \text{nmol ml}^{-1}$) and MDA ($61.18 \, \text{nmol ml}^{-1}$) were formed from cod liver oil. Figure 5.5 shows the structures of selected aldehydes generated during peroxidation of polyunsaturated dietary fatty acids.

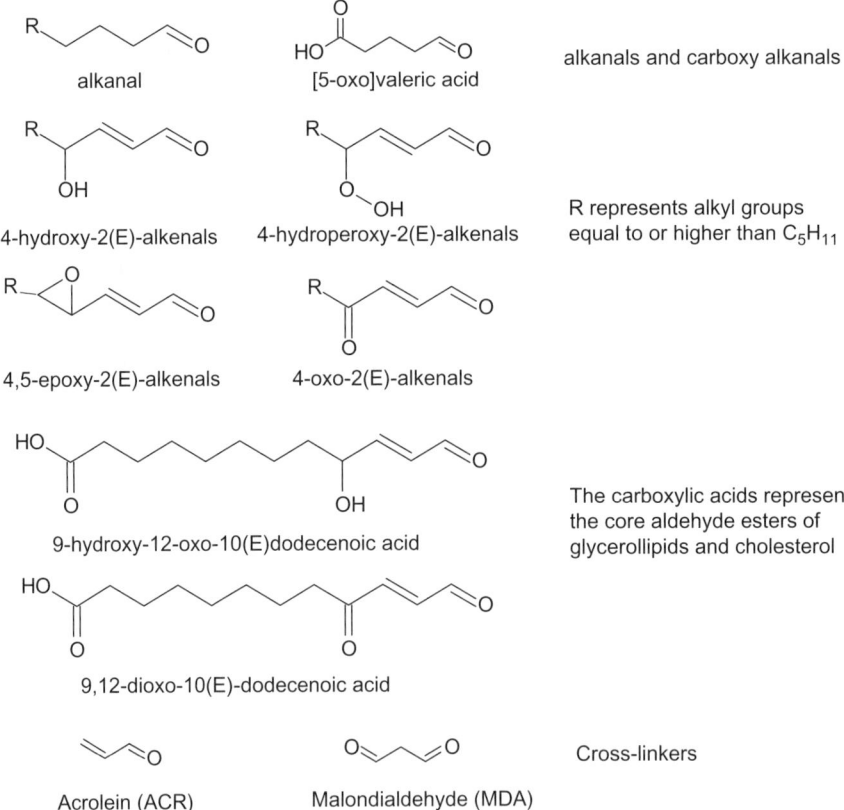

alkanal

[5-oxo]valeric acid

alkanals and carboxy alkanals

4-hydroxy-2(E)-alkenals

4-hydroperoxy-2(E)-alkenals

R represents alkyl groups equal to or higher than C_5H_{11}

4,5-epoxy-2(E)-alkenals

4-oxo-2(E)-alkenals

9-hydroxy-12-oxo-10(E)dodecenoic acid

The carboxylic acids represent the core aldehyde esters of glycerollipids and cholesterol

9,12-dioxo-10(E)-dodecenoic acid

Acrolein (ACR)

Malondialdehyde (MDA)

Cross-linkers

Figure 5.5   Structures of selected bioreactive bifunctional alkenals and core aldehydes. Representative saturated aldehydes derived from both methyl and carboxyl terminals of the fatty acids are included.

5.3.2.2.3 Core Aldehydes. It is frequently overlooked that, for every molecule of short-chain alkenal or hydroxyalkenal released from an oxo-lipid ester, there remains an equivalent of aldehyde bound to the Gro or Ch molecule. Oxidation of oleic acid esters yields a mixture of 8-, 9-, 10- and 11-oxo esters by cleavage on the ester side of the hydroperoxides.[6] These high molecular weight non-volatilc aldehyde esters are potentially more useful predictors of the quality of edible oils than the volatile aldehydes, which tend to evaporate.

Similar core aldehydes are generated by metal-catalyzed homolytic cleavage of the hydroperoxides of linoleic acid, ARA and DHA esters. Specifically, the monohydroperoxide of linoleic acid yields [9-oxo]nonanoate, while ARA yields the [5-oxo]valerate but DHA the [4-oxo]butyrate, which are retained by their parent ester molecules. The less stable hydroperoxy, hydroxy and epoxy core aldehydes—like the bifunctional low molecular weight analogues—are degraded at elevated temperatures to saturated core aldehydes and acids.[110,111]

Specific identification of core aldehydes among the peroxidation products of model TAGs and corn oil were first reported following *tert*-butyl hydroperoxide oxidation.[63] For identification, the core aldehydes were converted into the DNPH derivatives,[110] which were resolved by TLC using a neutral lipid solvent system. Later, representative C_9 and C_5 core aldehydes of glyceryl esters were synthesized to serve as model compounds for the LC-SI-MS identification of core aldehydes in lipoproteins, vegetable oils and animal fats.[64]

Retention of chain breakage products by the OOO molecule following losses of volatile fragments has been observed at frying temperatures.[68] A protonated molecular ion at m/z 747.7 appeared along with an acylium ion, RCO^+, at m/z 127.0. Other chain-shortened OOO species appeared as a result of losing C_8 (m/z 493.4), C_9 (m/z 479.4), C_{10} (m/z 465.3), C_{11} (m/z 451.3) and C_{12} (m/z 437.3) volatile short-chain fragments, which corresponded to core aldehydes anticipated from the chain cleavage of the various monohydroperoxides of OOO identified earlier among the *tert*-BHP oxidation products of both synthetic and natural TAGs containing oleic acid residues.[63,64] The latter methods,[63,64] were also used for the identification of mixed lipid ester hydroperoxides and core aldehydes, including C_4-core aldehydes, from autoxidized TAGs of Baltic herring oil.[76]

Likewise, TAGs containing short-chain aldehydes have been recognized among the oxidation products of vegetable oils using normal phase LC-MS.[19,20] A separation was obtained for oxo-TAGs into classes of oxidation products such as epoxy-, oxo-, hydroperoxy-, hydroxy- and core aldehydes ("21/2 glycerides"), which were identified by SIM.[19,20] TAGs containing chain-shortened core aldehydes were also reported among the peroxidation products from genetically modified canola oils using APCI-MS.[69,70]

The most extensive identification and quantification of TAG core aldehydes was performed during a rapid oxidation of corn and sunflower oils with *tert*-BHP/Fe^{2+}.[71,72] The core aldehydes were isolated by TLC as DNPH derivatives and identified by reversed phase HPLC with online ESI-MS and by reference to standards. A total of 113 species of TAG core aldehydes were specifically identified, accounting for 32–53% of the DNPH-reactive material of high molecular weight, representing 25–33% of the total oxidation products. The major core aldehyde species (50–60% of total TAG core aldehydes) were the mono [9-oxo]nonanoyl- and mono-[12-oxo]-9,10-epoxy dodecenoyl- or [12-oxo-]-9-hydroxy-10,11-dodecenoyl-DAGs.

These techniques were subsequently utilized in a systematic identification of TAG core aldehydes in air-oxidized sunflower seed oil.[73] The major species of core aldehydes were identified as 9-oxononanoyl (70%)-, 12-oxo-9,10-epoxydodecenoyl (10%)- and 13-oxo-9,11-tridecadienoyl (5%)-containing acylglycerols, plus smaller amounts of simple and mixed chain-length dialdehydes, and hydroxyl and epoxy monoaldehyde-containing acylglycerols (15% of total). Quantitatively, the core aldehydes made up $2–12 \, \mathrm{g \, kg^{-1}}$ of oil by UV detection and $2–9 \, \mathrm{g \, kg^{-1}}$ by ESI-MS detection, whereas the hydroperoxides measured in the unreduced state by HPLC with evaporative light scattering detection (ELSD) were estimated at $200 \, \mathrm{g \, kg^{-1}}$ after 18 days of autoxidation. The authors have provided[72,73] extensive tabulations of the oxo-TAGs (including the core aldehydes) of autoxidized corn and sunflower seed oils. In contrast to the samples of sunflower oil oxidized by *tert*-BHP, the air oxidized oil contained 12 : 1 and 13 : 2 core aldehydes. Subsequent LC-ESI-MS studies have identified chain-shortened TAG molecules in pig plasma chylomicrons following feeding of peroxidized dietary fats[112–114] (see Section 5.4.2.2).

Reference core aldehydes of acyl and alkyl glycerols have been prepared by chemical oxidation and pancreatic lipase hydrolysis of oxo-acylglycerols derived from bovine milk fat and shark liver oil.[115]

The presence of core aldehydes in frying oils has been demonstrated by HPLC,[116] as has been the formation of short-chain glycerol-bound aldehydes during thermo-oxidation in model systems containing single fatty acid TAGs.[117,118] The compounds formed from OOO and LnLnLn were identified as [8-oxo]octanoate, [9-oxo]nonanoate, octadionate and nonadionate. Recently, the core aldehyde formation in thermo-oxidized olive and sunflower oils has been investigated with the aid of MS/MS with similar results.[119]

Phospholipids containing core aldehydes at the *sn*-2-position were first detected in copper oxidized plasma lipoproteins.[88,120] The major core aldehydes from oxidized LDL and high density lipoprotein (HDL) were identified as 1-palmitoyl(stearoyl)-2-(9-oxo)-nonanoyl- and 1-palmitoyl(stearoyl)-2-(5-oxo)valeroyl-*sn*-glycerols after phospholipase C digestion of the DNPH derivatives of the oxidized glycerophospholipids. High levels of such aldehydes are formed when glycerophospholipids containing unsaturated fatty acids are oxidized *in vitro*.[121,122] Glycerophosphocholine (GroPCho) core aldehydes have been detected in atherosclerotic plaques, aged erythrocytes, apoptotic cells and in the plasma of smokers.[122,123] In another study,[124] oxidized LDL was found to contain *sn*-2-azelaic (Az) acid esterified to GroPCho (an ester not detected by DNPH derivatization), which induces apoptosis at low concentrations. Multiple reaction monitoring (MRM) at 22 eV collision energy and a transition from the molecular cation $[M + H]^+$ of *m/z* 652 to daughter ion at *m/z* 184 allowed the identification of hexadecyl/azelaoyl GroPCho

(HAzGroPCho). Recently the copper-catalyzed lipid oxidation products of LDL (20 h) was examined by MS/MS with Orbitrap.[28] One class of molecules recognized by peaks in m/z range 594–666 were identified as truncated (core aldehyde) phosphatidylcholine (PtdCho). It has been suggested that cell membranes "grow whiskers" as phospholipids undergo peroxidation and that many of their oxidized fatty acids protrude at the surface.[125]

The chain-shortened oxo-fatty acid esters of cholesterol were first recognized during copper-catalyzed oxidation of LDL.[120,126] The aldehydes were identified by reference to standards in GC-MS with ammonia and in LC-MS with NCI following preparation of DNPH and methoxime (MOX/TMS) derivatives. The components were Ch-[5-oxo]valerate (m/z 664) and Ch-[9-oxo]nonanoate (m/z 720), with much smaller amounts of Ch-[4-oxo]butyrate (m/z 650), Ch-[6-oxo]hexanoate (m/z 678), Ch-[7-oxo]heptanoate (m/z 692) and Ch-[10-oxo]decanoate (m/z 734), as indicated by SIM chromatograms. Similar core aldehydes were later isolated from copper-oxidized HDL.[127] Significantly, there were no unsaturated core aldehydes recognized, although monounsaturated derivatives were identified following *tert*-BHP oxidation of Ch-18 : 2 and Ch-20 : 4.[79] The oxidation in the absence of copper ions of Ch-18 : 2 and Ch-20 : 4 yielded readily detectable amounts of monounsaturated core aldehyde esters such as Ch-[7-oxo]heptenoate (m/z 690), with much smaller amounts of Ch-[6-oxo]hexenoate (m/z 676), [Ch-9-oxo]nonadienoate (m/z 716) and Ch-[10-oxo]decadienoate (m/z 730), as well as the earlier identified saturated core aldehydes. The formation of Ch-[9-oxo]nonanoate and Ch-[5-oxo]valerate as the major products in Cu^{2+} treated LDL has been independently confirmed.[128,129] The peak of aldehyde formation at 12 h was followed by a significant decrease in ChE core aldehyde proportion, which was accompanied by an increase in the [7-keto]ChE core aldehyde proportion.[129] The formation of the highly toxic 7-hydroperoxy ChE core aldehydes during autoxidation of Ch-18 : 2 has been noted, as has been the formation of Ch-[11-oxo]undecanoate (m/z 746), Ch-[12-oxo]dodecanoate (m/z 780) and Ch-[9-oxo]nonanoate (m/z 720).[130]

ChE core aldehydes in human atherosclerotic plaques have been detected using essentially identical LC-MS methods, including preparation of DNPH derivatives.[88,131] The core aldehydes produced by oxidation of ChE have been analyzed in detail by others, involving chemical synthesis of reference standards.[128,132] Phorbol esters were shown to stimulate the production of ChE core aldehydes by macrophages, which were also able to internalize externally added core aldehydes.[128]

5.4 Bioanalysis of Lipid Oxidation Products

5.4.1 Lumenal Degradation and Reduction of Oxo-lipids

The dietary oxo-lipids undergo further transformation in the mouth, stomach and small intestine before becoming exposed to the processes of fat absorption. Only a few of these alterations have been investigated and the fate of the products assessed. The ingested hydroperoxides are subject to degradation by reduction and lipolysis. In view of the high degree of oxygenation of the stomach and the anticipated peroxidation of dietary fat there, it has been referred to as a "bioreactor",[133] which can be accessed by an *in vitro* model.[134] It has been suggested that myoglobin and phenolic antioxidants participate in these transformations. An *in vitro* digestion model has been employed to assess the bioaccessibility of HNE and related aldehydes present in oxidized oils rich in ω-6 acyl groups.[135] It was found, that despite the great reactivity of the oxygenated α,β-aldehydes, study of the head space and the fluid matrix of the various digestion phases confirmed that HNE, ONE and EDE persist after digestion and are susceptible to absorption, and therefore could reach the systemic circulation. This study also showed that the antioxidant butylated hydroxyl toluene (BHT), when added to foods, can evolve to give toxic metabolites that are bio-accessible after digestion. The *in vitro* model for fat digestion has been recently adopted for an LC-/ESI-MS/MS examination of the lumenal digestion of oxidized rapeseed oil (M. Tarvainen, J-P. Suomela, A. Kuksis and H. Kallio, *Lipids*, submitted).

5.4.1.1 Lipases, Phospholipases and Esterases

The digestion of oxidized dietary fats and oils starts in the mouth, where lingual lipase has been shown to attack the short chain and oxo-fatty acids present in the *sn*-3-position of acylglycerol molecules. Pregastric esterase is a major fat-digesting enzyme in the calf as well as in man.[136] The preferential lipolysis of oxo-fatty acids continues in the stomach under the influence of gastric lipase which, like the lingual lipase, has shown preference for hydrolysis of the *sn*-3-position.[136] The pregastric lipases do not possess a free thiol group which could complex with the aldehydes, unlike the gastric lipases originating from the stomach.[137] Inhibition studies with enantiomeric phosphonate inhibitors have shown that the TAG molecule completely fills the active site crevice of dog gastric lipase, in contrast to what is observed with other lipases such as pancreatic lipase, which has a shallower and narrower active site.[138] The shallower and narrower active site may account for the lower stereoselectivity of pancreatic lipase and more extensive hydrolysis of the oxo-TAGs.[139]

Extensive studies of hydroperoxide digestion have shown that, at low oral doses, trilinoleoyl glycerol hydroperoxide (LLL-OOH) is soon broken down to linoleic hydroperoxide (L-OOH) and linoleic hydroxide (L-OH), probably by pregastric lipases; whereas, at high doses, LLL-OOH is retained in the stomach.[140,141] Neither LLL-OOH nor L-OOH reached the intestine in significant amounts, though unoxidized lipids moved to the intestines. When L-OOH was given intragastrically, the lipids decomposed under the acidic conditions of the stomach and linoleic acid hydroxides, hexanal, [9-oxo]nonanoic acid, and two novel compounds were detected 30 min after treatment. The two novel compounds were identified as epoxyketones: [11-oxo]-12,13-epoxy-9- and [11-oxo]-9,10-epoxy-12-octadecenic acid.[140]

The lipolysis of any TAGs unhydrolyzed in the stomach is continued in the small intestine, where pancreatic lipase releases indiscriminately both oxidized and non-oxidized fatty acids from the primary positions of acylglycerols.[60,64] The core aldehydes (as well as other oxo-lipids) occupying the *sn-2*-positions of the glycerol would be expected to reach the small intestine and be absorbed and re-esterified *via* the monoacylglycerol (MAG) pathway.

The dietary oxo-glycerophospholipids are hydrolyzed by mammalian low molecular weight phospholipase A_2 (also known as PAF acetyl hydrolase).[142] Oxo-PtdCho is a specific substrate for the enzyme, which releases an oxidized short-chain fatty acid and lyso-PtdCho. Recently, PAF acetyl hydrolase-like lipoprotein associated phospholipase A_2 (Lp-PLA$_2$) has been localized in necrotic cores and inflammatory areas of coronary vulnerable plaques.[143] Reports of phospholipid core aldehyde hydrolysis by paraoxonase[144] have been recently attributed to a contamination of the paraoxonase preparations with PAF acetyl hydrolase.[145] The glycerophospholipids containing the core aldehyde as well as their DNPH derivatives are subject to hydrolysis by phospholipase C (*Bacillus cereus*), which has been employed in analytical work.[120]

Cholesteryl esters (ChEs) are known to be hydrolyzed by acid ChE hydrolase (AChEH), although esters with fewer than 16 acyl carbons are attacked at a much lower rate.[146] Oxo-ChE are subject to hydrolysis by pancreatic ChE ester hydrolase.[147] However, the C_5 core aldehyde ester of cholesterol has been found to be resistant to hydrolysis by mouse peritoneal macrophage lysosomes.[128] It was speculated that the resistance of the ChE core aldehydes to AChEH and neutral ChE hydrolase (NChEH) could be due to the ability of ChE core aldehydes to bind to proteins, or to form dimers and thereby become unavailable for enzymatic degradation. Ch [5-oxo]valerate tends to undergo aldol condensation upon storage giving a dimer of *m/z* 969 (Ravandi and Kuksis, 1995, unpublished); bacterial (*Pseudomonas*) ChE hydrolases, however, readily hydrolyze all ChE core aldehydes.[79]

5.4.1.2 GSH/GSSG and Other Reduction Systems

The process of luminal degradation and reduction of oxo-lipids continues in the enterocyte. The factors involved in mucosal peroxide transport, metabolism and oxidative susceptibility have been reviewed,[148,149] and it has been concluded that the appearance of lipid hydroxides, which are the reduction products of lipid peroxides in mesenteric lymph, indicates a role for mucosal glutathione (GSH) and GSH peroxidases in enterocyte metabolism of lumenal lipid peroxides. There is good evidence to suggest that GSH is a key participant in the elimination of peroxides by the intestine.[150–152] GSH is a naturally occurring cellular reductant and antioxidant[153,154] which functions to detoxify reactive oxygen metabolites of endogenous or exogenous origin. GSH is present in high concentrations in tissues;[151,153] in normal tissues, GSH homeostasis is maintained by *de novo* synthesis from cysteine, from regeneration of glutathione disulfide (GSSG), and from GSH uptake from exogenous sources *via*

Na$^+$-dependent transport systems. Dietary GSH and biliary GSH output is the major source of luminal GSH.

A conscious rat lymph fistula model[149,151] predicts that, at a given peroxide concentration, the amount of peroxide (ROOH) recovered from the intestinal lumen and lymph is governed by the mucosal GSH status. This mechanism of GSH-dependent "metabolic trapping" highly favours peroxide disposal in the small intestine. During GSH deficiency, in contrast, decreased intracellular peroxide catabolism impairs mucosal ROOH absorption, increases luminal ROOH retention, and promotes lymphatic ROOH transport. These predictions have been confirmed experimentally.[148,150] Exogenous GSH supplements either from diet or bile increase mucosal GSH concentrations and promote metabolism of lipid peroxides, thereby increasing luminal peroxide uptake into enterocytes. That biliary GSH is a major contributor to the luminal GSH pool is evidenced by studies showing that diversion of biliary GSH significantly decreases intestinal peroxide metabolism and results in elevated peroxide recoveries in lumen and lymph.[148,150] In view of the findings with lipid peroxide administration[140,141] and the role of the GSH–GSSG system,[148,149] it is clear that the quantities of dietary fatty acid hydroperoxides available for intestinal uptake would be greatly reduced due to the degradative processes described above.

5.4.2   Absorption of Oxo-lipids

The intestine is a primary site of dietary fat absorption and a critical defence barrier against dietary-derived peroxides, mutagens and carcinogens. Early studies in the 1960s and 1970s investigated the absorption and lymphatic transport of fatty acid hydroperoxides, but the results of these studies had remained equivocal because a clear distinction was not made between the absorption of primary and secondary products of lipid oxidation, and between micellar and lymphatic pathways of absorption. More recent work[140,141,154] would appear to exclude extensive direct absorption of hydroperoxidized fatty acids unless the reducing potential of the enterocytes has been exhausted. The original reports of lymphatic absorption of lipid hydroperoxides were apparently compromised by the inability to distinguish clearly between lipid hydroperoxides and their degradation products, which are readily absorbed (see below). Nevertheless, differences in mucosal GSH status under the various dietary and experimental conditions could also explain the lack of general agreement in the literature and provide a mechanism for luminal peroxide handling by the small intestine.[148]

5.4.2.1   Micellar Absorption

There is evidence that the volatile aldehydes generated during lipid peroxidation can diffuse within, or even escape, the cell and attack targets far from the original site of initiation. Therefore, it has been suggested that they are not only end-products of lipid peroxidation but also act as mediators of oxidative stress.[108] It is likely that other water-soluble products of lipid oxidation, including MAG core aldehydes and acids, would also be taken up in the micellar form. Lipid uptake from micellar phase is a generally accepted mechanism for intestinal transport of free fatty acids.[155,156]

5.4.2.2   Lymphatic Absorption

The monoacylglycerol (MAG) pathway represents the major route of dietary fat absorption. It involves the formation of acyl coenzyme A (CoA) esters of free fatty acids and a transfer of the activated acids to *sn*-2-monoacylglycerols to yield *sn*-1,2(2,3)-diacylglycerols, which are further acylated by the activated acids to form TAGs.[139] During GSH deficiency, increased luminal retention of the peroxides of both free fatty acids and 2-monoacylglycerols would promote lymphatic transport of oxo-lipids. The MAG pathway would be expected to be involved in the mucosal uptake of *sn*-2-MAG hydroperoxides, *sn*-2-MAG-hydroxides and *sn*-2-MAG core aldehydes. Alternatively, the dietary fatty acids could be absorbed *via* the phosphatidic acid pathway, which involves ester-ification of *sn*-3-glycerophosphate by acyl-CoA, dephosphorylation and acy-lation of the released DAGs to TAGs.[139] The formation of oxo-acyl-CoA esters by the intestine or other tissues does not appear to have been reported. Although oxo-TAGs have been recovered from lymph following feeding of oxo-lipids to pigs, the mechanism of their formation has not been estab-lished;[112–114] the products consisted of TAG molecules with a single hydroxyl, epoxy, keto or aldehyde function. On the basis of peak intensities, the hydro-xides were the most abundant. The aldehydes were identified as DNPH deri-vatives. The isolation of oxo-TAGs from pig lymph finds support in studies with conscious rat lymph model[149,150] and in peroxide feeding studies,[157,158] showing that rat chylomicrons contain oxo-lipids at concentrations determined by the lipid peroxide content of the diet.

Others have shown that the TAG-OOH appearance in lymph and chylo-microns was very low to undetectable as long as ascorbate and/or ubiquinols were present.[159] A limited of lipid hydroperoxide absorption has also been demonstrated in Caco-2 cell cultures, where the hydroperoxide transport was prevented (attenuated) by a gastrointestinal peroxidase.[160,161] It was shown that [1-^{14}C]-13-HPODE, up to a nontoxic concentration of 100 M, did not cross the Caco-2 cell monolayer unreduced when applied to the luminal side.[161] The [1-^{14}C]HPODE derived radioactivity was preferentially recovered from intra-cellular and released DAGs, glycerophospholipids (PL) and oxo-ChE. A similar distribution pattern was obtained with 13-HODE. In contrast, linoleic acid was preferentially incorporated into TAGs, ChEs and PL (but mainly released as TAGs). It was concluded that food-borne hydroperoxy fatty acids are instantly reduced by the gastrointestinal glutathione peroxidase. The hydroperoxy fatty acids, however, disturbed the intestinal lipid metabolism by being esterified as hydroxyl fatty acids into complex lipids, which could have rendered the resulting lipoproteins susceptible to further oxidative modification.

Heating oils and fats leads to cyclization of polyunsaturated fatty acids, and cyclohexenyl and cyclopentenyl fatty acids have been identified in some edible oils. The absorption of cyclic fatty acid monomers are greatly influenced by their positioning within the TAG, but the structure of the cyclic fatty acids influences their lymphatic recovery only when they are absorbed as free fatty acids (phosphatidic acid pathway).[162]

5.4.3 Oxo-lipids in Plasma, Tissues and Urine

Indirect evidence for the transport of dietary lipid peroxides into the circulation is provided by independent measurements of oxo-lipids in plasma, tissues and urine, although such measurements do not exclude the endogenous origin of these lipids. The general inadequacy of the experimental methods used to distinguish lipid peroxides and their degradation products, however, has not been always overcome. In addition, the formation of lipid hydroperoxides and their degradation products during the isolation and storage of the samples has not always been rigorously excluded.

5.4.3.1 Chylomicrons and VLDL

A few studies have specifically documented the presence of lipid peroxides in the lymph,[163] isolated chylomicrons,[112–114,158] blood plasma[164] and tissues[165] after administration of a lipid peroxide containing diet. Some studies[158] have suggested that rat lymph chylomicrons contain lipid hydroperoxides at concentrations that are determined by the lipid peroxide content in the diet. In view of other reports that have documented the presence in the rat of efficient GSH-dependent intestinal systems that detoxify dietary lipid hydroperoxides, this could have happened only because these animals had received large amounts of peroxidized lipid or their GSH-defence mechanism had been compromised.[148,150,151]

Recent studies[112–114] have identified oxidized and chain-shortened TAG molecules in pig plasma chylomicrons. The identified products consisted of TAG molecules with a single hydroxyl, epoxy, keto or aldehyde function. On the basis of peak intensities, the hydroxides were the most abundant. The aldehydes were identified as DNPH derivatives.

Although plasma is endowed with an array of antioxidant defence mechanisms such as plasma ascorbate and, in a site-specific manner, bilirubin (which are much more effective in protecting lipids from peroxidative damage by aqueous oxidants than all the other endogenous antioxidants[166]), the presence of fatty acid peroxides (FAOOHs) and fatty acid hydroxides (FAOHs) in plasma has been clearly demonstrated[167] using HPLC and UV detection; up to 13 distinct regioisomers of FAOOH and FAOH were identified, the analytical range being 0.033–1.6 μmol L^{-1} for each compound. Analysis of human plasma exposed to artificial oxidation with Cu^{2+} ion and hydrogen peroxide (a free radical generation reaction) showed marked increase in hydroxy and hydroperoxy PUFA. Furthermore, feeding of DHA increases hydroperoxide levels in plasma, liver and kidney, whereas the α-tocopherol level was reduced concomitantly.[168] Consistent with these results, rats fed DHA-containing oils had more thiobarbituric acid reactive substances in these organs than controls. These results confirm previous findings,[169] which showed that a fish oil diet increased the phospholipid hydroperoxide (PLOOH) concentration of liver and plasma more than safflower oil, perilla oil and olive oils, supporting the idea that a diet rich in DHA oils promotes lipid peroxidation *in vivo*.

Others have presented evidence to indicate that consumption of repeatedly heated soy oil increases the serum thiobarbituric acid reactive substances (TBARS) and LDL compared to fresh oil.[170] It was concluded that the anti-atherosclerotic effect of soy oil is lost when repeatedly heated. Still other work[171] has shown that CLA given in a butter diet results in increased lipid peroxidation measured as an 83% higher 8-iso-prostaglandin $F_{2\alpha}$ concentration compared with the control in healthy young men. A mixture of *cis*-9,*trans*-11- and *trans*-12,*cis*-10-CLA was given as part of a diet rich in butter, but no change in plasma total, LDL and HDL Ch and TAG concentrations, or in inflammatory and hemostatic risk markers was found, nor was there any change in fasting insulin and glucose levels. Clearly, the oxo-fatty acid levels of these men increased after a PUFA "precursor" loading.

5.4.3.2 Tissues and Atheroma

Lipid hydroperoxides and isoprostanes are present in tissues other than intestine and their levels vary with the intake of polyunsaturates. The proinflammatory F_2-isoprostanes are produced *in vivo* by free radical catalyzed oxidation of ARA. These products accumulate in plasma and tissues in response to dietary supplementation with ARA containing diets.[51]

Arachidonoyl GroPCho became readily autoxidized to Iso-PGE_2/D_2 and Iso-PGF_2 GroPChos during storage of plasma at $-20\,^\circ$C and could reach 1000–$4000\,pg\,ml^{-1}$ after several months; this was about 50 fold higher than the levels in fresh plasma. Incubation of plasma at $37\,^\circ$C yields both unesterified and esterified isoprostanes in readily measurable amounts. Figure 5.6 demonstrates the presence of the palmitoyl/stearoyl-isoprostanoyl-GroPChos in human plasma HDL.[172]

EPA is the most abundant PUFA in fish oil. Recent studies suggest that the beneficial effects of fish oil are due, in part, to the generation of various bioactive oxidation products. The oxidation of EPA *in vitro* yields a series of F_3-IsoPs, which are virtually undetectable at baseline *in vivo*, but supplementation of animals with EPA markedly increases the quantities—up to $27.4\,ng\,g^{-1}$ of heart tissue of mice.[56,76]

The anti-inflammatory mediators A_4/J_4-NPs derived from DHA by free radical oxidation possess anti-inflammatory effects in macrophages in a dose dependent manner, as shown by addition of synthetic standards.[56,57] It was suggested that, in some cases, the susceptibility of DHA to oxidation may be a virtue rather than a vice. Since the bulk of the main ω3 fatty acid, DHA, is

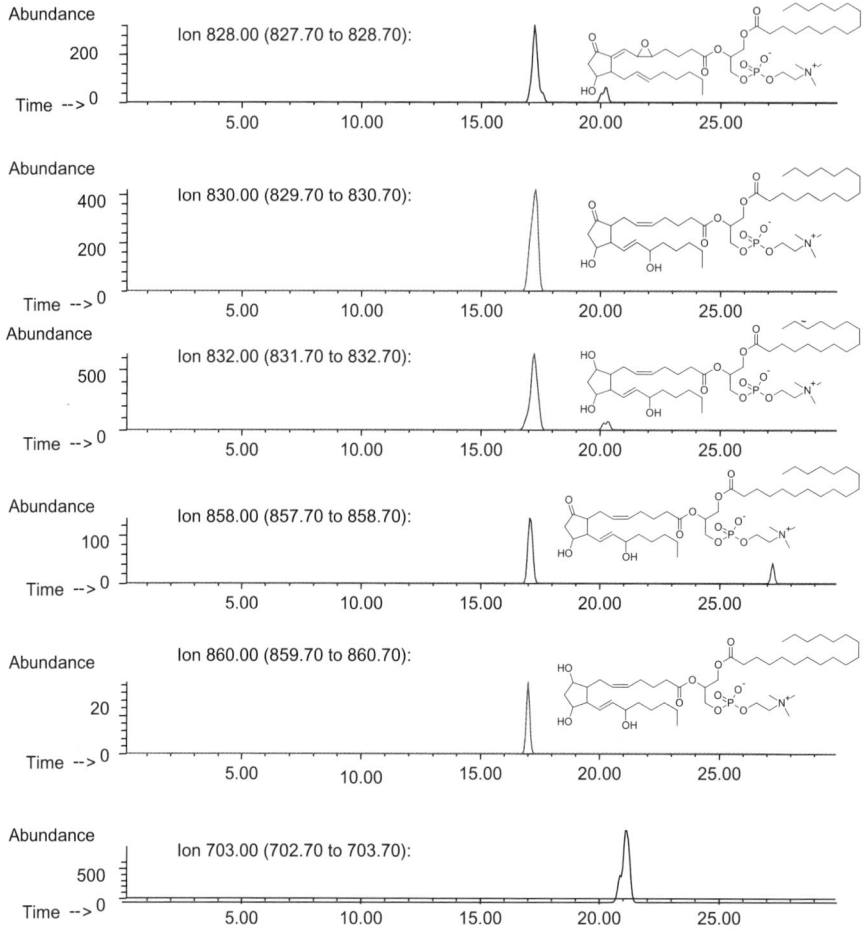

Figure 5.6   Single ion mass chromatograms of isoprostane GroPChos isolated from human HDL. The proposed identities of the peaks are shown by the structural formulae: m/z 828, 1-palmitoyl-2-[5,6-epoxy]-Iso-PGE$_2$GroPCho; m/z 830, 1-palmitoyl-2-Iso-PGE$_2$/D$_2$GroPCho; m/z 832, 1-palmitoyl-2-Iso-PGF$_2$GroPCho; m/z 858, 1-stearoyl-2-Iso-PGE$_2$/D$_2$GroPCho; m/z 860, 1-stearoyl-2-IsoPGF$_2$GroPCho. Modified from ref. 172.

located in the cerebrovascular system, an appreciation of brain lipid peroxidation can be gained better by measuring 4-hydroxy-hexanoic acid (4-HHA) by GC-MS after derivatization into pentafluorobenzyl and heptabutyryl esters.[173]

Several authors have observed the presence of mercapturic acid and/or glutathione conjugate diastereoisomers in various metabolic profiles from rat injected with labelled HNE.[174,175] The formation of these diastereoisomers has been explained[176] by suggesting that, under physiological pH conditions (pH 7.5), the racemic HNE may react spontaneously with glutathione, resulting in four diastereoisomers (two from *R*-HNE and two from *S*-HNE). Additionally, ring closure of the conjugated acids yields another chiral carbon, so that theoretically eight diastereomers of HNE-SG conjugates can be formed. Both *R*-HNE and *S*-HNE are substrates for GSTs, even though *S*-HNE was preferentially detoxified by rat GSTs.[177] Recent *in vitro* studies have confirmed that hepatic cytosols at pH 7.5 form two major HNE-GSH diastereomers, representing each 4% of the racemic HNE dose (one from *R*-HNE and another one from *S*-HNE formed by the GST-catalyzed reaction).[178]

Recently, the stereochemical configuration of 4-hydroxy-2-nonenal-cystein adducts and their stereoselective formation in a redox-regulated protein has been determined.[179] Authentic (*R*)-HNE and (*S*)-HNE-cysteine adducts were prepared and characterized by HPLC and NMR. These adducts in proteins were identified using a pyridylamination-based approach, which enabled the analysis of individual (*R*)-HNE- and (*S*)-HNE-cysteine adducts by LC-MS following acid hydrolysis. Analysis of HNE-cysteine adducts from human thioredoxin showed the HNE preferentially modifies Cys^{73} and to a lesser extent, the active site Cys^{32}. While the (*R*)-HNE- and (*S*)-HNE-cysteine adducts were equally formed at Cys^{73}, Cys^{32} exhibited a remarkable preference for (*R*)-HNE. The present method provides a platform for the chemical analysis of protein *S*-associated aldehydes *in vitro* and *in vivo*.

In brain, isoprostanes are formed as stable metabolites of ARA and are enriched in glial cells, whereas the neuroprostanes derived from DHA are highly concentrated in neuronal membranes.[180–182] These compounds are sensitive and specific biomarkers of oxidative stress in *post mortem* adult human brain. Selective vulnerability of preterm white matter to oxidative damage has been reported to be due to increased F_2-isoprostanes.[181] F_2 -and F_4-neuroprostanes are measured by NCI-MS with SIM.[182] Brain tissue has also been found to contain various short chain aldehydes such as HNE and ONE.[183–185] An essential role of extracellular signal-regulated kinase (ERK) has been demonstrated in ONE-mediated cytotoxicity in SH-SY5Y human neuroblastoma cells.[186]

Oral intake of excess polyunsaturated fatty acids (*i.e.* EPA and DHA) in a fish oil diet has been shown to lead to accelerated membrane lipid peroxidation, resulting in red blood cell (RBC) senescence linked to lowering of immune response of spleen cells, and that supplementation with α-tocopherol as antioxidant does not effectively prevent such oxidative degeneration.[187]

Oxidized membrane vesicles and blebs from apoptotic cells of patients with arteriosclerosis have been shown to contain biologically active oxidized phospholipids (palmitoyl/[5-oxo]-valeroyl-GroPCho) that induce monocyte–endothelium interactions.[188]

Oxidized lipid accumulation in the presence of α-tocopherol has also been shown in atherosclerosis.[189] Specifically, *cis,trans* isomers of Ch-18 : 2-OOH are formed predominantly during plasma and lesion lipoprotein oxidation in the presence of endogenous α-tocopherol. Furthermore, since *cis,trans*-Ch-18 : 2-OOH are the primary products found in human lesions and in a rabbit model of arterial injury, it was suggested that lipid peroxidation *in vivo* occurs in the presence rather the absence of α-tocopherol.

It has been demonstrated that complex ChE hydroperoxides are biologically active components of minimally oxidized LDL (mmLDL) and that similar polyoxygenated ChEs, as found in mmLDL, are also present in murine atherosclerotic lesions.[190] Polyoxygenated ChE hydroperoxides, found *in vivo* in atherosclerotic tissue, are believed to be generated by exposure to 12/15-LO-containing cells. Detailed analyses of the oxo-phospholipid composition in atherosclerotic plaques from different stages of development have shown remarkable similarity, which suggests continuous oxo-lipid formation and breakdown.[191] More recently, the distribution of intact oxidized glycerophospholipid species in lipid extracts of carotid artery plaques from patients with arthrosclerosis has been determined by ion-trap-Orbitrap MS.[28]

5.4.3.3   Urine

Numerous studies have shown that the urinary excretion of oxo-lipids increases with the oxidative stress of the system. The level of urinary oxo-lipids is also related to the dietary intake of unsaturates, which are required for the formation of the oxo-lipid markers, and which may have been in part responsible for the increased oxidative stress.

The degradation of hydroxy-alkenals takes place in the liver and leads to different metabolites including acids, which are excreted in the urine.[192,193] Their GC-MS measurement permits a non-invasive discrimination between the two PUFA oxidation pathways because 4-hydroxy-hexanoic acid (4-HHA) and 4-hydroxy-nonanoic acid (4-HNA) reflect the peroxidation of ω3 and ω6 PUFA families, respectively. MDA is commonly measured in biological fluids (*e.g.* urine and plasma) as well as in isolated cells after reaction with thiobarbituric acid (TBA) or diethyl TBA, which are purified by HPLC and measured by absorbance at 532 nm or by fluometry.[97]

The biotransformation of HNE *in vivo* involves reduction/oxidation of the aldehyde group and conjugation to endogenous glutathione leading to mercapturic acid (MA) conjugates in urine. After intravenous administration of radioactive HNE to the rat, 67% appeared in the urine in 48 h.[194] The end products were isolated by HPLC and identified by MS as HNE-MA, 1,4-dihydroxynonene-MA, 4-hydroxynonenoic-MA and the corresponding lactone. The HNA was characterized by FAB-MS giving a pseudo molecular ion $[M - H]^-$ at *m/z* 171. The HNA-MA gave $[M - H]^-$ at *m/z* 334, while the HNA-lactone-MA gave $[M - H]^-$ at *m/z* 316. The study showed that hepatic GSH plays a major role in determining HNE metabolic fate, as well as in a potential unbalance of the GSH/GSSG system.

1,4-Dihydroxynonane (DHN) is a product of alcohol dehydrogenase reduction of HNE and is best detected as the unmodified DHN-MA product. DHN-MA is also excreted in human urine and has been proposed as an improved non-invasive biomarker for lipid oxidation.[195] Positive ESI produced an intense $[M + H]^+$ ion at *m/z* 322. When subjected to MS/MS with collision induced activation in the ion trap, the selected m/z 322 parent ion decomposed to mainly m/z 304 ion, corresponding to loss of water from the $[M + H]^+$ species, and to a lesser extent, the m/z 164 fragment ion, which corresponded to the N-acetylcysteine moiety of DHN-MA. All the chromatograms from rat and human urine showed that DHN-MA generated *in situ* is excreted in the urine as a mixture of at least two stereoisomers. This result was in accordance with that of another study of HNE metabolism, which showed that DHN-MA was excreted in rat urine as a pair of diastereoisomers,[196] and by other authors[197] who showed that HNE formed *in vivo* is a racemic mixture of 4-R and 4-S isomers.

Assuming that all amino acids making up the glutathione have an L configuration, Michael addition of glutathione at the double bond would lead to an additional chiral centre at the C_3 position. Therefore, the excreted DHN-MA could be composed theoretically of four isomers. The metabolism in rats of the two enantiomers with radiolabelled R- and S-HNE markers yielded results[178] comparable to those obtained previously with a racemic mixture.[192] Urinary excretion of radioactivity was estimated and urinary metabolites were identified using FAB-MS.[194] There was evidence that the major GST isoforms isolated from guinea-pig or rat liver catalyzed the preferential conjugation of the S-HNE enantiomers.[178]

Lipid peroxidation also yields the 4-oxo-2-nonenal (ONE) along with HNE as products of LOX oxidation and therefore conjugates of HNE and ONE metabolites have value as markers of *in vivo* oxidative stress and lipid peroxidation. Thus, carbon tetrachloride (CCl_4) treatment of rats—a widely accepted animal model of acute oxidative stress—results in a significant increase in the urinary excretion of DHN-MA, HNA-MA lactone, ONE-MA and ONA-MA.[197] The conjugates of the volatile aldehydes along with MS fragmentation products were identified by reversed phase HPLC interfaced with a hybrid triple quadrupole/linear ion trap MS with a TurboV ESI source. Various scanning techniques, all run in negative ion mode, were used for the characterization and detection of lipid peroxidation (LPO) MA products.

In contrast to a general lack of data for absorption of conjugated hydroperoxydienes, there is evidence for the absorption of complex secondary products of lipid autoxidation.[198] Thus, although about 50% of an oral dose of autoxidized [^{14}C]-labelled linoleic acid given to rats was excreted in the faeces, the majority (75%) of the remaining activity was absorbed and excreted in the urine, or as CO_2. No indication of the nature of the absorbed materials was obtained, although fractionation of the products into aldehyde-containing and high (endoperoxide-containing) molecular mass fractions confirmed that both classes of compounds were absorbed. It was demonstrated subsequently[199] that typical *trans*-2-alkenal compounds known to be produced from the thermally induced autoxidation of PUFAs are readily absorbed from the gut into the systemic circulation *in vivo*, metabolized (primarily *via* the addition of glutathione across their electrophilic carbon–carbon double bonds), and excreted in the urine as C_3-mercapturate conjugates in rats. Figure 5.7 shows the structures of the proposed intermediates.[199]

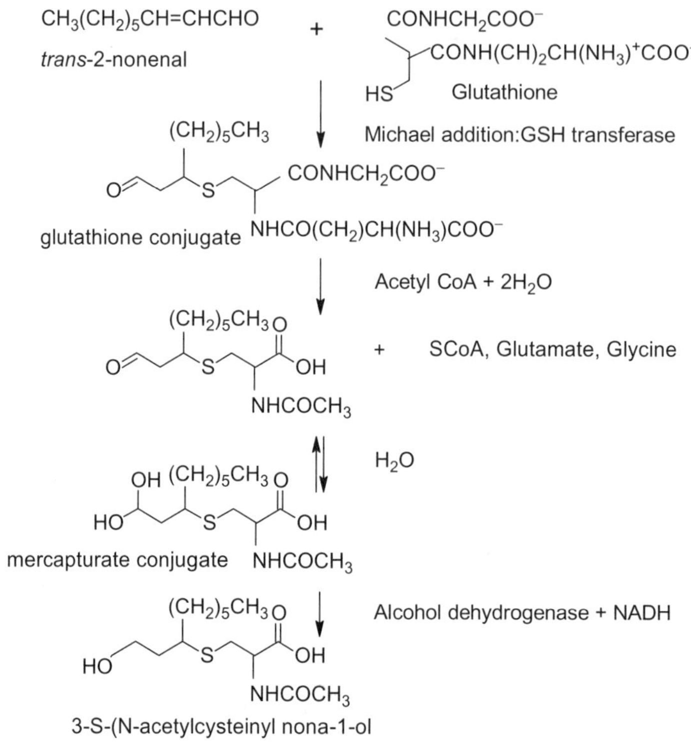

Figure 5.7 Mechanism for the generation of *trans*-2-nonenal mercapturate conjugates *in vivo*. Modified from ref. 199.

An investigation of the effect of CLA on lipid peroxidation in humans showed a significant increase in 8-iso-PGF$_{2\alpha}$ and 15-keto-dihydro-PGF$_{2\alpha}$ in urine after three months of daily CLA intake (4.2 g day^{-1}) compared to the control group. Likewise, a diet rich in CLA and butter increased lipid peroxidation in healthy young men as indicated by 83% higher urinary excretion of 8-iso-PGF$_{2\alpha}$.[200] However, an *in vivo* oxidation and urinary excretion of oxo-CLA itself was not reported by the radioimmunoassays employed.

Isoprostanes possess potent biological activity as inflammatory mediators that augment the perception of pain.[51] Among these IsoPs, 8-iso-PGF$_{2\alpha}$ (also known as isoPGF$_{2\alpha}$-III and 15-F$_{2t}$-IsoP), which is excreted in human urine, is the most thoroughly investigated and its measurement is claimed to be the most reliable approach for assessing the oxidative stress status.[201] However, these authors observed no statistical difference between the urinary levels of F$_{2\alpha}$-isoprostanes obtained from treated and control animals. A new technique that allows a higher sample throughput substitutes ethyl acetate extraction for HPLC initial sample isolation. The analysis is completed by conventional derivatization and mass spectrometry with stable isotope dilution.[202] The wisdom and conclusions based on the measurement of only one isomer of

F$_2$-IsoP family as a reflection of the global oxidative stress has been challenged[34] because measurement in plasma or urine of IsoPs and NPs merely reflects the oxidation of 20 : 4ω6 and 22 : 6ω3, respectively.

5.4.3 Biological Activity of Oxo-lipids

The biological activity of the oxo-lipids parallels their chemical reactivity and varies with the nature of the oxo-lipid, the hydroxy aldehydes being the most reactive. Below the biological activity is considered according to the cellular target—amino lipids, amino acids and proteins, and nucleic acids. The interactions involve aldol condensation, Michael condensation and Schiff's base reaction.

5.4.4.1 Lipid Targets

Both short chain and core aldehydes are known to undergo self-condensation and polymerization, which account for a significant part of the discrepancy between the moles of substrate destroyed and the amount of aldehyde recovered during peroxidation. HNE and MDA have been shown to undergo self-condensation and polymerization.[104,110,183] More recently, PtdCho core aldehydes were found to undergo self-condensation and to react with aminoglycerophospholipids.[203,204]

Self-condensation among core aldehydes of GroPCho has also been observed and dimer, trimer and tetramer units have been identified.[205] The formation of polymers was confirmed by demonstrating a change in the ratio between the aldehyde hydrogen and the hydrogens of the trimethylamine of the head group, yielding 1 : 6 for the dimer and 1 : 7 for the trimer and tetramer. TAG and MAG core aldehydes also undergo self-condensation and react with aminoglycerophospholipids such as phosphatidylethanolamine (PtdEtn) and phosphatidylserine (PtdSer).[206] Likewise, the MAG C$_9$ core aldehyde reacted readily with aminophospholipids to form Schiff's bases. MDA reacts with the amine groups of amino phospholipids.[207]

The formation of N-(hexanoyl)PtdEtn (HEPE)—a novel PtdEtn adduct— during the oxidation of erythrocyte membrane and LDL has been demonstrated by MS to result from a reaction of PtdEtn with lipid peroxides.[208] Upon reaction of egg PtdEtn with 13-hydroperoxy-octadecadienoic acid (HPODE) or other oxidized polyunsaturated acids followed by phospholipase D-mediated hydrolysis, the formation of *N*-(hexanoyl)ethanolamine (HEEA), a head group of HEPE, was confirmed by isotope-dilution LC-MS/MS. Moreover, increasing amounts of HEEA were detected in the hydrolysates of oxidized erythrocyte ghosts and LDL with increasing lipid peroxidation levels.

Dopamine adducts of simple aliphatic aldehydes have been synthesized and demonstrated to occur in brain homogenates.[209] Specifically, the succinyl, hexanoyl, propanoyl and glutaroyl adducts of dopamine were identified by LC-MS/MS. The hexanoyl dopamine, an adduct derived from arachidonic acid, caused severe cytotoxicity in human dopaminergic neuroblastoma SH-SY5Y cells, whereas the other adducts showed only slight effects. Dopamine adduct formation was suggested as a possibility for a novel mechanism responsible for the pathogenesis in Parkinson disease.

Early work with pure Ch-[5-oxo]valerate showed that it tends to dimerize upon storage giving a m/z value of 969, while mixtures of Ch-[5-oxo]-valerate and Ch-[9-oxo]nonanoate yielded m/z values corresponding to C_5–C_5 and C_9–C_9, as well as mixed C_5–C_9 combinations.[126,127]

Among the lipid targets of oxo-lipids may be included such fat-soluble antioxidants as the tocopherols. Free radical scavengers such as tocopherols, butylated hydroxytoluene (BHT) and plant phenolics inhibit lipid oxidation by reducing peroxyl and alkoxyl radicals into stable compounds. There is evidence that free radical scavenging capacity as measured by hydrogen atom transfer, electron transfer or oxygen radical absorbance capacity (ORAC) has limited value in predicting the antioxidant activity in complex foods or tissues.[210] Thus, the coexistence of oxidized lipids and α-tocopherol in all lipoprotein fractions isolated from advanced human atherosclerotic plaques have been known for some time.[211]

Recently, the decomposition of 13-HPODE was shown not to be affected by α-tocopherol, several other antioxidants or antioxidant enzymes.[212] In fact, the inclusion of α-tocopherol during the decomposition of 13-HPODE resulted in an accumulation of aldehydes. In contrast, the 13-HPODE was converted to 13-hydroxylinoleic acid by undergoing decarboxylation into acetate. Furthermore, human clinical trials using α-tocopherol as one of the major antioxidant ingredients were mostly negative and failed to show any evidence of protection against cardiovascular diseases.[213]

5.4.4.2   Amino Acid and Protein Targets

Short chain aldehydes have been long known to react with various amino acids, peptides and proteins.[104,110,183] MDA is often the most abundant aldehyde resulting from lipid peroxidation and occurs in biological materials in various covalently bound forms. MDA primarily forms adducts with lysine residues of proteins.[214] The major reaction of MDA is its addition to lysine, generating N-(2-propenal) lysine (N-propenal-lysine),[215,216] which is the major form in which endogenous MDA is excreted in urine of rats and humans.[216] MDA also forms fluorescent products such as dihydropyridine (DHP) lysine, a model of fluorescent components in lipofuscin.[217] Food products contain protein-bound MDA which, in the course of digestion, is broken down to N-ε-(2-propenal)lysine,[218] a molecule found to be absorbed by the gut and which reaches many organs.[219] Since N-ε-(2-propenal) lysine is an α,β-unsaturated aldehyde, it could possibly react further with plasma proteins, such as albumin, and apoB-48 and apoB-100.[220] Interestingly, red vine polyphenols have been shown to prevent the absorption of the lipotoxin MDA in humans.[220]

The MDA-modification sites on human serum albumin (HSA) have been identified using MS/MS techniques.[207] Thus, when HSA was exposed to MDA and the modified proteins were digested with V8 protease and analyzed by LC-ESI-MS/MS, six peptides were identified. These peptides contained *N*-propenal adducts at Lys[136], Lys[174], Lys[240], Lys[281], Lys[525] and Lys[541], and revealed that Lys[525] is the most reactive residue for MDA modification.

ACR, a highly toxic lipid peroxidation aldehyde, is known to form four different types of adducts, namely aldimine-, propanal-, methylpyridinium (MP)- and formyl-dehydropiperidino (FDP) type adducts. Nucleophilic groups in a peptide sequence may undergo Michael reactions with the electrophilic double bond of ACR, or give rise to nucleophilic additions on the carbonyl group to form Schiff bases.[221] ACR is strong cross-linking agent of cellular components such as proteins. The ACR-induced cross-links in chain B from bovine insulin have been characterized by MS in order to identify the cross-linking sites. The ACR-treated peptide was digested with a protease and the resulting peptides were analyzed by reversed phase LC-ESI-MS/MS. Inter- and intra-molecular cross-linking adducts were identified between amino groups and the side chain of histidine in the peptide. The results indicated that the ACR-induced cross-links were accompanied by two reactions, namely Michael addition and Schiff's base formation.[222]

A proteonomic profiling of rat lung epithelial cells has recently been performed following exposure to 20 µM ACR for 24 h.[38] There were 34 proteins that showed changes from control cells after ACR treatment; the expression of 18 proteins was increased and the expression of 16 proteins was decreased. It was speculated that some of the proteins affected by ACR may be involved in the development of lung diseases. Others have investigated the activation of anti-apoptosis survival factors in relation to the induction of cell death by apoptosis, following exposure to low doses of ACR.[223]

The protein adducts generated from interaction with HNE and ONE derived from lipid oxidation have also been extensively investigated.[34] Reversibly formed adducts, such as the HNE-Lys Michael adduct, could be found on proteolytic peptides only if a sodium borohydride (NaBH$_4$) reduction step was used prior to proteolysis. Among the adducts specifically identified were HNE-Lys-derived 2-pentylpyrrole, ONE-Lys-derived 4-ketoamide, ONE-derived His-Lys-pyrrole cross-link and a Lys-derived 3-formyl-4-pentyl-pyrrole, which results from a combined action of ONE and ACR.[34] The NaBH$_4$ quench achieved reductive "cementing" of Michael adducts present at equilibrium in solution at the time of the quench. Incubation of proteins with LOX-derived aldehydes in the presence of sodium cyano-borohydride (NaBH$_3$CN) selectively reduced protonated Schiff bases (iminium species) that were in equilibrium with free amine and free carbonyl compounds. Thus, an analysis of adducts formed using NaBH$_3$CN rather than NaBH$_4$ did not provide a true reflection of what was present in solution[224] (Figure 5.8).

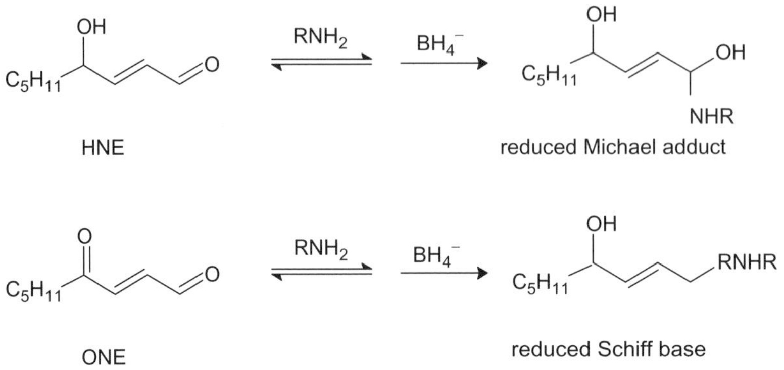

Figure 5.8 Major products isolated from reductive quenching of initial reactions of HNE and ONE with amines. Modified from ref. 34.

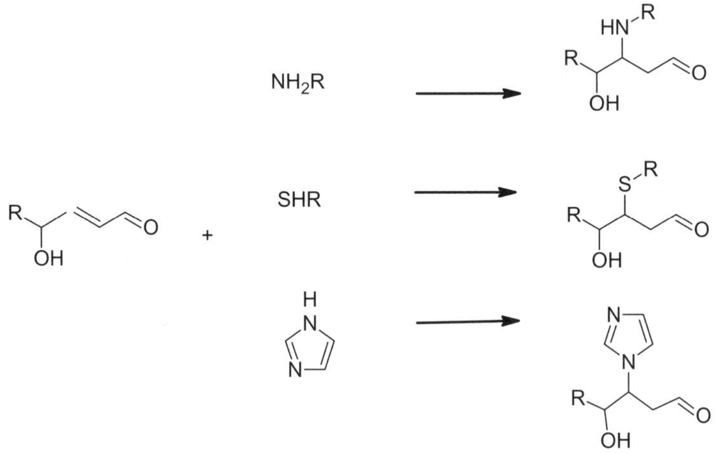

Figure 5.9 Formation of Michael reaction compounds by addition of amino, imino and thiol groups to the double bond of oxygenated α,β-unsaturated aldehydes. Modified from ref. 32.

A variety of simple Michael and Schiff base adducts were formed initially, but only some of these adducts, such as the HNE and ONE-derived Michael adducts on Cys and His residues, were found to survive the conditions of proteolysis and LC-MS.[34] Figure 5.9 shows the Michael addition products obtained between 4-hydroxy-2-nonenal and various amino, imino and sulfhydryl groups. In other studies,[225] HNE has been shown to induce mitochondrial uncoupling by specific and inhibitable interactions with the uncoupling proteins UCP1, UCP2 and UCP3, and with the adenine nucleotide translocase.

The *trans*-4-hydroxy-2-hexenal (HHE) derived from docosahexaenoic acid (22 : 6 n-3) is toxic to primary cultures of cerebral cortical neurons.[226] Using an antibody raised against HHE–protein adducts, it was observed that HHE modified specific proteins of 75, 50 and 45 kDa in concentration- and time-dependent manner. This study showed that HHE reacts with specific proteins rather than modifying proteinaceous nucleophilic groups non-discriminately.

Amino acid and peptide binding of HNE, ONE and other low molecular weight aldehydes has been recently investigated by (MALDI-ToF-MS).[102] For this purpose, the α-amino groups of histidine and lysine were derivatized with *p*-carboxybenzyltriphenylphosphonium to form pseudo dipeptides, which could be sensitively detected by MALDI-ToF-MS. The pseudo dipeptides were excellent surrogates for His- or Lys-containing peptides in model reactions with reactive electrophiles (HNE and ONE). Amino acid and protein binding has also been demonstrated[77] for highly reactive γ-ketoaldehydes, which form lactams and hydroxylactams with lysine, and for ketoaldehyde (neuroketal, NK) lysyl adducts from brain synaptosomes.[78] It was earlier demonstrated[227] that HNE treatment of glyceraldehyde-3-phosphate dehydrogenase (GAPDH) results in the modification of amino acid residues primarily located at the surface of the GAPDH molecule.

Large differences have been demonstrated in the reduction and *retro*-Michael conversion steps in the metabolism between the *R*- and *S*-4-hydroxy-2-nonenal conjugates of glutathione in the rat.[194] HNE has a chiral centre at C_4 and is produced by lipid peroxidation as racemic mixture of *R*- and *S*-enantiomers.

The eicosanoids containing α,β-unsaturation carbonyl moieties including 15-A_{2t}-IsoP and A_4/J_4-NPs, readily adduct GSH in the presence of GST *in vitro*,[228,229] while levuglandin E_2, a cytotoxic seco prostanoic acid co-generated with prostaglandins by non-enzymatic rearrangement of the cycloxygenase-derived endoperoxide, prostaglandin H_2, has been reported to avidly bind proteins.[230]

Glycerophospholipid core aldehydes react readily with amino acids and polypeptides.[203,204] Specifically, adducts of [5-oxo]valeroyl- and [9-oxo]nona-noyl GroPCho and free lysine, valine, isoleucine and the methyl ester of lysine were isolated after reduction with $NaBH_3CN$ and identified by LC-ESI-MS. Likewise, adducts of [5-oxo]valeroyl- and [9-oxo]nonanoyl GroPCho and myoglobin were isolated and identified. Myoglobin treated with PtdCho core aldehydes shows the presence of three series of multicharged ions, one corresponding to the original horse skeletal muscle apomyoglobin (MW 16,949.89), a second corresponding to the $NaBH_3CN$ treatment product (MW 17,030.74) and the third corresponding to the reduced PtdCho core aldehyde Schiff base adduct containing two molecules of the 16 : 0/9 : 0 core aldehyde (MW 18,218.14). Other incubations gave evidence of the formation of Schiff bases with the mono-16 : 0/9 : 0 Ald and mono-18 : 0/9 : 0 Ald, and the di-16 : 0/9 : 0 Ald, and the mixed 16 : 0/9 : 0 and 18 : 0/9 : 0 Ald adducts.[204] Subsequently, oxidized PtdCho–protein adducts were prepared and identified using mono-clonal autoantibodies.[231] Others complexed oxo-PtdCho derived from perox-ynitrite oxidation (SIN-1) with apoprotein A1.[232] The structures of these adducts, however, were not established.

The most detailed investigation to date has been reported on protein binding of biotinylated core aldehydes of glycerophospholipids.[39] At least 20 different biotinylated human aortic endothelial cell (HAEC) proteins were recognized to which 1-palmitoyl-2-[5-oxo]valeroyl GroPEtn-*N*-biotin (Ox-PAPE-*N*-biotin) was covalently bound. Such adducts were not detected during treatment with unoxidized 16 : 0/20 : 4 GroPEtn-*N*-biotin. Figure 5.10 shows the preparation of 16 : 0/[5-oxo]valeroyl GroPEtn-*N*-biotin and the isolation of its adducts with HAEC cell proteins represented by R. The biotin derivatives permitted a selective isolation of the adducted proteins using avidin binding beads and SDS-PAGE. The 16 : 0/[5-oxo]valeroylGroPEtn-*N*-biotin bound peptides were characterized by MS/MS following trypsin digestion, reduction and alkylation.

Binding of ChE core aldehydes to amino acids, peptides and aminophospholipids has also been reported, as has been the binding of radiolabelled ChE core aldehydes to mouse peritoneal macrophages and serum proteins.[128] A monoclonal antibody prepared to lysine-bound oxidized ChE showed that it recognized exclusively protein bound ChE core aldehydes, including Ch-[9-oxo]nonanoyl-lysine and Ch-[5-oxo]valeroyl-lysine, in atherosclerotic plaques.[131] Due to the reversible nature of such unconjugated Schiff bases, these aldehydes have received little attention as the causative agent for modification of nucleophilic biomolecules. However, it has recently been shown that irreversible covalent protein modification by such aldehydes may be obtained in which hydrogen peroxide and alkyl hydroperoxides mediate the binding resulting in a protein *N*-acylation.[233]

MAG core aldehydes also readily react with amino acids and peptides,[206] as demonstrated by reversed phase LC-ESI-MS of the $NaBH_3CN$-reduced Schiff's base of 2-MAG-C_9-ALD and glycine-histidine-lysine (GHK). In the total negative ion mass spectrum of the Schiff's base, the deprotonated molecular ions for the reduced base of GHK and 2-MAG-C_9-ALD were seen at m/z 569 and the unreacted peptide at m/z 339; the adducts of glycyl-glycyl-histidine (GGH) and glycyl-glycyl-glycine (GGG) eluted at 2.0 and 2.2 min, respectively. Similar LC-ESI-MS methods were used to demonstrate the formation of the 2-MAG-C_9-ALD adducts of valine, lysine, as well as PtdEtn, in addition to the three tripeptides.[206]

A Schiff's base is also readily formed between Ch-[9-oxo]nonanoate and the ε-amino group of N-BOC-Lys.[128] The base was stabilized by $NaBH_4$ reduction, which increased mass to m/z 771 from the $[M + 1]^+$ value of 769. Increasing the CapEx from 120 to 300 V gave fragment ions supporting the Schiff base structure. Evidence was also obtained for a covalent interaction of $[^3H]ChE$ core aldehydes with serum proteins.[128] Likewise, 7-keto-ChE core aldehydes were demonstrated to bind to lysine and to LDL apoB. As seen from the LC-MS of the major product eluted at 21 min, a pseudomolecular ion peak at m/z 832.1 $[M + H]^+$ was formed corresponding to the reduced form of Ch-[9-oxo]nonanoate. The binding of the C_9 aldehyde of ChE to apoB protein was confirmed using a monoclonal antibody raised against the C_9 core aldehyde modified protein. Among the aldehydes tested, Ch-[9-oxo]nonanoate was the only source of immunoreactive material generated with the protein. It was

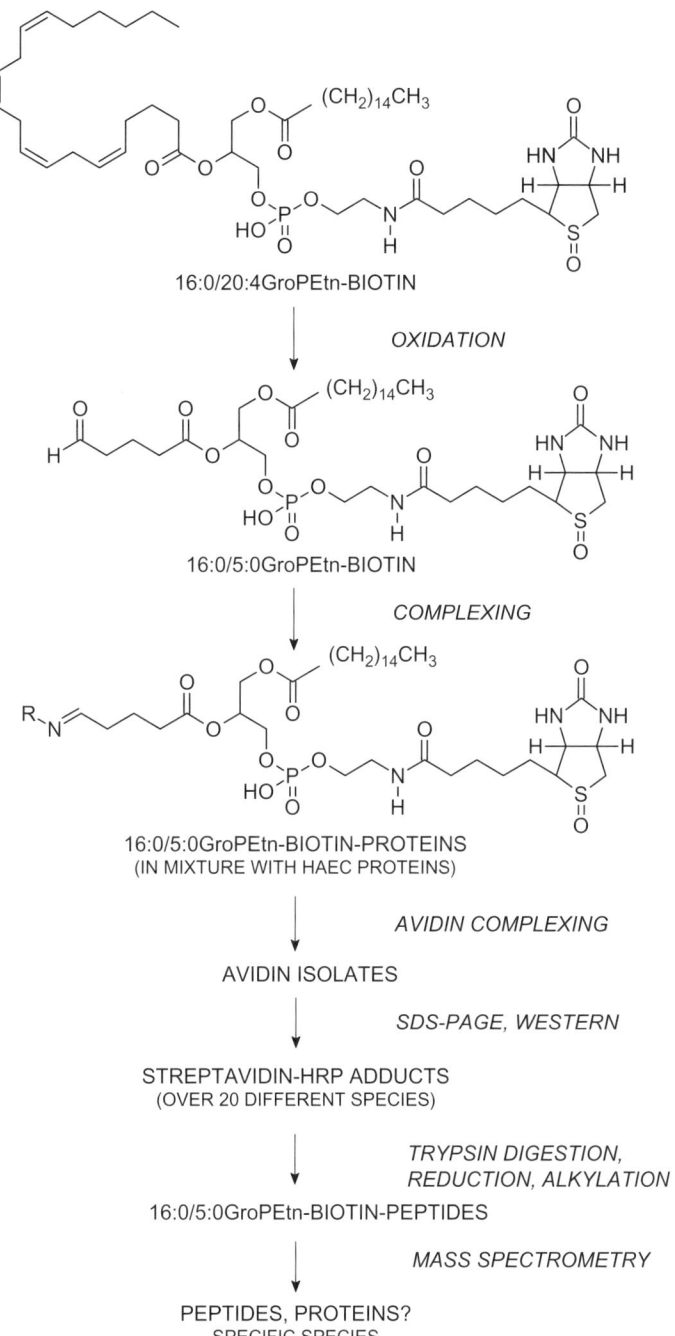

16:0/20:4GroPEtn-BIOTIN

OXIDATION

16:0/5:0GroPEtn-BIOTIN

COMPLEXING

16:0/5:0GroPEtn-BIOTIN-PROTEINS
(IN MIXTURE WITH HAEC PROTEINS)

AVIDIN COMPLEXING

AVIDIN ISOLATES

SDS-PAGE, WESTERN

STREPTAVIDIN-HRP ADDUCTS
(OVER 20 DIFFERENT SPECIES)

*TRYPSIN DIGESTION,
REDUCTION, ALKYLATION*

16:0/5:0GroPEtn-BIOTIN-PEPTIDES

MASS SPECTROMETRY

PEPTIDES, PROTEINS?
SPECIFIC SPECIES

Figure 5.10 Oxidative formation of 1-palmitoyl-2-[5-oxo]valeroyl GroPEtn-*N*-biotin from 1-palmitoyl-2-arachidonoyl GroPEtn-*N*-biotin and its addition to membrane proteins. Modified from ref. 39.

suggested that the binding of ChE core aldehydes to LDL might represent a process common to the oxidative modification of LDL and atherosclerosis.[128] Interestingly, unsaturated lipid peroxidation-derived aldehydes have been reported to activate autophagy in vascular smooth-muscle cells.[234]

Truncated phospholipid hexadecyl/azelaoyl GroPCho (Az-LPAF), derived from the fragmentation of *sn*-2-linoleoyl residues, depolarized mitochondria of intact cells and allowed NADH loss, but did not interfere with complex 1 function.[235] Suppression of mitochondrial function by oxidatively truncated phospholipids was shown to be reversible, aided by Bid (a member of the Bcl-2 family, which promotes mitochondrial-dependent apoptosis) and Bcl-X$_L$ (a member of the Bcl-2 family, which obstructs apoptosis). In other studies, OKL38, an oxidative stress response gene, was found to be stimulated by oxidized phospholipids.[236] The oxidative signal was induced by oxidized 1-palmitoyl-2-arachidonoyl GroPCho and its derivative 1-palmitoyl-2-epoxy-isoprostane E$_2$-GroPCho.

5.4.4.3   Nucleic Acid Adducts

Numerous DNA lesions resulting from exposure to reactive oxygenated species (ROS) have been identified. All four bases and nucleosides undergo modification as a result of exposure to ROS.[237] Most of the low molecular weight aldehydes were found to react with the exocyclic amino groups of DNA bases, giving rise to adducts. Thus, products of DNA treated with 4-hydroxy-2-nonenal have been isolated and identified by HPLC-MS/MS methods, as have the reaction products of 4-hydroperoxy-2-nonenal and 2′-deoxyguanosine.[238] Figure 5.11 shows the formation of etheno DNA and heptanone-etheno DNA adducts through homolytic decomposition of lipid hydroperoxides, along with the base excision repair (BER) products appearing in the urine. Others have shown that RNA is modified by the same bifunctional reactive electrophiles derived from lipid peroxidation that covalently modify DNA.[239]

Recently, a 5-LOX mediated endogenous DNA damage has been reported.[240] Chiral lipidomics analysis showed that a 5-LOX-derived lipid hydroperoxide was responsible for endogenous DNA–adduct formation. Stable isotope dilution chiral LC-electron capture (EC) APCI-MS/RM-MS demonstrated a concomitant formation of lipid hydroperoxides in the presence of different enzyme stimulators or inhibitors. It was concluded that a targeted lipidomics approach provides a powerful tool to study pathways of cellular COX-1-, 5-LOX- and ROS-mediated lipid peroxidation and to examine their relationship to cellular DNA damage. The bifunctional electrophiles are genotoxic, with 4-ONE being a particularly potent genotoxin, which covalently modifies 2′-deoxyguanosine, 2′-deoxy and 2′-deoxycytidine residues on DNA.[241] In hepatocellular carcinoma, the *trans*-4-hydroxy-2-nonenal preferentially forms DNA adducts at carbon 249 of human p53 gene—a unique mutation hotspot.[242] The formation of HNE and other enal-derived cyclic DNA adducts from ω-3- an ω-6-PUFA, and their role in DNA repair and human p53 gene mutation, have been reviewed.[243]

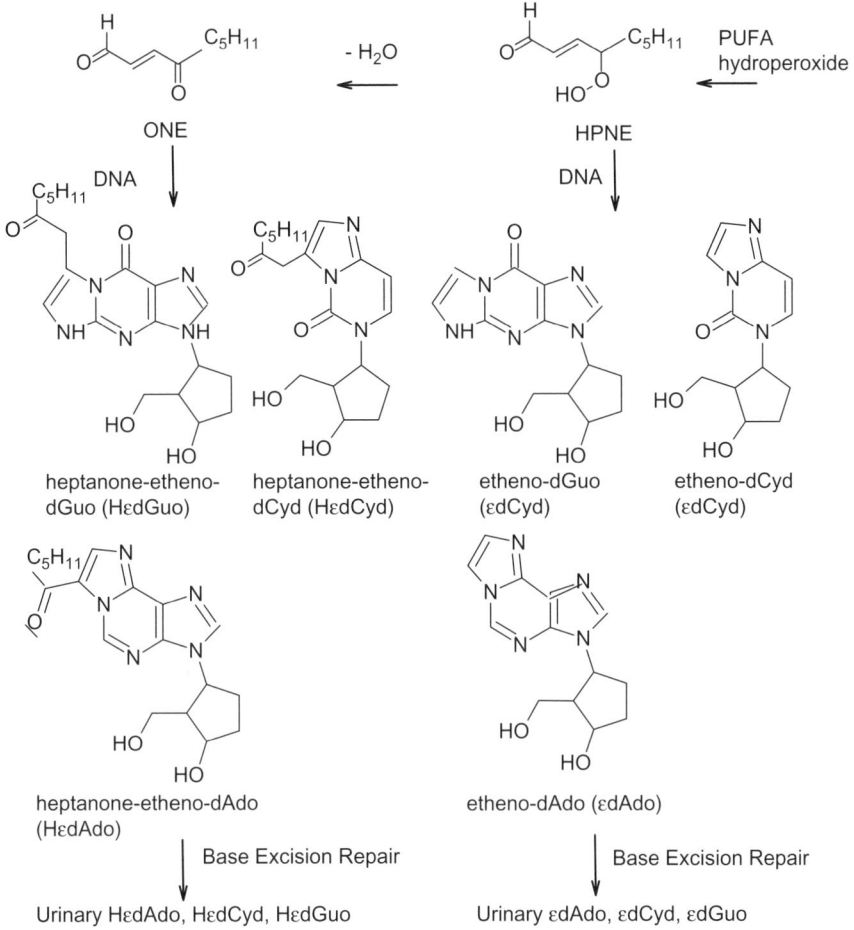

Figure 5.11 Formation of etheno DNA and heptanone-etheno DNA adducts through homolytic decomposition of lipid hydroperoxides, and of urinary excretion products through base excision repair. Redrawn from ref. 238.

5.4.4.4 Other Effects

In addition to specifically identified covalent binding, biologically active lipid oxidation products are also involved in less well-defined interactions with carrier and receptor proteins, as well as with cell membranes. Thus, a deep-fried fat was shown to cause peroxisome proliferator-activated receptor (PPAR) α activation in the liver of pigs as a non-proliferating species.[244] Upregulation of sterol regulatory element-binding protein (SREBP) and its target genes in liver and small intestine suggests that the oxidized fat could stimulate synthesis of cholesterol and TAG in these tissues. It was concluded that deep-fried fats could exert similar effects in man.

A selective vulnerability of preterm white matter to oxidative damage defined by F_2-isoprostanes has also been shown.[181] A novel family of oxidized choline glycerophospholipids (oxPC$_{CD36}$) that mediate thrombospondin receptor (CD36)-dependent recognition of LDL oxidized by various pathways has been identified[245] and the structural basis for the recognition of oxPC$_{CD36}$ by CD36 has been determined[246] using a combination of site-directed mutagenesis and ligand-binding analyses of various GST-CD36 fusion proteins bound to glutathione-Sepharose beads. Others reported that oxPC$_{CD36}$ serves as a ligand for another member of the scavenger receptor (SR) class B, type 1 (SR-B1).[247]

Another study showed that oxo-alkylGroPCho in atheroma induces platelet aggregation and inhibits endothelium-dependent arterial relaxation.[248] The 1-O-alkyl-2-[ω-oxo]acyl-*sn*-glycerols in shark liver oil and human milk have been recognized as a potential source of PAF mimics and of γ-hydroxy butyric acid.[115] The C_4 core aldehydes of both ChE and glycerolipids after reduction may be released as the neurologically active street drug, gamma hydroxy butyrate (GHB).[249] Furthermore, incubation of minimally modified LDL (mmLDL) containing 12/15-LOX oxidized ChE activated macrophages induced membrane ruffling and cell spreading, activated extra-cellular signal regulated kinase (ERK 1/2) and Akt signalling, and promoted the secretion of proinflammatory cytokines.[190] Many of these activities were diminished by ebselen, a reducing agent.

5.4.5 Exhaustion of Reducing Equivalents

There is good evidence that various antioxidant systems function to control lipid hydroperoxide formation, degradation and absorption. GSH peroxidase and GSSG reductase act as an enzyme pair in the reduction of peroxides, with concomitant oxidation of GSH and the regeneration of GSH with reducing equivalents donated from NADPH.[148,149] This system is responsible for much of the protection against intestinal absorption of hydroperoxides.

The supply of NADPH is functionally coupled to the pentose phosphate pathway[250] and is regulated by glucose availability and glucose 6-phosphate dehydrogenase activity.[251] The responsiveness of the gut to glucose suggests that nutrient availability is an important contributing factor in the intestinal catabolism of luminal peroxides. It has been shown that TAG-OOH appearance in lymph and chylomicrons is very low or undetectable as long as ascorbate and/or ubiquinols are present.[157]

Experiments suggest that it is the efficiency of lipid peroxide absorption from the lumen[150] and the availability of cellular reductants (GSH and NADPH)[149–152] that largely dictate the kinetics and extent of intracellular peroxide detoxification by the intestine. Studies in animal models support a paradigm of redox modulation of oxidative susceptibility of the intestinal epithelium *in vivo*. Rats placed on chronic lipid peroxide diets exhibited suppressed mucosal proliferative apoptotic activity in accordance with oxidized tissue GSH/GSSG redox balance.[252,253] This state of cytostasis created by chronic peroxide challenge was ameliorated by supplemental GSH that restored tissue GSH–GSSG redox balance.[252–255] Thus, physiological or pathological factors that induce peroxide generation, decrease luminal GSH transport, or promote intestinal GSH oxidation are expected to impact intestinal turnover characteristics and organ survival. Toxicity of aldehydes can frequently be prevented by replenishing the exhausted thiol scavengers.[226]

Neuronal mitochondrial toxicity has been recently described along with inhibitory effects on respiratory function and enzyme activities in rat brain mitochondria.[256] Mitochondrial aldehyde dehydrogenase 2 ($ALDH_2$) detoxifies HNE by oxidizing its aldehyde group.[257] It was shown that transgenic mice, which expressed a dominant-negative form of $ALDH_2$ in the brain, had decreased ability to detoxify HNE in their cortical neurons and accelerated accumulation of HNE in the brain. Consequently, the life span was shortened and age-dependent neurodegeneration and hyperphosphorylation of tau-protein (a hallmark of Alzheimer's) were observed.

Other antioxidant systems, based on ascorbate, ubiquinol and α-tocopherol have also been found to provide an efficient defence mechanism against formation of lipid peroxides in rat intestine and mesenteric lymph.[158] However, α-tocopherol[21] has been shown to be sometimes ineffective as an antioxidant, while ascorbic acid[185] has been found to actively decompose linoleic hydroperoxides *via* one electron reduction, yielding a variety of reactive aldehyde and ketone products. The gastrointestinal GSH peroxidase isoenzyme represents a first line of defence against ingested hydroperoxides: GSH reduced the amount of hydroperoxides transported from the rat or human gut into lymph.[258]

Aldo-keto reductase (AKR) family 1member 10 (AKR1B10) promotes cell survival by regulating lipid synthesis and eliminating carbonyls.[259] AKR1B10 is a monomeric enzyme that efficiently catalyzes the reduction to corresponding alcohols of a range of aromatic and aliphatic aldehydes and ketones, including highly electrophilic α,β-unsaturated carbonyls and anti-tumor drugs containing carbonyl groups, with NADPH as a co-enzyme. AKR1B10 is primarily expressed in the normal human colon and small intestine, but overexpressed in liver and lung cancer. AKR1B10 silencing increases the levels of α,β-unsaturated carbonyls, leading to a 2–3 fold increase of cellular lipid peroxides. In any event, exhaustion of endogenous or exogenous antioxidants clearly exposes the tissues to the adverse effects of lipid peroxidation.[259]

5.4.5.1   *Overdosing with Polyunsaturates*

Epidemiological studies have shown an inverse relation between the dietary consumption of fish containing EPA/DHA and mortality from coronary heart disease (CHD); a RBC EPA/DHA (called ω-3 index) has been proposed as a new risk factor for death from CHD.[260,261] The cardioprotective effects of EPA/DHA are widespread and appear to act independently of blood cholesterol reduction, and are mediated by diverse mechanisms. Their overall effects include anti-arrhythmia, TAG lowering, anti-thrombosis, anti-inflammation and endothelial relaxation. Current intakes of EPA/DHA in North America and elsewhere are well below those recommended by the American Heart Association for the management of patients with coronary heart disease.[262]

However, a review of human clinical studies that evaluated the effects of dietary antioxidants, vitamins, monounsaturated fatty acids (MUFA), PUFA and specific flavonoid-rich foods on LDL particle oxidation failed to clearly support this notion.[263] While some studies reported no difference in susceptibility of LDL particles to oxidation with fish oil supplementation, others found an increase[264–266] in the susceptibility of LDL particles to oxidation. Difference in design of studies and supplementation dosage may explain in part these inconsistencies in results. Unfortunately, these comparisons failed to include measurements of oxidative stress by appropriate indicators.

It is generally recognized that the replacement of saturated fats by vegetable and fish oils containing high levels of PUFAs renders individuals susceptible to lipid peroxidation and potential development of cardiovascular lesions.[267–270] PUFA serves as a major substrate for the peroxidation of dietary and tissue lipids, and overdosing with PUFA promotes this process and potentially overwhelms the defence mechanisms. There would appear to be no doubt that diet-derived aldehydes are readily absorbed[106,194,199] and there is good evidence that many chronic diseases represent states of heightened oxidative stress characterized by tissue lipid oxidation.[32,34,43]

Due to the presence of multiple double bonds and methylene carbons, EPA/DHA are much more susceptible to peroxidation than linoleic acid, linolenic acid and ARA, as shown by animal studies. Feeding of DHA to rats (90 mg per 100 g of rat mass) caused increased hydroperoxide accumulation in plasma (191–192% of controls), liver (170–230%) and kidney (250–340%), with a concomitant reduction in α-tocopherol levels (21–73% of controls).[168] Likewise, oral intake of excess polyunsaturated fatty acids (EPA/DHA) in a fish oil diet (15% fish oil) led to an acceleration of membrane lipid peroxidation in mice resulting in RBC senescence linked to lowering of immune response of spleen cells, which was not effectively prevented by supplementation with α-tocopherol.[187]

Lipid peroxidation induced by DHA enrichment modifies paracellular permeability of Caco-2 cells.[271] Incubation with 100 μM DHA increased lipid peroxidation and paracellular permeability, in parallel with a redistribution of the tight junction proteins occludin and ZO-1. Taurine, which acted as an antioxidant, partially prevented all these effects. Supplementation with DHA induced dose- and time-dependent oxidative changes in C6 glioma cells.[272] Low doses (25 μM) of DHA strengthened the cellular antioxidant defence system, as shown by a rise in glutathione peroxidase (GPX) and catalase activity, and decreased levels of lipid peroxidation (ROS). The opposite effect was observed with high doses (50 and 75 μM) of DHA and increased time of exposure (72 h and longer). Acute consumption of fish oil improved postprandial very low density lipoprotein (VLDL) without affecting LDL and HDL profiles in healthy men aged 50–65 years without evidence of lipid peroxidation.[273] This effect, however, does not rate as a long time overdosing.

Flax oil, which contains over 50% 18 : 3 ω-3 (ALA), and flax seed oil is not recommended for use for frying, pan heating, or preparation of food when heat is involved.[274] In small amounts, ALA is effectively metabolized to long chain ω-3 in plasma and tissues,[275] but overdosing may cause disease (see below).

The increasing trend to enrich foods with PUFA entails the potential risk of simultaneously enriching the food with the oxygenated unsaturated aldehydes. Thus, variable amounts of toxic HHE and HNE have been found in baby food and baby formula.[276] Obviously, overdosing with PUFAs (peroxidized or not) runs the risk of exhausting the protective mechanisms against both exogenous and endogenous lipid peroxidation products, and potentially precipitates disease. Consequently, a search for new antioxidants and new analytical approaches in antioxidant evaluation has been undertaken.[44]

5.4.5.2 Aging and Disease

It was postulated more than 50 years ago that aging could be partly due to lipid peroxidation and an accumulation of molecular and cellular damage induced by free radicals.[277] Lipid peroxides are now recognized as major sources of dietary oxidants of mutagenic or carcinogenic potential which are of nutritional and toxicological importance.[278,279] Figure 5.12 summarizes the metabolic effects demonstrated so far to result from *in vitro* and *in vivo* peroxidation of polyunsaturated dietary fatty acids in man and experimental animals.

By virtue of their ability to generate oxyradicals, lipid peroxides are capable of initiating degenerative processes and promote digestive system disorders such as inflammation and cancer.[280,281] Dietary intake of highly unsaturated fats is an important contributor to lipid peroxides in the intestinal lumen, and an estimated daily intake of 1.4 mmol lipid peroxides per day could occur in humans consuming an average of 84 g of fat daily.[282] In animal studies, luminal peroxide contents of 0.4 μmol have been detected after experimental duodenal infusion of peroxidized lipids;[151] levels that have been shown to alter intestinal cell responses *in vitro* (0.2–10 μM).[283,284] Early studies demonstrated that the toxicity of dietary polyunsaturated oils correlated directly with their peroxide content and that excessive consumption of lipid peroxides was cytotoxic *in vivo*.[285–287] In rats, administration of hydroperoxy and hydroxyl fatty acids stimulated DNA synthesis and induced ornithine decarboxylase activity consistent with enhanced cellular proliferation.[288] Moreover, mucosal hypertrophy of the colon developed in rats given peroxidized ethyl linoleate.[289]

Oxidation of unsaturated fatty acids of lipoproteins and lipids forms alkoxyl and peroxyl radicals that decompose into several metastable carbonyl end products, including ACR.[290] MDA is a product of oxidative damage to lipids, amino acids and DNA, and accumulates with aging and disease.

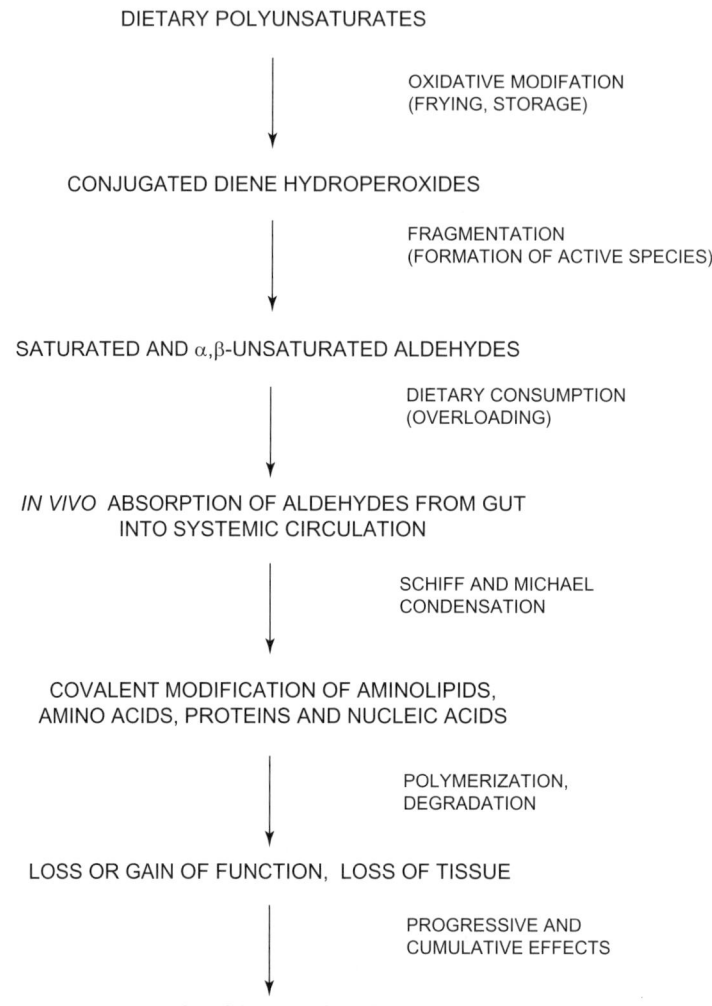

DIETARY POLYUNSATURATES

OXIDATIVE MODIFATION
(FRYING, STORAGE)

CONJUGATED DIENE HYDROPEROXIDES

FRAGMENTATION
(FORMATION OF ACTIVE SPECIES)

SATURATED AND α,β-UNSATURATED ALDEHYDES

DIETARY CONSUMPTION
(OVERLOADING)

IN VIVO ABSORPTION OF ALDEHYDES FROM GUT
INTO SYSTEMIC CIRCULATION

SCHIFF AND MICHAEL
CONDENSATION

COVALENT MODIFICATION OF AMINOLIPIDS,
AMINO ACIDS, PROTEINS AND NUCLEIC ACIDS

POLYMERIZATION,
DEGRADATION

LOSS OR GAIN OF FUNCTION, LOSS OF TISSUE

PROGRESSIVE AND
CUMULATIVE EFFECTS

DISEASE AND AGEING

Figure 5.12 Overall summary of the *in vitro* and *in vivo* demonstrated effects of peroxidized dietary PUFA on ageing and disease.

Oxidative mitochondrial damage is a key contributor to aging and age-associated degenerative diseases such as Parkinson's and Alzheimer's.[291] MDA depressed mitochondrial membrane potential and also showed a dose-dependent inhibition of mitochondrial complex I and complex-I-linked respiration. MDA significantly elevated mitochondrial reactive oxygen species and protein carbonyls at 0.2 and $0.002\,\mu\text{mol}\,\text{mg}^{-1}$, respectively.[292] These results suggest that MDA-induced neuronal mitochondrial toxicity may be an important factor in brain ageing and neurodegenerative diseases.

Reactive oxygen species and insulin signalling in the adipose tissue are critical determinants of ageing and age-related diseases, but it is not known if they represent independent factors or if they are linked mechanically. P66Shc is a redox enzyme that generates mitochondrial ROS and promotes ageing in mammals. Insulin activates the redox enzyme activity of p66Shc in adipocytes. Deletion of p66Shc resulted in increased mitochondrial uncoupling and reduced TAG accumulation in adipocytes. These results suggested that intracellular oxidative stress might accelerate ageing by favouring fat deposition and fat-related disorders.[293]

A considerable diminution of protein synthesis takes place when organisms age. There is no consensus about the mechanism underlying this decline with age because protein synthesis is a complex process requiring the participation of many molecules. Recent work has shown that treatment of young rats with cumene hydroperoxide yielded the same adduct formation between MDA and HNE with elongation factor-2 (eFF-2).[294] This suggests that ageing could result from a gradual increase in the levels of adduct formation with these aldehydes secondary to lipid peroxidation. MDA and HNE total protein adducts were determined by fluorescence (356/460 nm), while adducts with eFF-2 were determined by polyclonal goat anti-MDA and anti-HNE antibody.

α-Linolenic acid (18 : 3 n3) (ALA) has been associated inconsistently with an increased risk of prostate cancer. Flax oil has been shown to produce the strongest response in the genotoxicity of vegetable cooking oils in the *Drosophila* Wing Spot test when fed a medium containing 6% and 12% of each of the oils.[295] It was argued that the genotoxicity was probably due to the fatty acid composition of the oils, which after peroxidation can form specific DNA-adducts. A recent comperhensive review[296] concluded that high ALA intakes or high blood and adipose tissue concentrations of ALA may be associated with a small increased risk of prostate cancer. However, these conclusions were qualified because of the heterogeneity across the studies and the likelihood of publication bias.

Several disease states are associated with enhanced lipid peroxidation that leads to accumulation of ACR and ACR–protein and ACR–DNA adducts. Increased levels of ACR have been measured in plasma of patients with renal failure[297] and in Alzheimer's disease brain tissue.[298] Moreover, increased ACR–protein adducts have been detected in vascular tissues of patients with diabetes,[299] Alzheimer's disease[300] and atherosclerosis.[106,108,301] Both dietary and endogenous sources of ACR, including some enzymatic processes, have been recently reviewed.[302]

Chronic gut pathological states such as inflammatory bowel diseases are associated with chronic inflammation of the intestine,[303–305] suggesting a role for GSH–GSSG redox in disease etiology. A specific linkage between the oxidation of mucosal GSH, intestinal inflammation and disease activity in human ulcerative colitis has been shown.[306,307] The active disease state was resolved with normalization of mucosal redox balance by therapeutic intervention with azosulfapyridine, an effective agent for ulcerative colitis and a free radical scavenger.[308]

Oxidatively truncated phospholipids are substrates for lipoprotein-associated phospholipase A_2(Lp-PLA$_2$) which promotes formation of putative pro-inflammatory and pro-apoptotic products.[309] Initial descriptions of the enzyme focused on its similarity to PAF acetyl hydrolase to degrade PAF, but subsequent work has shown that LpPLA$_2$ has a much broader specificity than PAF acetyl hydrolase.[310] The LpPLA$_2$ activity is inhibited by darapladib (Glaxo-SmithKline), which greatly reduces the release of the highly atherogenic short chain aldehydes and lysoPtdCho.[311]

Cokacayne syndrome complementation group B (CSB) protein is engaged in transcription-coupled repair of UV-induced DNA damage and its deficiency leads to progressive multisystem degeneration and premature ageing.[312] HNE–DNA adducts block *in vitro* transcription by T7 RNA polymerase. When these lesions are not removed from the transcribed DNA strand due to CSB gene mutation or CSAB protein inactivation by high pathological HNE concentrations, they may contribute to accelerated premature ageing.[312]

The deleterious effects of repeatedly heated soy oil on the development of atherosclerosis have been explored using overiectomized rats, which represent an estrogen-deficient state.[170] Repeatedly heated soy oil caused significant increases in TBARS and LDL compared with fresh oil. Consumption of fried food has been suggested to promote obesity, but this association has been seldom studied. A recent study reports a positive association between fried food and general and central obesity among subjects in the highest quintile of energy intake from fried food.[313] Frying food absorbs degradation products of the frying oils such as polymers and polar compounds. These products have been associated with different types of cancer, endothelial dysfunction and hypertension.[314]

A recent review[315] has highlighted the role of oxidative stress in Alzheimer's, Parkinson's and Huntington's diseases, as well as amyotrophic lateral sclerosis and multiple sclerosis, which are recognized as neurodegenerative and neuro-inflammatory disorders where there is evidence for a primary contribution of oxidative stress in neuronal death. The authors conclude with a call not to ignore the contributions of diet and lifestyle to oxidative stress.

5.5   Conclusions

It is well established that unsaturated dietary fats and oils are subject to rapid peroxidation upon exposure to air during isolation, storage and food preparation. Polyunsaturated fatty acids also provide major substrates for endogenous peroxidation. Overdosing with polyunsaturates results in the accumulation of various lipid peroxides in cell membranes and lipoproteins, exhibiting a complexity and biochemical activity comparable to that seen for the autoxidation products of dietary fats and oils. While antioxidants appear to be effective in model systems, their effectiveness in treating disease has yet to be shown—though they would be expected to slow down the disease processes.

Of particular interest has been the formation and accumulation of 4-oxo-2-nonenal from peroxidation of ω-6-polyunsaturates and of 4-oxo-2-hexenal from peroxidation of ω-3-polyunsaturates. The bifunctional aldehydes are freely water-soluble and readily penetrate cell membranes, and can enter the systemic circulation. The presence of these highly reactive oxo-lipids in plasma and cells results in extensive protein and nucleic acid modification leading to either loss or gain of function, both changes having been associated with alterations in membrane structure and properties, mitochondrial uncoupling and subcellular messengers in gene regulatory and signal transduction pathways. All these effects have been associated either directly or indirectly with chronic disease such as Parkinson's, Alzheimer's, Huntington's, inflammation, atherosclerosis, diabetes and cancer.

The various unsaturated oxo- aldehydes exhibit a common mechanism of action characterized by formation of irreversible Schiff and Michael adducts with proteins and nucleic acids when the normally protective reducing equivalents (thiolate pool) have been exhausted. In this respect, the tissue damage accompanying lipid peroxidation may not differ from that caused by the free radicals arising from smoking.

Although the danger of overloading with polyunsaturates has been suggested before, clear evidence for it has been lacking due to inadequate methodology for recognizing and quantifying endogenous lipid oxidation. The various indicators so far proposed have dealt only with individual products of lipid peroxidation, which fail to reflect the scope or scale of the problem. The use of mass spectrometric methods to identify and quantify the entire spectrum of products along with the loss of substrate has greatly advanced the field, which has now progressed from a hypothesis-driven speculation to a systematic collection of well-documented analytical evidence.

The extension of mass spectrometric lipidomics to protein and nucleic acid carbonylation on a proteome and nucleome-wide scale expands our knowledge of the actual sites of aldehyde attack on the protein and nucleic acid chains, which affect the function of specific enzymes, cellular receptors and genes, and the mechanisms that promote disease and ageing.

Acknowledgments

The advice and assistance of my collaborators Drs J. J. Myher, H. Kamido, A. Ravandi, L. Marai, K. Hartvigsen, J-P. Kurvinen, O. Sjovall and Professor H. Kallio is acknowledged and greatly appreciated. The studies conducted in the author's laboratory were supported by the Heart and Stroke Foundation of Ontario, Toronto, and the Medical Research Council of Canada, Ottawa, Canada. I wish to thank Dr Inga Kuksis for reading and commenting on the final copy of the manuscript.

References

1. M. Morita and M. Tokita, *Lipids*, 2006, **41**, 91.
2. L. J. Marnett, S. E. Rowlinson, D. C. Goodwin, A. S. Kalgutkar and C. A. Lanzo, *J. Biol. Chem.*, 1999, **274**, 22903.

3. N. A. Porter, S. E. Caldwell and K. A. Mills, *Lipids*, 1995, **30**, 277.
4. E. N. Frankel, *J. Am. Oil Chem. Soc.*, 1984, **61**, 1908.
5. D. A. Forss, *Prog. Lipid Res.*, 1972, **13**, 177.
6. E. N. Frankel, *Lipid Oxidation,* The Oily Press, Dundee Scotland, 1998, p. 23–41.
7. H. R. Rawls and P. A. Van Santen, *J. Am. Oil Chem. Soc.*, 1970, **47**, 121.
8. E. N. Frankel, *Prog. Lipid Res.*, 1983, **22**, 1.
9. W. E. Neff, E. N. Frankel and D. Weisleder, *Lipids*, 1981, **16**, 439.
10. C. Schneider, N. A. Porter and A. R. Brash, *J. Biol. Chem.*, 2008, **283**, 15539.
11. H. W-S. Chan and D. T. Coxon, in *Autoxidation of unsaturated lipids*, ed. H. W.-S. Chan, Academic Press, London, 1987, p. 17–50.
12. U. Garscha, T. Nilsson and E. H. Oliw, *J. Chromatogr. B*, 2008, **872**, 90.
13. T. Miyazawa, H. Kunika, K. Fujimoto, Y. Endo and T. Kaneda, *Lipids*, 1995, **30**, 1001.
14. B. Halliwell and J. M. C. Gutteridge, *Free Radicals in Biology and Medicine,* Clarendon Press, Oxford, 1989.
15. H. W. Gardner, *Biochim. Biophys. Acta*, 1991, **1084**, 221.
16. D. Marchand, V. Grossi, A. Hirschler-Rea and J.-F. Rotani, *Lipids*, 2002, **37**, 541.
17. H. Kuhn, T. Schewe and S. M. Rapoport, *Adv. Enzymol.*, 1986, **58**, 273.
18. C. Schneider, D. A. Pratt, N. A. Porter and A. R. Brash, *Chem. Biol.*, 2007, **14**, 473.
19. L. Steenhorst-Slikkerveer, A. Louter, H.-G. Janssen and C. Bauer-Plank, *J. Am. Oil Chem. Soc.*, 2000, **77**, 837.
20. C. Bauer-Plank and L. Steenhurst-Slikkerveer, *J. Am. Oil Chem. Soc.*, 2000, **77**, 477.
21. J. M. Upston, J. Neuziol and R. Stocker, *J. Lipid Res.*, 1996, **37**, 2650.
22. C. N. Serhan, C. B. Clish, J. Brannon, S. P. Colgan, N. Chang and K. Gronert, *J. Exp. Med.*, 2000, **192**, 1197.
23. B. Dangi, M. Obeng, J. M. Nauroth, M. Teymourlouei, M. Needham, K. Raman and L. M. Arterburn, *J. Biol. Chem.*, 2009, **284**, 14744.
24. K. E. Furse, D. A. Pratt, C. Schneider, A. R. Brash, N. A. Porter and T. P. Lybrand, *Biochemistry*, 2006, **45**, 3206.
25. A. Kuksis, *Front. Biosci.*, 2007, **12**, 3203.
26. A. Kuksis, J.-P. Suomela, M. Tarvainen and H. Kallio, *Mol. Biotechnol.*, 2009, **42**, 224.
27. A. Kuksis, J.-P. Suomela, M. Tarvainen and H. Kallio, in *Lipidomics. Volume 2: Methods and Protocols*, ed. D. Armstrong, Humana Press, New York, 2010, Methods of Molecular Biology vol. 580, pp. 39–91.
28. B. Davis, G. Koster, L. J. Douet, M. Scigelova, G. Woffendin, J. M. Ward, A. Smith, J. Humphries, K. G. Burnand, C. H. Macphee and A. D. Postle, *J. Biol. Chem.*, 2008, **283**, 6428.
29. M. Careri, F. Bianchi and C. Corradini, *J. Chromatogr.*, 2002, **970**, 3.
30. H. Kim, G. P. Page and S. Barnes, *Nutrition*, 2004, **20**, 155.

31. J. Astle, J. T. Ferguson, J. B. German, G. H. Harrigan, N. L. Kelleher, T. Kodadek, B. A. Parks, M. J. Roth, K. W. Singletary, C. D. Wenger and G. B. Mahady, *J. Nutr.*, 2007, **137**, 2787.

32. M. D. Guillén and E. Goicoechea, *Crit. Rev. Food Sci. Nutr.*, 2008, **48**, 119.

33. M. Kussmann, M. Affolter, K. Nagy, B. Holst and L. B. Fay, *Mass Spectrom. Rev.*, 2007, **26**, 727.

34. L. M. Sayre, D. Lin, Q. Yuan, X. Zhu and X. Tang, *Drug Metab. Rev.*, 2006, **38**, 651.

35. H.-G. Schmarr, W. Sang, S. Gass, U. Fischer, B. Kopp, C. Schulz and T. Potouridis, *J. Sep. Sci.*, 2008, **31**, 3458.

36. E. K. Long, I. Smoliakova, A. Honzatko and M. J. Picklo Sr., *Lipids*, 2008, **43**, 765.

37. S. Uchiyama, Y. Inaba, M. Matsumoto and G. Suzuki, *Anal. Chem.*, 2009, **81**, 485.

38. P. Sarkar and B. E. Hayes, *Life Sci.*, 2009, **85**, 188.

39. B. G. Gugiu, K. Mouillesseaux, V. Duong, T. Herzog, A. Hekimian, L. Koroniak, T. M. Vondriska and A. D. Watson, *J. Lipid Res.*, 2008, **49**, 510.

40. M. Keeney, I. Katz and D. P. Schwartz, *Biochim. Biophys. Acta*, 1962, **62**, 615.

41. J. H. Weihrauch, C. R. Brewington and D. P. Schwartz, *Lipids*, 1974, **9**, 883.

42. D. P. Schwartz, A. H. Rady and S. Castaneda, *J. Am. Oil Chem. Soc.*, 1994, **71**, 441.

43. M. Guichardant and M. Lagarde, *Eur. J. Lipid Sci. Technol.*, 2009, **111**, 75.

44. M. Laguerre, L. J. Lopez-Giraldo, J. Lecomte and P. Villeneuve, *INFORM*, 2009, **20**, 328.

45. N. A. Porter, K. A. Mils and R. L. Carter, *J. Am. Chem. Soc.*, 1994, **116**, 6690.

46. H. W. S. Chan and G. Levett, *Lipids*, 1977, **12**, 99.

47. M. P. Yurawecz, J. K. Hood, M. M. Mossoba, J. A. G. Ropach and Y. Ku, *Lipids*, 1995, **30**, 595.

48. N. A. Porter, R. A. Wolf, E. M. Yarbro and H. Weenen, *Biochem. Biophys. Res. Commun.*, 1979, **89**, 1058.

49. J.-P. Iliou, D. Jourd'heuil, F. Robin, B. Serkiz, P. Guivarc'h, J.-V. Volland and J.-P. Vilaine, *Lipids*, 1992, **27**, 959.

50. J. D. Morrow and C. J. Roberts, *Prog. Lipid Res.*, 1997, **36**, 1.

51. J. D. Morrow, T. M. Harris and L. J. Roberts, *Anal. Biochem.*, 1990, **184**, 1.

52. M. VanRollins, P. D. Frade and O. A. Carretero, *Biochim. Biophys. Acta*, 1988, **966**, 133.

53. M. VanRollins and R. C. Murphy, *J. Lipid Res.*, 1984, **25**, 507.

54. A.-M. Lyberg, E. Fasoli and P. Adlercreutz, *Lipids*, 2005, **40**, 969.

55. A.-M. Lyberg and P. Adlercreutz, *Lipids*, 2006, **41**, 67.

56. L. Gao, H. Yin, G. L. Milne, N. A. Porter and J. D. Morrow, *J. Biol. Chem.*, 2006, **281**, 14092.
57. H. Yin, E. S. Musiek, L. Gao, N. A. Porter and J. D. Morrow, *J. Biol. Chem.*, 2005, **280**, 26600.
58. M. Masoodi and A. Nicolaou, *Rapid Commun. Mass Spectrom.*, 2006, **20**, 3023.
59. A. A. Christy, *Lipids*, 2009, **44**, 1105.
60. W. E. Neff, E. N. Frankel and K. Miyashita, *Lipids*, 1990, **25**, 33.
61. E. N. Frankel, W. E. Neff and K. Miyashita, *Lipids*, 1990, **25**, 40.
62. K. Miyashita, E. N. Frankel, W. E. Neff and R. A. Awl, *Lipids*, 1990, **25**, 48.
63. A. Kuksis, J. J. Myher, L. Marai and K. Geher, in *LIPIDFORUM*, ed. Y. Malkki and G. Lambertson, Bergen, Norway, 1993, 230.
64. O. Sjovall, A. Kuksis and L. and J. J. Myher, *Lipids*, 1997, **32**, 1211.
65. C. Wijesundera, C. Ceccato, P. Watkins, P. Fagan, B. Fraser, N. Thienthong and P. Perlmutter, *J. Am. Oil Chem. Soc.*, 2008, **85**, 543.
66. T. Miyazawa, H. Kunika, K. Fujimoto, Y. Endo and T. Kaneda, *Lipids*, 1995, **30**, 1001.
67. W. E. Neff and W. C. Byrdwell, *J. Chromatogr. A*, 1998, **818**, 169.
68. W. C. Byrdwell and W. E. Neff, *J. Chromatogr. A*, 1999, **852**, 417.
69. W. C. Byrdwell and W. E. Neff, *J. Chromatogr.*, 2001, **905**, 85.
70. W. C. Byrdwell and W. E. Neff, *Rapid Commun. Mass Spectrom.*, 2002, **16**, 300.
71. O. Sjovall, A. Kuksis and H. Kallio, *J. Chromatogr. A.*, 2001, **905**, 119.
72. O. Sjovall, A. Kuksis and H. Kallio, *Lipids*, 2002, **37**, 81.
73. O. Sjovall, A. Kuksis and H. Kallio, *Lipids*, 2003, **38**, 1179.
74. F. Giufrida, F. Distaillats, L. H. Skibsted and F. Dionisi, *Chem. Phys. Lipids*, 2004, **131**, 41.
75. L. Fauconnot, J. Hau, J. M. Aeschlimann, L. B. Fay and F. Dionisi, *Rapid Commun. Mass Spectrom.*, 2004, **18**, 214.
76. H. Yin, J. D. Brooks, L. Gao, N. A. Porter and J. D. Morrow, *J. Biol. Chem.*, 2007, **282**, 29890.
77. C. J. Brame, R. G. Salomon, J. D. Morrow and L. J. Roberts, *J. Biol. Chem.*, 1999, **274**, 13139.
78. N. Bernoud-Hubac, S. S. Davies, O. Boutaud, T. J. Montine and L. J. Roberts II, *J. Biol. Chem.*, 2001, **276**, 30964.
79. H. Kamido, A. Kuksis, L. Marai and J. J. Myher, *Lipids*, 1993, **28**, 331.
80. M. L. Lenz, H. Hughes, J. R. Mitchell, D. P. Vis, J. R. Guyton, A. A. Taylor, A. M. Gotto and C. V. Smith, *J. Lipid. Res.*, 1990, **31**, 1043.
81. T. Wang, Y. Wengui and W. S. Powell, *J. Lipid Res.*, 1992, **33**, 525.
82. L. Kritharides, W. Jessup, J. Gifford and R. T. Dean, *Anal. Biochem.*, 1993, **213**, 79.
83. J. A. Kenar, C. M. Havrilla, N. A. Porter, J. R. Guyton, S. A. Brown, K. F. Kemp and E. Selinger, *Chem. Res. Toxicol.*, 1996, **9**, 737.
84. V. A. Folcik and M. K. Cathcart, *J. Lipid Res.*, 1994, **35**, 1570.

85. C. M. Havrilla, D. L. Hachey and N. A. Porter, *J. Am. Chem. Soc.*, 2000, **122**, 8042.

86. H. Yin, C. M. Havrilla, J. D. Morrow and N. A. Porter, *J. Am. Chem. Soc.*, 2002, **124**, 7745.

87. H. Yin, J. D. Morrow and N. A. Porter, *J. Biol. Chem.*, 2004, **279**, 3766.

88. H. Kamido, A. Kuksis, L. Marai, J. J. Myher and H. Pang, *Lipids*, 1992, **27**, 645.

89. P. Spiteller, W. Kern, J. Reiner and G. Spiteller, *Biochim. Biophys. Acta*, 2002, **1531**, 188.

90. Y. Deng and R. G. Saloman, *J. Org. Chem.*, 2000, **65**, 6660.

91. H. Han and A. S. Csallany, *J. Am. Oil Chem. Soc.*, 2009, **86**, 253.

92. C. M. Seppanen and A. S. Csallany, *J. Am. Oil Chem. Soc.*, 2004, **81**, 1137.

93. S. H. Lee and I. A. Blair, *Chem. Res. Toxicol.*, 2000, **13**, 698.

94. N. A. Porter and W. A. Pryor, *Free Radic. Biol. Med.*, 1990, **8**, 541.

95. S. Bacot, N. Bernould-Hubac, N. Baddas, B. Chantegrel, C. Deshayes, A. Doutheau, M. Lagarde and M. Guichardant, *J. Lipid Res.*, 2003, **44**, 917.

96. F. J. van Kuijk, A. N. Siakotos, L. G. Fong, R. J. Stephens and D. W. Thomas, *Anal. Biochem.*, 1995, **224**, 420.

97. M. Guichardant, L. Valette-Talbi, C. Caradini, G. Crozier and M. Berger, *J. Chromatogr.*, 1994, **655**, 112.

98. M. Guichardant, N. Bernoud-Hubac, B. Chantegrel, C. Deshayes and M. Lagarde, *Prostaglandins Leukot. Essent. Fatty Acids*, 2002, **67**, 147.

99. D. L. Carbone, J. A. Doorn, Z. Kiebler and D. R. Petersen, *Chem. Res. Toxicol.*, 2005, **18**, 1324.

100. B. J. Steward, J. A. Doorn and D. R. Petersen, *Chem. Res. Toxicol.*, 2007, **20**, 1111.

101. R. M. LoPachin, T. Gavin, D. R. Petersen and D. S. Barber, *Chem. Res. Toxicol.*, 2009, **22**, 1499.

102. X. Zhu, V. E. Anderson and L. M. Sayre, *Rapid Commun. Mass Spectrom.*, 2009, **23**, 2113.

103. D. Lin, H.-G. Liu, G. Perry, M. A. Smith and L. M. Sayre, *Chem. Res. Toxicol.*, 2005, **18**, 1219.

104. H. Esterbauer, in *Free Radicals, Lipid Peroxidation and Cancer*, ed. D. C. H. McBrien and F. T. Slater, Academic Press, New York, 1982, p. 101–105.

105. M. Hecker, M. Haurand, V. Ullrich, U. Diczfalausy and S. Hammarstrom, *Arch. Biochem. Biophys.*, 1987, **254**, 124.

106. K. Uchida, M. Kanematsu, Y. Morimitsu, T. Osawa, N. Noguchi and E. Niki, *J. Biol. Chem.*, 1998, **273**, 16058.

107. K. Umano, K. J. Dennis and T. Shibamoto, *Lipids*, 1988, **23**, 811.

108. K. Uchida, *Trends Cardiovasc. Med.*, 1999, **9**, 109.

109. T. Miyatake and T. Shibamoto, *Food Chem. Toxicol.*, 1996, **34**, 1009.

110. H. Esterbauer, H. Zollner and R. J. Schaur, in *Membrane Lipid Oxidation,* ed. C. Vigo-Pelfrey, CRC Press, Boca Raton, FL, 1989, **vol. 1**, p. 239–268.

111. H. Esterbauer and K. H. Cheeseman, *Methods Enzymol.*, 1990, **186**, 407.

112. J.-P. Suomela, M. Ahotupa, O. Sjovall, J.-P. Kurvinen and H. Kallio, *Lipids*, 2004, **39**, 639.

113. J.-P. Suomela, M. Ahotupa and H. Kallio, *Lipids*, 2005, **40**, 349.

114. J.-P. Suomela, M. Ahotupa and H. Kallio, *Lipids*, 2005, **40**, 437.

115. K. Hartvigsen, A. Ravandi, R. Harkewicz, H. Kamido, K. Bukhave, G. Holmer and A. Kuksis, *Lipids*, 2006, **41**, 679.

116. M. C. Dobarganes and G. Marquez-Ruiz, in *Deep Frying*, ed. M. D. Erickson, AOCS Press, Urbana, IL, 2007, p. 87–110.

117. O. Berdeaux, J. Velasco, G. Marquez-Ruiz and C. Dobarganes, *J. Am. Oil Chem. Soc.*, 2002, **79**, 279.

118. O. Berdeaux, G. Marquez-Ruiz and C. Dobarganes, *J. Chromatogr. A*, 1999, **863**, 171.

119. J. Velasco, S. Marmesat, O. Berdeo, G. Marquez-Ruiz and C. Dobarganes, *J. Agric. Food Chem.*, 2005, **53**, 4006.

120. H. Kamido, A. Kuksis, L. Marai, J. J. Myher and H. Peng, *Lipids*, 1992, **27**, 645.

121. A. D. Watson, N. Leitinger, M. Navab, K. F. Faull, S. Horkko, J. L. Witztum, W. Palinski, D. Schwenke, R. G. Salomon, W. Sha, G. Subbanagounder, A. M. Fogelman and J. A. Berliner, *J. Biol. Chem.*, 1997, **272**, 13597.

122. G. Subbanagounder, N. Leitinger, D. CSchwenke, J. W. Wong, H. Lee, C. Rizza, A. D. Watson, K. F. Faull, A. M. Fogelman and J. A. Berliner, *Arterioscler. Thromb. Vasc. Biol.*, 2000, **20**, 2248.

123. T. M. McIntyre, G. A. Zimmerman and S. M. Prescott, *J. Biol. Chem.*, 1999, **274**, 25189.

124. R. Chen, L. Yang and T. M. McIntyre, *J. Biol. Chem.*, 2007, **282**, 24842.

125. M. E. Greenberg, X.-M. Li, B. G. Gugiu, X. Gu, J. Qin, R. G. Salomon and S. L. Hazen, *J. Biol. Chem.*, 2008, **283**, 2385.

126. H. Kamido, A. Kuksis, L. Marai and J. J. Myher, *FEBS Lett.*, 1992, **304**, 269.

127. H. Kamido, A. Kuksis, L. Marai and J. J. Myher, *J. Lipid Res.*, 1995, **36**, 1876.

128. B. Karten, H. Bochzelt, P. M. Abuja, M. Mittelbach and W. Sattler, *J. Lipid Res.*, 1999, **40**, 1240.

129. Y. Kawai, A. Saito, N. Shibata, M. Kobayashi, S. Yamada, T. Osawa and K. Uchida, *J. Biol. Chem.*, 2003, **278**, 21040.

130. D. Herrera, A. Ravandi and A. Kuksis, in *Abstracts, 98th Annual Meeting & Expo of AOCS*, Quebec City, Quebec, Canada, May 13–16, 2007, p. 117.

131. G. Hoppe, A. Ravandi, D. Herrera, A. Kuksis and H. F. Hoff, *J. Lipid Res.*, 1997, **38**, 1347.

132. B. Karten, H. Boechzelt, P. M. Abuja, M. Mittelbach, KI. Oettl and W. Sattler, *J. Lipid Res.*, 1998, **39**, 1508.
133. J. Kanner and T. Lapidot, *Free Radic. Biol. Med.*, 2001, **31**, 1388.
134. T. Lapidot, R. Granit and J. Kanner, *J. Agric. Food Chem.*, 2005, **53**, 3391.
135. E. Goicoechea, E. K. Van Twillert, M. Dutts, E. D. F. A. Brandon, P. R. Kootstra, M. H. Blockland and M. Guillen, *J. Agric. Food Chem.*, 2008, **56**, 8475.
136. F. Paltauf, F. Esfandi and A. Holasek, *FASEB Lett.*, 1974, **40**, 119.
137. M. Y. J. Timmermans, G. Reekmans, H. J. H. Teuchy and L. P. M. Kuipers, *Biochem. J.*, 1996, **314**, 931.
138. N. Miled, A. Rousel, C. Busetta, L. Berti-Dupuis, M. Riviere, G. Buono, R. Verger, C. Cambillau and S. Canaan, *Biochemistry*, 2003, **42**, 11587.
139. A. Kuksis and R. Lehner, in *Intestinal Lipid Metabolism*, ed. C. M. Mansbach II, P. Tso and A. Kuksis, Kluwer Academic/Plenum Publishers, New York, 2001, p. 185–213.
140. K. Kanazawa and H. Ashida, *Biochim. Biophys. Acta*, 1998, **1393**, 336.
141. K. Kanazawa and H. Ashida, *Biochim. Biophys. Acta*, 1998, **1393**, 349.
142. K. E. Stremler, D. M. Stafforini, S. M. Prescott and T. M. McIntyre, *J. Biol. Chem.*, 1991, **266**, 11095.
143. F. D. Kolodgie, A. P. Burke, K. S. Skorija, E. Ladich, R. Kutys, A. T. Makuria and R. Virmani, *Arterioscler. Thromb. Vasc. Biol.*, 2006, **26**, 2523.
144. Z. Ahmed, S. Babaei, G. F. Maguire, D. Draganov, A. Kuksis, B. N. La Du and P. W. Connelly, *Cardiovasc. Res.*, 2003, **57**, 225.
145. P. W. Connelly, D. Draganov and G. F. Maguire, *Free Radic. Biol. Med.*, 2005, **38**, 164.
146. A. Negre, R. Salvayre, P. Rogalle, Q. Q. Dang and L. Douste-Blazy, *Biochim. Biophys. Acta*, 1987, **918**, 76.
147. P. Lechene de la Porte, N. Abouakil, H. Lafont and D. Lombardo, *Biochim, Biophys. Acta*, 1987, **920**, 237.
148. T. Y. Aw, *Toxicol. Appl. Pharmacol.*, 2005, **204**, 320.
149. T. S. LeGrand and T. Y. Aw, in *Intestinal Lipid Metabolism*, ed. C. H. Mansbach II, P. Tso and A. Kuksis, Kluwer Academic/Plenum Publishers, New York, 2001, p. 351–366.
150. T. Y. Aw and M. W. Williams, *Am. J. Physiol.*, 1992, **263**, G665.
151. T. Y. Aw, M. W. Williams and L. Gray, *Am. J. Physiol.*, 1992, **262**, G99.
152. D. P. Kowalski, R. M. Feeley and D. P. Jones, *J. Nutr.*, 1990, **120**, 1115.
153. N. Kaplowitz, T. Y. Aw and M. Ookhtens, *Ann. Rev. Pharmacol. Toxicol.*, 1985, **25**, 715.
154. F. Ursini, A. Zamburlini, G. Cazzolato, M. Maiorino, G. B. Bon and A. Sevanian, *Free Radic. Biol Med.*, 1998, **25**, 250.
155. J. M. Johnston and B. Borgstrom, *Biochim. Biophys. Acta*, 1964, **84**, 412.
156. W. C. Breckenridge and A. Kuksis, *Can. J. Biochem.*, 1975, **53**, 1184.
157. I. Staprans, H. J. Rapp, X. M. Pan, K. Y. Kim and K. R. Feingold, *Arterioscler. Thromb. Vasc. Biol.*, 1994, **14**, 1900.

158. I. Staprans, H. J. Rapp, X. M. Pan, D. A. Hardman and K. R. Feingold, *Arterioscler. Thromb. Vasc. Biol.*, 1996, **16**, 533.
159. D. Mohr, Y. Umeda, T. G. Redgrave and R. Stocker, *Redox Rep.*, 1999, **4**, 79.
160. K. Wingler, C. Muller, K. Schmehl, S. Florian and R. Brigelius-Flohe, *Gastroenterology*, 2000, **119**, 420.
161. C. Mueller, B. Friedrichs, K. Wingler and R. Brigelius-Flohe, *Biol. Chem.*, 2002, **383**, 637.
162. J. C. Martin, C. Caselli, S. Broquet, P. Juaneda, M. Nour, J.-L. Sebedio and A. Bernard, *J. Lipid Res.*, 1997, **38**, 1666.
163. F. G. Bergen and H. H. Draper, *Lipids*, 1970, **5**, 976.
164. J. Glavind and C. Sylven, *Acta Chem. Scand.*, 1970, **24**, 3723.
165. K. Nagatsugawa and T. Kaneda, *J. Jpn. Oil Chem. Soc.*, 1983, **32**, 362.
166. B. Frei, R. Stocker and B. N. Ames, *Proc. Natl. Acad. Sci. U. S. A.*, 1988, **85**, 9748.
167. R. W. Browne and D. Armstrong, *Clin. Chem.*, 2000, **46**, 829.
168. J. H. Song, K. Fujimoto and T. Miyazawa, *J. Nutr.*, 2000, **130**, 3028.
169. T. Miyazawa, T. Suzuki and K. Fujimoto, *Lipids*, 1993, **28**, 789.
170. S. K. Adam, S. Das, I. N. Soelaiman, N. A. Umar and K. Jaarin, *Tohoku J. Exp. Med.*, 2008, **215**, 219.
171. M. Raff, T. Tholstrup, S. Basu, P. Nonboe, M. T. Sorensen and E. M. Straarup, *J. Nutr.*, 2008, **138**, 509.
172. W. Pruzanski, E. Stefanski, F. C. de Beer, M. G. de Beer, A. Ravandi and A. Kuksis, *J. Lipid Res.*, 2000, **41**, 1035.
173. E. S. Musiek, J. D. Brooks, M. Joo, E. Brunoldi, A. Porta, G. Zanoni, G. Vidari, T. S. Blackwell, T. J. Montine, G. L. Milne, B. McLaughlin and J. D. Morrow, *J. Biol. Chem.*, 2008, **283**, 19927.
174. P. J. M. Boon, H. S. Marinho, R. Oosting and G. J. Mulder, *Toxicol. Appl. Pharmacol.*, 1999, **159**, 214.
175. A. Laurent, J. Alary, L. Debrauwer and J. P. Cravedi, *Chem. Res. Toxicol.*, 1999, **12**, 887.
176. J. Alary, F. Gueraud and J.-P. Cravedi, *Mol. Aspects Med.*, 2003, **24**, 177.
177. A. Hiratsuka, K. Tobita, H. Saito, Y. Sakamoto, H. Nakano, K. Ogura, T. Nishiyama and T. Watabe, *Biochem. J.*, 2001, **355**, 237.
178. F. Gueraud, F. Crouzet, J. Alary, D. Rao, L. Debrauwer, F. Laurent and J.-P. Cravedi, *BioFactors*, 2005, **24**, 97.
179. C. Wakita, T. Maeshima, A. Yamazaki, T. Shibata, S. Ito, M. Akagawa, M. Ojika, J. Yodoi and K. Uchida, *J. Biol. Chem.*, 284, 28810. .
180. T. J. Montine and J. D. Morrow, *Am. J. Pathol.*, 2005, **166**, 1283.
181. S. A. Back, N. L. Luo, R. A. Mallinson, J. P. O'Malley, L. D. Wallen, B. Frei, J. D. Morrow, C. K. Petito, C. T. Roberts, G. H. Murdoch and T. J. Montine, *Ann. Neurol.*, 2005, **58**, 108.
182. T. J. Montine, M. F. Beal, M. Cudkowicz, H. O'Donnell, R. A. Margolin, L. McFarland, A. F. Bachrach, W. E. Zacker, L. J. Roberts and J. D. Morrow, *Neurology*, 1999, **52**, 562.

183. H. Esterbauer, R. J. Schaur and H. Zollner, *Free Radic. Biol. Med.*, 1991, **11**, 81.

184. S. H. Lee and I. A. Blair, *Chem. Res. Toxicol.*, 2000, **13**, 698.

185. S. H. Lee, T. Oe and I. A. Blair, *Science*, 2001, **292**, 2083.

186. S. H. Lee, T. Oe and I. A. Blair, *Chem. Res. Toxicol.*, 2002, **15**, 300.

187. M. Oarada, H. Furukawa, T. Majima and T. Miyazawa, *Biochim. Biophys. Acta*, 2000, **1487**, 1.

188. J. Huber, A. Vales, G. Mitulovic, M. Blumer, R. Schmid, J. L. Witztum, B. R. Binder and N. Leitinger, *Arterioscler. Thromb. Vasc. Biol.*, 2002, **22**, 101.

189. J. M. Upston, A. C. Terentis, K. Morris, J. F. Keaney Jr and R. Stocker, *Biochem. J.*, 2002, **363**, 753.

190. R. Harkewicz, K. Hartvigsen, F. Almazan, E. A. Dennis, J. L. Witztum and Y. I. Miller, *J. Biol. Chem.*, 2008, **283**, 10241.

191. A. Ravandi, S. Babaei, R. Leung, D. J. Stewart, G. Hoppe, H. Hoff, H. Kamido and A. Kuksis, *Lipids*, 2004, **39**, 1.

192. J. Alary, L. Debrauwer, Y. Fernandez, J.-P. Cravedi, D. Rao and G. Bories, *Chem. Res. Toxicol.*, 1998, **11**, 130.

193. E. Rathahao, G. Peiro, N. Martins, J. Alarey, F. Guereaud and L. Debrauwer, *Anal. Bioanal. Chem.*, 2005, **381**, 1532.

194. J. Alary, F. Bravais, J.-P. Cravedi, L. Debrauwer, D. Rao and G. Bories, *Chem. Res. Toxicol.*, 1995, **8**, 34.

195. L. L. De Zwart, R. C. A. Hermanns, N. P. E. Vermeulen and N. J. Van Sittert, *Xenobiotica*, 1996, **26**, 1087.

196. G. Bringmann, M. Gassen and R. Lardy, *Tetrahedron*, 1994, **50**, 10245.

197. H. C. Kuiper, C. L. Miranda, J. D. Sowell and J. F. Stevens, *J. Biol. Chem.*, 2008, **283**, 17131.

198. K. Kanazawa, E. Kanazawa and M. Natake, *Lipids*, 1985, **20**, 412.

199. M. Grootveld, M. D. Atherton, A. N. Sheerin, J. Hawkes, D. R. Blake, T. E. Richens, C. J. L. Silwood, E. Lynch and A. W. D. Claxson, *J. Clin. Invest.*, 1998, **101**, 1210.

200. M. Raff, T. Tholstrup, S. Basu, P. Nonboe, M. T. Sorensen and E. M. Straarup, *J. Nutr.*, 2008, **138**, 509.

201. J. A. Lawson, H. Li, J. Rokach, M. Adiyaman, S.-W. Hwang, S. P. Khanapure and G. A. FitzGerald, *J. Biol. Chem.*, 1998, **273**, 29295.

202. J. Nourooz-Zadeh, M. B. Cooper, D. Ziegler and D. J. Betteridge, *Biochem. Biophys. Res. Commun.*, 2005, **330**, 731.

203. A. Ravandi, A. Kuksis, J. J. Myher and L. Marai, *J. Biochem. Biophys. Methods*, 1995, **30**, 271.

204. A. Ravandi, A. Kuksis, N. Shaikh and G. Jackowski, *Lipids*, 1997, **32**, 989.

205. P. Friedman, S. Horkko, D. Steinberg, J. L. Witztum and E. A. Dennis, *J. Biol. Chem.*, 2002, **277**, 7010.

206. J.-P. Kurvinen, A. Kuksis, A. Ravandi, O. Sjovall and H. Kallio, *Lipids*, 1999, **34**, 299.

207. T. Ishii, S. Ito, S. Kumazawa, T. Sakurai, S. Yamaguchi, T. Mori, T. Nakayama and K. Uchida, *Biochem. Biophys. Res. Commun.*, 2008, **371**, 28.

208. K. Tsuji, Y. Kawai, Y. Kato and T. Osawa, *BioFactors*, 2004, **21**, 263.

209. X. Liu, N. Yamada, W. Maruyama and T. Osawa, *J. Biol. Chem.*, 2008, **283**, 34887.

210. J. Alamed, W. Chaiyasit, D. J. McClements and E. A. Decker, *J. Agric. Food Chem.*, 2009, **57**, 2969.

211. X. Niu, V. Zammit, J. M. Upston, R. T. Dean and R. Stocker, *Arterioscler. Thromb. Vasc. Biol.*, 1999, **19**, 1708.

212. A. Raghavemenon, M. Garelnabi, S. Babu, A. Aldrich, D. Litvinov and S. Partasarathy, *Antioxid. Redox Signal.*, 2009, **11**, 1237.

213. E. Lonn, J. Bosch, S. Yusuf, P. Heridan, J. Pogue, J. M. Arnold, C. Ross, A. Arnold, P. Sleight, J. Probstfield and G. R. Dagenais, *JAMA*, 2005, **293**, 1338.

214. M. E. Haberland, A. M. Fogelman and P. A. Edvards, *Proc. Natl. Acad. Sci., U. S. A.*, 1982, **79**, 1712.

215. K. Uchida, K. Sakai, K. Itakura, T. Osawa and S. Toyokuni, *Arch. Biochem. Biophys.*, 1997, **346**, 45.

216. L. G. McGirr, M. Hadley and H. H. Draper, *J. Biol. Chem.*, 1985, **260**, 15427.

217. K. Uchida, *Free Radic. Res.*, 2006, **40**, 1335.

218. L. A. Piche, P. D. Cole, M. Hadley, R. Van den Bergh and H. Draper, *Carcinogenesis*, 1988, **9**, 473.

219. J. Giron-Calle, M. Alaiz, F. Millan, V. Rutz-Gutierrez and E. Vioque, *J. Agric. Food Chem.*, 2002, **50**, 6194.

220. J. Kanner, *Mol. Nutr. Food Res.*, 2007, **51**, 1094.

221. E. R. Stadtman and R. L. Levine, *Ann. N.Y. Acad. Sci.*, 2000, **899**, 191.

222. T. Ishii, T. Yamada, T. Mori, S. Kumazawa, K. Uchida and T. Nakayama, *Free Radic. Res.*, 2007, **41**, 1253.

223. J. Roy, P. Pragathi, A. Bettaieb, A. Tanel and D. A. Averill-Bates, *Chem. Biol. Interact.*, 2009, **181**, 154.

224. K. S. Echtay, T. C. Esteves, J. L. Pakay, M. B. Jekabsons, A. J. Lambert, M. Portero-Otin, R. Pamplona, A. J. Vidal-Puig, S. Wang, S. J. Roebuck and M. D. Brand, *EMBO J.*, 2003, **22**, 4103.

225. E. K. Long, T. C. Murphy, L. J. Leiphon, J. Watt, J. D. Morrow, G. L. Milne, J. R. H. Howard and M. J. Picklo Sr, *J. Neurochem.*, 2008, **105**, 714.

226. F. Fenaille, P. A. Guy and J. C. Tabet, *J. Am. Soc. Mass Spectrom.*, 2003, **14**, 215.

227. T. Isahii, E. Tatsuda, S. Kumazawa, T. Nakayama and K. Uchida, *Biochemistry*, 2003, **42**, 3474.

228. Y. Chen, J. D. Morrow and L. J. Roberts II, *J. Biol. Chem.*, 1999, **274**, 10863.

229. P. A. Grimsrud, H. Xie, T. J. Griffin and D. A. Bernlohr, *J. Biol. Chem.*, 2008, **283**, 21837.

230. R. G. Saloman, W. Sha, C. Brame, K. Kaur, G. Subbanagounder, J. O'Neil, H. F. Hoff and L. J. Roberts II, *J. Biol. Chem.*, 1999, **274**, 20271.

231. S. Horkko, D. A. Bird, E. Miller, H. Itabe, N. Leitinger, G. Subbanagounder, J. A. Berliner, P. Friedman, E. A. Dennis, L. K. Curtis, W. Palinski and J. L. Witztum, *J. Clin. Invest.*, 1999, **103**, 117.

232. Z. Ahmed, A. Ravandi, G. F. Maguire, A. Kuksis and P. W. Connelly, *Cardiovasc. Res.*, 2003, **58**, 712.

233. K. Ishino, T. Shibata, T. Ishii, Y.-T. Liu, S. Toyokuni, X. Zhu, L. M. Sayre and K. Uchida, *Chem. Res. Toxicol.*, 2008, **21**, 1261.

234. B. G. Hill, P. Haberzettl, Y. Ahmed, S. Srivastava and A. Bhatnagar, *Biochem. J.*, 2008, **410**, 525.

235. R. Chen, A. E. Feldstein and T. M. McIntyre, *J. Biol. Chem.*, 2009, **284**, 26297.

236. R. Li, W. Chen, R. Yanes, S. Lee and J. A. Berliner, *J. Lipid Res.*, 2007, **48**, 709.

237. M. D. Evans, M. Dizdaroglu and M. S. Cooke, *Mutat. Res. Rev.*, 2004, **567**, 1.

238. I. A. Blair, *J. Biol. Chem.*, 2008, **283**, 15545.

239. P. Zhu, S. H. Lee, S. Wehrli and I. A. Blair, *Chem. Res. Toxicol. 2006*, **19**, *809*.

240. W. Jian, S. H. Lee, M. V. Williams and I. A. Blair, *J. Biol. Chem.*, 2009, **284**, 16799.

241. S. H. Lee, M. V. Williams and I. A. Blair, *Prostaglandins Other Lipid Mediat.*, 2005, **77**, 141.

242. W. Hu, Z. Feng, J. Eveleigh, G. Iyer, J. Pan, S. Amin, F. I. Chung and M. S. Tang, *Carcinogenesis*, 2002, **23**, 1781.

243. F. L. Chung, J. Pan, S. Chouhury, R. Roy, W. Hu and M. S. Tang, *Mutat. Res.*, 2003, **531**, 25.

244. S. Luci, B. Konig, B. Giemsa, S. Huber, G. Haouse, H. Kluge, G. I. Stangl and K. Eder, *Br. J. Nutr.*, 2007, **97**, 872.

245. X. M. Li, R. G. Salomon, J. Qin and S. L. Hazen, *Biochemistry*, 2007, **46**, 5009.

246. N. S. Kar, M. Z. Ashraf, M. Valiyaveettil and E. A. Podrez, *J. Biol. Chem.*, 2008, **283**, 8765.

247. M. Z. Ashraf, N. S. Kar, X. Chen, J. Choi, R. G. Salomon, M. Bebbraio and E. A. Podrez, *J. Biol. Chem.*, 2008, **283**, 10408.

248. H. Kamido, H. Eguchi, H. Ikeda, T. Imaizumu, K. Yamana, K. Hartvigsen, A. Ravandi and A. Kuksis, *J. Lipid Res.*, 2002, **43**, 158.

249. M. Mamelak and A. Kuksis, *INFORM*, 2006, **17**, 607.

250. I. R. Brude, C. A. Drevon, I. Hjermann, I. Seljeflot, S. Lund-Katz and K. Saarem, *et al., Arterioscler. Thromb. Vasc. Biol.*, 1997, **17**, 2576.

251. L. V. Eggleston and H. A. Krebs, *Biochem. J.*, 1974, **138**, 425.

252. M. D. Mesa, R. Buckley, A. M. Minihane and P. Yaqoob, *Atherosclerosis*, 2004, **175**, 333.

253. S. Tsunada, R. Iwakiri, K. Fujimoto and T. Y. Aw, *Dig. Dis. Sci.*, 2003, **48**, 2333.

254. S. Tsunada, R. Iwakiri, T. Noda, K. Fujimoto, J. A. Fuseler and T. Y. Aw, *Dig. Dis. Sci.*, 2003, **48**, 210.

255. S. Tsunada, R. Iwakiri, H. Ootani, T. Y. Aw and K. Fujimoto, *Scand., J. Gastroenterol.*, 2003, **38**, 1002.

256. J. Long, C. Liu, L. Sun, H. Gao and J. Liu, *Neurochem. Res.*, 2009, **34**, 786.

257. R. Flohe-Brigelius, *Free Radic. Biol. Med.*, 1999, **27**, 951.

258. I. Ohsawa, K. Nishimaki, Y. Murakami, Y. Suzuki, M. Ishikawa and S. Ohta, *J. Neurosci.*, 2008, **28**, 6239.

259. C. Wang, R. Yan, D. Luo, K. Watanabe, D.-F. Liao and D. Cao, *J. Biol. Chem.*, 2009, **284**, 26742.

260. W. S. Harris and C. Van Schacky, *Prev. Med.*, 2004, **39**, 212.

261. W. S. Harris, *Pharmacol. Res.*, 2007, **55**, 217.

262. B. M. Anderson and D. W. Ma, *Lipids Health Dis.*, 2009, **8**, 33.

263. A. Lapointe, C. Couillard and S. Lemieux, *J. Nutr. Biochem.*, 2006, **17**, 645.

264. E. C. Leigh-Firbank, A. M. Minihane, D. S. Leake, J. W. Wright, M. C. Murphy, B. A. Griffin and C. M. Williams, *Br. J. Nutr.*, 2002, **87**, 435.

265. Y. E. Finnegan, A. M. Minihane, E. C. Leigh-Firbank, S. Kew, G. W. Meijer, R. Muggli, P. C. Calder and C. M. Williams, *Am. J. Clin. Nutr.*, 2003, **77**, 783.

266. M. D. Mesa, R. Buckley, A. M. Minihane and P. Yaqoob, *Atherosclerosis*, 2004, **175**, 333.

267. S. Higgins, Y. L. Carroll, S. N. McCarthy, B. M. Corridan, H. M. Roche, J. M. Wallace, N. O'Brien and P. A. Morrissey, *Br. J. Nutr.*, 2001, **85**, 23.

268. A. Piolot, D. Blache, L. Boulet, L. J. Fortin, D. Dubreuil, C. Marcoux, J. Davignon and S. Lussier-Cacan, *J. Lab. Clin. Med.*, 2003, **141**, 41.

269. A. Catala, *Chem. Phys. Lipids*, 2009, **157**, 1.

270. G. S. Oostenbrug, R. P. Mensink, M. R. Hardeman, T. De Vries, F. Brouns and G. Hornstra, *E. J. Appl. Physiol.*, 1997, **83**, 746.

271. S. Roig-Perez, F. Guardiola, M. Moreto and R. Ferrer, *J. Lipid Res.*, 2004, **45**, 1418.

272. F. Leonardi, L. Attorri, R. Di Benedetto, A. Di Biase, M. Sanchez, F. P. Tregno, M. Nardini and S. Salvati, *Free Radic. Res.*, 2007, **41**, 748.

273. G. Burdge, J. Powell, T. Dadd, D. Talbot, J. Civil and P. Calder, *Br. J. Nutr.*, 2009, **102**, 160.

274. W. S. Choo, E. J. Birch and J.-P. Dufour, *J. Am. Oil Chem. Soc.*, 2007, **84**, 735.

275. S. Lee, Master's Thesis, Seoul National University, Seoul, South Korea, 2004.

276. A. J. Sinclair, N. M. Attar-Bashi and D. Li, *Lipids*, 2002, **37**, 1113.

277. D. Harman, *J. Gerontol.*, 1956, **11**, 298.

278. R. A. Floyd, M. West and K. Hensley, *Exp. Gerontol.*, 2001, **36**, 619.

279. J. Liu and A. Mori, *Neurochem. Res.*, 1999, **24**, 14769.

280. J. M. Huerta, S. Gonzalez, S. Fernandez, A. M. Patterson and C. Lasheras, *Free Radic. Res.*, 2006, **40**, 571.
281. M. B. Grisham and D. N. Granger, *Dig. Dis. Sci.*, 1998, **33**, 6S.
282. S. P. Wolf and J. Nourooz-Zadeh, *Atherosclerosis*, 1996, **119**, 261.
283. Y. Gotoh, T. Noda, R. Iwsaki, K. Fujimoto, C. A. Rhoads and T. Y. Aw, *Cell Prolif.*, 2002, **35**, 221.
284. T. G. Wang, Y. Gotoh, M. H. Jennings, C. A. Rhoads and T. Y. Aw, *FASEB J.*, 2000, **14**, 1567.
285. J. S. Andrews, W. H. Griffith, J. F. Mead and R. A. Stein, *J. Nutr.*, 1960, **70**, 199.
286. T. Kaneda, H. Sakai and S. Ishiii, *J. Biochem.*, 1955, **42**, 561.
287. T. Kimura, K. Iida and Y. Takei, *J. Nutr. Sci. Vitaminol.*, 1984, **30**, 125.
288. A. W. Bull, N. D. Nigro, W. A. Golembieski, J. D. Crissman and L. J. Marnett, *Cancer Res.*, 1984, **44**, 4924.
289. H. Hara, K. Miyashita, S. Ito and T. Kasai, *J. Nutr.*, 1996, **126**, 800.
290. W. Grosch, in *Autoxidation of Unsaturated Lipids*, ed. W. H. S. Chan, Academic Press, London, 1987, p. 95–140.
291. J. Liu and B. N. Ames, *Nutr. Neurosci.*, 2005, **8**, 67.
292. J. Long, C. Liu, L. Sun, H. Gao and J. Liu, *Neurochem. Res.*, 2009, **34**, 786.
293. I. Berniakovich, M. Trinei, M. Stendardo, E. Migliaccio, S. Minucci, P. Bernardi, P. G. Pelicci and M. Giorgio, *J. Biol. Chem.*, 2008, **283**, 34283.
294. S. Arguelles, A. Machado and A. Ayala, *Free Radic. Biol. Med.*, 2009, **47**, 324.
295. M. Rojas-Molina, J. Campos-Sanchez, M. Analla, A. Munoz-Serrano and A. Alonso-Moraga, *Environ. Mol. Mutagen.*, 2005, **45**, 90.
296. J. A. Simon, Y.-H. Chen and S. Bent, *Am. J. Clin. Nutr.*, 2009, **89**(suppl), 1558S.
297. K. Sakata, K. Kashiwagi, S. Sharmin, S. Ueda, Y. Irie, N. Murotani and K. Igarashi, *Biochem. Biophys. Res. Commun.*, 2003, **305**, 143.
298. M. A. Lovell, C. Xie and W. R. Markesbery, *Neurobiol. Aging*, 2001, **22**, 187.
299. M. Daimon et al., K. Sugiyama, W. Kameda, T. Saitoh, T. Oizumi, A. Hirata, H. Yamaguchi, H. Ohnuma, M. Igarashi and M. Kato, *Endocr. J.*, 2003, **50**, 61.
300. N. Y. Calingasan, K. Uchida and G. E. Gibson, *J. Neurochem.*, 1999, **72**, 751.
301. B. Shao, K. D. O'Brien, T. O. McDonald, X. Fu, J. F. Oram, K. Uchida and J. W. Heinecke, *Ann. N. Y. Acad. Sci.*, 2005, **1043**, 396.
302. D. J. Conklin, A. Bhatnagar, H. R. Cowley, G. H. Johnson, R. J. Wiechmann, L. M. Sayre, M. B. Trent and P. J. Boor, *Toxicol. Appl. Pharmacol.*, 2006, **217**, 277.
303. C. Fiocchi, *Gastroenterology*, 1998, **115**, 182.
304. M. B. Grisham and D. N. Granger, in *Inflammatory Bowel Disease*, ed. J. B. Kirsner, W. B. Saunders, Philadelphia, PA, 2000, p. 55–64.
305. F. Powrie, *Immunity*, 1995, **3**, 171.

306. A. Keshavarzian, S. Sedghi, J. Kanofsky, T. List, C. Robinson, C. Ibrahim and D. Winship, *Gastroenterology*, 1992, **103**, 177.
307. N. J. Simmons, R. E. Allen, T. R. J. Stevens, R. Niall, M. Van Someren, D. R. Blake and D. S. Rampton, *Gastroenterology*, 1992, **103**, 186.
308. M. B. Grisham, *Lancet*, 1994, **344**, 859.
309. A. Zalewski and C. Mcphee, *Arterioscler. Thromb. Vasc. Biol.*, 2005, **25**, 923.
310. J. H. Min, C. Wilder, J. Aoki, H. Arai, K. Inoue, L. Paul and M. H. Gelb, *Biochemistry*, 2001, **40**, 4539.
311. E. R. Mohler III, C. M. Ballantyne, M. H. Davidson, M. Hanefeld, L. M. Ruilope, J. L. Johnson and A. Zalewski, *J. Am. Coll. Cardiol.*, 2008, **51**, 1632.
312. L. Maddukuri, E. Speina, M. Christansen, D. Dudzinska, J. Zaim, T. Obtulowicz, S. Kabaczyk, M. Komisarski, Z. Bokowy, J. Szczegielniak, A. Wojcik, J. T. Kusmierek, T. Stevsner, V. A. Bohr and B. Tudek, *Mutat. Res.*, 2009, **666**, 23.
313. P. Guallar-Castillon, F. Rodriguez-Artalejo, N. Schmid Formes, J. R. Banegas, P. A. Etxezarreta, E. Ardanaz, A. Barricarte, M.-D. Chiraque, M. D. Iraeta, N. L. Larranaga, A. Losada, M. Mendez, C. Martinez, J. R. Quiros, C. Navarro, P. Jakszyn, M. J. Sanchez, M. J. Toromo and C. A. Gonzalez, *Am. J. Clin. Nutr.*, 2007, **86**, 198.
314. F. Sorriguer, G. Rojo-Martinez, M. C. Dobarganes, J. M. Garcia Almeida, I. Esteva, M. Beltran, M. S. Ruiz de Adana, F. Tinahones, J. M. Gomez-Zumaquero, E. Garcia-Fuentes and S. Gonzalez-Romero, *Am. J. Clin. Nutr.*, 2003, **78**, 1092.
315. L. Sayre, G. Perry and M. A. Smith, *Chem. Res. Toxicol.*, 2008, **21**, 172.

Mass Spectrometry in Phytonutrient Research

JEAN-LUC WOLFENDER,[a] AUDE VIOLETTE[a] AND
LAURENT B. FAY[b]

[a] School of Pharmaceutical Sciences, EPGL, University of Geneva,
University of Lausanne, 30, quai Ernest-Ansermet, CH-1211 Geneva 4,
Switzerland; [b] Nestlé Research Centre, Vers-chez-les-Blanc, PO Box 44,
CH-1000, Lausanne 26, Switzerland

6.1 Introduction

Phytonutrients, also called phytochemicals, cover a broad range of chemically
and physiologically diverse, low molecular weight, secondary metabolites.
These natural products (NPs) are often non-essential food components which
are found in vegetables and fruits, but they are also found in spices and traditional ingredients. Whether as food, spices, traditional ingredients or medicinal plants, a large number of NPs are consumed by individuals in their
normal, everyday lives. All of these compounds have the potential to provide
either beneficial or toxic effects. These compounds exhibit a wide range of
physiochemical properties and need specific analytical methods for their profiling, identification and quantification in their original matrices or in body
fluids.

Foods rich in bioactive food components have been termed functional foods;
they include fruits, vegetables and whole grains. Although there is no uni-
versally accepted definition of a functional food, the International Food
Information Council defines "functional foods" as those foods that provide

RSC Food Analysis Monographs No. 9
Mass Spectrometry and Nutrition Research
Edited by Laurent B. Fay and Martin Kussmann
© The Royal Society of Chemistry 2010
Published by the Royal Society of Chemistry, www.rsc.org

health benefits beyond basic nutrition. The American Dietetic Association states that "functional foods, including whole foods and fortified, enriched, or enhanced foods, are foods that have a potentially beneficial effect on health when consumed as part of a varied diet on a regular basis, at effective levels".[1] Many bioactive food components are plant secondary metabolites (phytonutrients), but these components may also be found in foods of animal origin (zoonutrients), as well as in fungal and bacterial products. The exact concentration of bioactive components in a given food can be determined only by analysis because, as with vitamins, content varies with the genotype of the plant or animal, growing conditions and processing methods. Unlike whole foods, extracts or concentrates sold as dietary supplements may have specific analytical information, but such labelling is not currently required.[2]

A list of main phytonutrients and their food source is provided in Table 6.1. In this chapter, not all the phytonutrients listed in this table are discussed and the focus is on those for which bioavailability studies have been well documented and where mass spectrometry is commonly used for their analysis.

In addition to their health-beneficial properties in food, bioactive NPs have provided the inspiration for most of the active ingredients in medicines. The reason for this success in drug discovery can probably be explained by their high chemical diversity, the effects of evolutionary pressure to create biologically active molecules and/or the structural similarity of protein targets across many species.[4] Around 80% of medicinal products up to 1996 were either directly derived from naturally occurring compounds or were inspired by a natural product: more recent analysis confirms the continuing importance of

Table 6.1 Examples of phytonutrients, their main sources and their claimed health benefits. Adapted from ref. 3.

Phytonutrient	Food of remarkably high content	Health benefit
Allyl sulfide	Onion, garlic	1
Isoflavones	Soybeans and other legumes	3
Quercetin	Onion, red grapes, citrus fruit, broccoli, Italian yellow squash	2, 3, 4
Lycopene	Tomatoes and tomato products	3
Isothiocyanates	Cruciferous vegetables	1
Resveratrol	Grapes (skin), red wine	2
β-Carotene	Citrus fruit, carrots, squash, pumpkin	3
Carnosol	Rosemary	1
Catechins	Teas, berries	1
Curcumin	Tumeric	1, 4
Ellagic acid	Grapes, strawberries, raspberries, walnuts	1, 3
Anthocyanins	Red wine, berries	1, 3
Lutein, zeaxanthine	Kale, collards, spinach, corn, eggs, citrus	3
Tannins	Wine, tea	2, 3

Key: 1 = anticancer activity; 2 = positive influence on blood lipid profile; 3 = antioxidant; 4 = antiinflammatory activity.

natural products for drug discovery.[5] Phytopreparations and herbal drugs have also been more and more successful for the public at large, and these types of natural remedies require better control in terms of efficacy and safety. Moreover, some of these plant extracts can be used to prepare functional foods and this opens questions related to their correct usage.

Bioactive NPs are derived from different sources. Depending on the countries they might reach, they could be marketed as phytomedicines (used for curing disease), dietary supplements (that can bear health claims) or as part of nutritional intake in the form of functional food. Of course, the claims of safety and efficacy for these various classes of products are different.

6.1.1 Traditional Ingredients (Functional Foods)

Functional foods or neutraceuticals are foods or food components for which a health benefit beyond basic nutrition is claimed. They can be found in a wide range of foods, from conventional to fortified or enriched foods, and also as dietary supplements. However, despite the growing interest in herbal products, there is still no unified regulation. Indeed, phytonutrients from herbs and botanicals can be provided in different ways that could result in different regulation. On one hand, they can be consumed as capsules, powders or another form of pills, and they would then be regulated as dietary supplements. On the other hand, they can be ingested as additives in conventional foods (*e.g.* teas, juices, chips, energy bars) and would then be regulated as such.[6]

Since the legislation concerning the use of phytopharmaceuticals and traditional ingredients varies considerably from country to country, different issues regarding their quality control and standardization requirements remain open and the differences between phytopharmaceuticals and nutraceuticals are not always well-defined. A striking example is that a preparation containing *Hypericum perforatum* (Saint John's wort) is available as a natural antidepressant only in pharmacy shops in Switzerland because of severe problems related to drug interactions,[7] while similar extracts can be present in chocolate energy bars in the United States.

6.2 Main Classes of Natural Products with Health Benefits

In plants or vegetables, two mains categories of NPs are present: primary and secondary metabolites. Secondary metabolites are organic compounds that are not directly involved in the normal growth, development or reproduction of organisms. Unlike primary metabolites, the absence of secondary metabolites does not result in immediate death, but rather in long-term impairment of the organism's survivability, fecundity or aesthetics, or perhaps in no significant change at all. In particular, they are synthesized by the plants to protect themselves against pathogenic agents, predators and ultraviolet (UV) light, and to attract pollinators and seed-dispersing insects. In addition, phytochemicals give plants their characteristic colour and flavour and play an important role as signalling molecules for nitrogen-fixing bacteria to form root nodules. Secondary metabolites are often restricted to a narrow set of species within a phylogenetic group.

Phytonutrients thus encompass a very large group of structurally and physiologically diverse plant compounds[8,9] Those that are claimed for a given health benefit are summarized in Table 6.1 and are detailed in this chapter according to their biosynthetic origin.

6.2.1 Polyphenols

Polyphenols are secondary plant metabolites having at least one aromatic ring with one or more hydroxyl groups attached. More than 8000 phenolic structures are currently known. Their structural complexity is enormous, from simple phenols, phenolic acids or flavonoids to complex high molecular weight compounds such as tannins and proanthocyanidins (Figure 6.1).[10] In plants, they arise biogenetically from two main synthetic pathways: the shikimate pathway and the acetate pathway,[11] and are commonly found conjugated to sugars and organic acids.

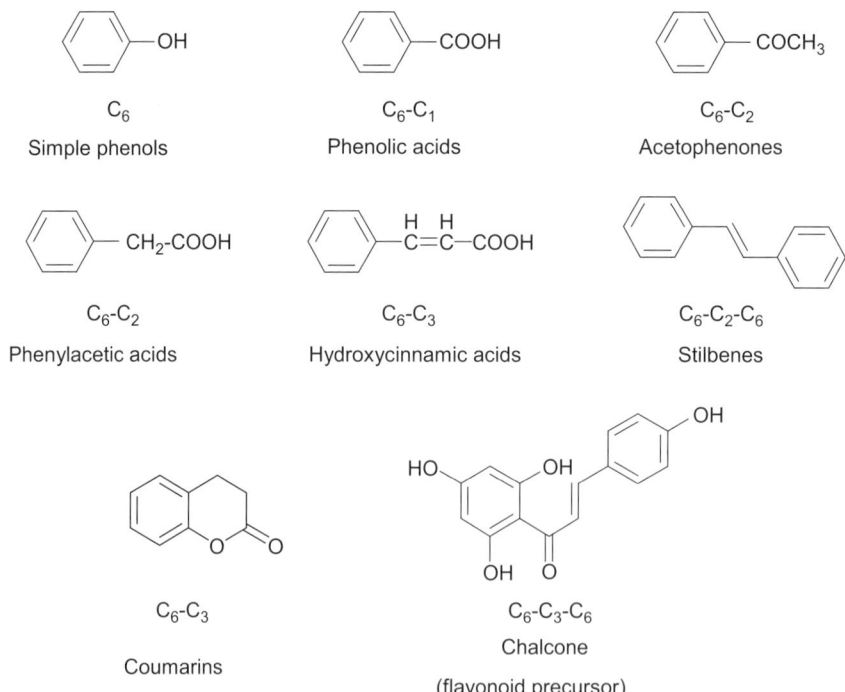

Figure 6.1   Main classes of phenolic compounds.

During the last 15 years, research on the effects of dietary polyphenols on human health has developed considerably, generating significant evidence of a supporting role for polyphenols in the prevention of degenerative diseases, particularly cardiovascular disease and cancer. Polyphenols have strong anti-oxidant properties and can scavenge free radicals in cell oxidation processes, but it has become clear that the mechanisms of action of polyphenols go beyond the modulation of oxidative stress.[12] Polyphenol contributions to the antioxidant capacity of the human diet are much larger than that of any vitamins. Indeed, their total dietary intake could be as high as $1\,g\,d^{-1}$, which is ten times higher than the intake of vitamin C and 100 times higher than the intakes of vitamin E and carotenoids.[13,14]

The main dietary sources of polyphenols are fruits and plant-derived beverages such as fruit juices, tea, coffee and red wine. Vegetables, cereals, chocolate and dry legumes also contribute to the total polyphenol intake. Below we briefly review the different classes of plant phenolic compounds.

6.2.1.1 Simple Phenols

Simple C_6-phenols (see Figure 6.1) such as phenol, cresol, thymol or resorcinol are widely present in the plant kingdom. Two major subgroups of phenolic acids are also widely represented and consist of classes of compounds such as hydroxybenzoic and hydroxycinnamic acids. Hydroxybenzoic acids include gallic, *p*-hydroxybenzoic, protocatechuic, vanillic and syringic acids, which have the C6–C1 structure (phenolic acids and aldehydes) in common. Hydroxycinnamic acids, on the other hand, are aromatic compounds with a three-carbon side chain (phenylpropanoid derivatives, C6–C3), with caffeic, ferulic, *p*-coumaric and sinapic acids being the most common. Phenylacetic acids and acetophenones (C6–C2) are less described.

6.2.1.2 Stilbene Derivatives, Resveratrol

More complex phenols appear in the form of C6–C2–C6 units, as is the case for stilbene derivatives. These compounds are derived from the same polyketone intermediate, which yields chalcones, the precursors of flavonoids (see below). The main representative of this category is resveratrol (3,5,4′-trihydroxy-stilbene)—a stilbene-derived polyphenol present in fruits and, in particular, in red grapes and cranberries. Being an important component of the Mediterranean diet and supposedly responsible for the "French paradox", it has been extensively studied and several beneficial pharmacological activities have been demonstrated such as antioxidant and anticancer activity, cardioprotection, inhibition of platelet aggregation and anti-inflammatory activity.[15]

6.2.1.3 Flavonoids

Flavonoids (C6–C3–C6) are present in high concentrations in the epidermis of leaves and the skin of fruits, and have important and varied roles as secondary metabolites. They are found in fruits, vegetables, tea, wine, grains, roots, stems and flowers, and are therefore regularly consumed by humans.

The main subclasses of flavonoids are flavones, flavonols, flavan-3-ols, iso-flavones, flavanones and anthocyanidins (Figure 6.2). Other flavonoid groups are the dihydroflavones, flavan-3,4-diols, coumarins, chalcones, dihy-drochalcones and aurones, which are thought to be less important from a dietary perspective. The basic flavonoid skeleton can have numerous con-stituents, although hydroxyl groups are usually present at the 4-, 5- and 7-positions. The majority of flavonoids exist naturally as glycosides. The presence of both sugars and hydroxyl groups increases water solubility, but other con-stituents such as methyl or isopentyl groups render flavonoids lipophilic.

Flavonoids are benzo-γ-pyrones that can be grouped according to the pre-sence of different types of substituents and the degree of benzo-γ-pyrone ring saturation. They are formed by a series of condensation reactions between hydroxycinnamic acid (B-ring) and malonyl residues (A-ring); carbons 2, 3 and 4 of the C-ring belong to the hydroxycinnamic acid residue. Flavonoids *per se* are compounds in which the B-ring is at position 2 of the C-ring, while iso-flavonoids contain the B-ring in position 3. In addition, the C ring can be either γ-pyrone (flavones, flavonols) or its dihydroderivative (flavanones, flavanolols) (Figure 6.2).

In plants, flavonoids occur in various forms. The core structures presented in Figure 6.1 can be hydroxylated, methylated and are sometimes substituted by aromatic and aliphatic acids, sulfate, prenyl, methylenedioxyl or isoprenyl groups. Flavonoids can also be found as dimers (biflavonoids) that can be linked by either C– or C–O–C bonds at different positions of the aglycones.

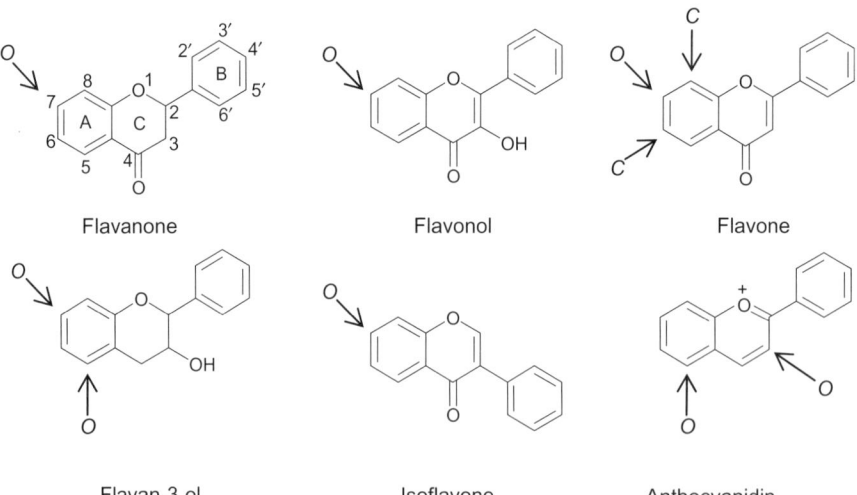

Figure 6.2   Structure of flavonoids. Arrows indicate the most common positions of *O*- and *C*-glycosylation.

However, the most encountered modification is glycosylation in the form of either *O*-glycosides or *C*-glycosides. Glycosylations render the flavonoids more polar and represent an efficient way for plants to store them in cell vacuoles. In *O*-glycosides, the aglycone is linked to the sugar by an acid-labile hemiacetal bond, while in *C*-glycosides, the linkage of the sugar to the aglycone occurs as an acid-resistant C–C bond. Biosynthetically in *O*-glycosides, the specific hydroxyl positions that are most often glycosylated are the 7-hydroxyl group in flavones, flavanones and isoflavones, the 3- and 7-hydroxyls in flavonols and flavanols, and the 3- and 5-hydroxyls in anthocyanidins. The 5-hydroxyl group participates in hydrogen bonding with the carbonyl group in most flavonoids and thus its glycosylation is rare. C-glycosylation has been reported to occur only at the C-6 and C-8 positions on the B-ring.

The sugar found in flavonoid glycosides is mainly glucose, but galactose, rhamnose, xylose and arabinose are also rather frequent, while mannose, fructose and glucuronic and galacturonic acids are rare. Disaccharides such as rutinose [rhamnosyl-(α1->6)-glucose] and neohesperidose [rhamnosyl-(α1->2)-glucose] are also often found. Higher oligosaccharides may also occur, but do so more rarely. The sugars can be modified by acetylation or linkage with malonyl residues.

6.2.1.4 Flavanones

The flavanones are the first flavonoid products of the flavonoid biosynthetic pathway and are therefore the immediate precursors of other flavonoids. They are characterized by the presence of a chiral centre at C-2 and the lack of the C2–C3 double bond, and they can thus exist as 2*S* or 2*R* stereoisomers. Nevertheless, the enzymatic reaction converting chalcones into flavanones is stereospecific, so most of the flavanones isolated from plants have the 2*S* configuration. The flavanone structure is highly reactive and hydroxylation, glycosylation and *O*-methylation reactions have been widely reported. Flavanones bearing more complex substituents such as prenyl and benzyl groups have also been described.

Flavanones are present in high levels in citrus fruits, with the most common glycoside known as hesperidin (hesperetin-7-*O*-rutinoside), which is present in citrus peel. Interestingly, flavanone rutinosides are tasteless, whereas the flavanone neohesperidoside conjugates (*e.g.* neohesperidin) from bitter orange and naringenin (naringenin-7-*O*-neohesperidoside) from grapefruit peel have an intensely bitter taste.

6.2.1.5 Flavonols

Flavonols are 3-hydroxyflavone derivatives. Their distribution and structural variations are extensive and have been well-documented. Extensive information on the different flavonols present in commonly consumed fruits, vegetables and drinks is available. However, there is wide variability in the levels present in specific foods, in part due to seasonal changes and varietal differences. The most common flavonol aglycones are kaempferol, quercetin, isorhamnetin and myricetin and conjugation mostly occurs at the positions 3, 7 and 4'.[16]

6.2.1.6 Flavones

Flavones have a close structural relationship to the flavonols (no hydroxyl in position 3), but unlike flavonols, they are not widely distributed in plants. The only significant occurrences in plants are in celery, parsley and a few other herbs, and they predominantly occur as 7-*O*-glycosides (*e.g.* luteolin and apigenin). In addition, polymethoxylated flavones have been found in citrus fruits (*e.g.* nobiletin and tangeretin).

6.2.1.7 Flavan-3-ols, Catechins

Flavan-3-ols, often referred to as flavanols, are the most complex class of the flavonoids because they range from simple monomers (catechin and its isomer epicatechin) to the oligomeric and polymeric proanthocyanidins, which are also known as condensed tannins. Tannins are indeed highly hydroxylated molecules and can form insoluble complexes with carbohydrates and proteins. This property of plant tannins is responsible for the astringency of tannin-rich foods because of the precipitation of salivary proteins.

Plant tannins can be subdivided into two major groups: hydrolysable and condensed tannins (Figure 6.3). Condensed tannins, or proanthocyanidins, are high molecular-weight polymers. The monomeric unit is catechin or epicatechin. Oxidative condensation occurs between carbon C-4 of the heterocycle and carbons C-6 or C-8 of adjacent units. Hydrolysable tannins consist of gallic acid and its dimeric condensation product hexahydroxydiphenic acid, esterified to a polyol, which is mainly glucose. These metabolites can oxidatively condense to other galloyl or hexahydroxydiphenic molecules, and form high molecular weight polymers. As their name indicates, these tannins are easily hydrolysed with acid, alkali and hot water and by enzymatic action, yielding polyhydric alcohols and phenylcarboxylic acids.[10]

6.2.1.8 Isoflavones

Isoflavones are flavonoids, but they are also called phytoestrogens because of their estrogenic activity. Structurally, they exhibit a similarity to mammalian estrogens and bind to estrogen receptors α and β. Apart from basic structural similarities, the key to their estrogenic effect is the presence of the hydroxyl groups on the A and B rings. They are classified as estrogen agonists, but also as estrogen antagonists since they compete with estrogen for their receptor. They have also been demonstrated to exert effects that are independent of the estrogen receptor.

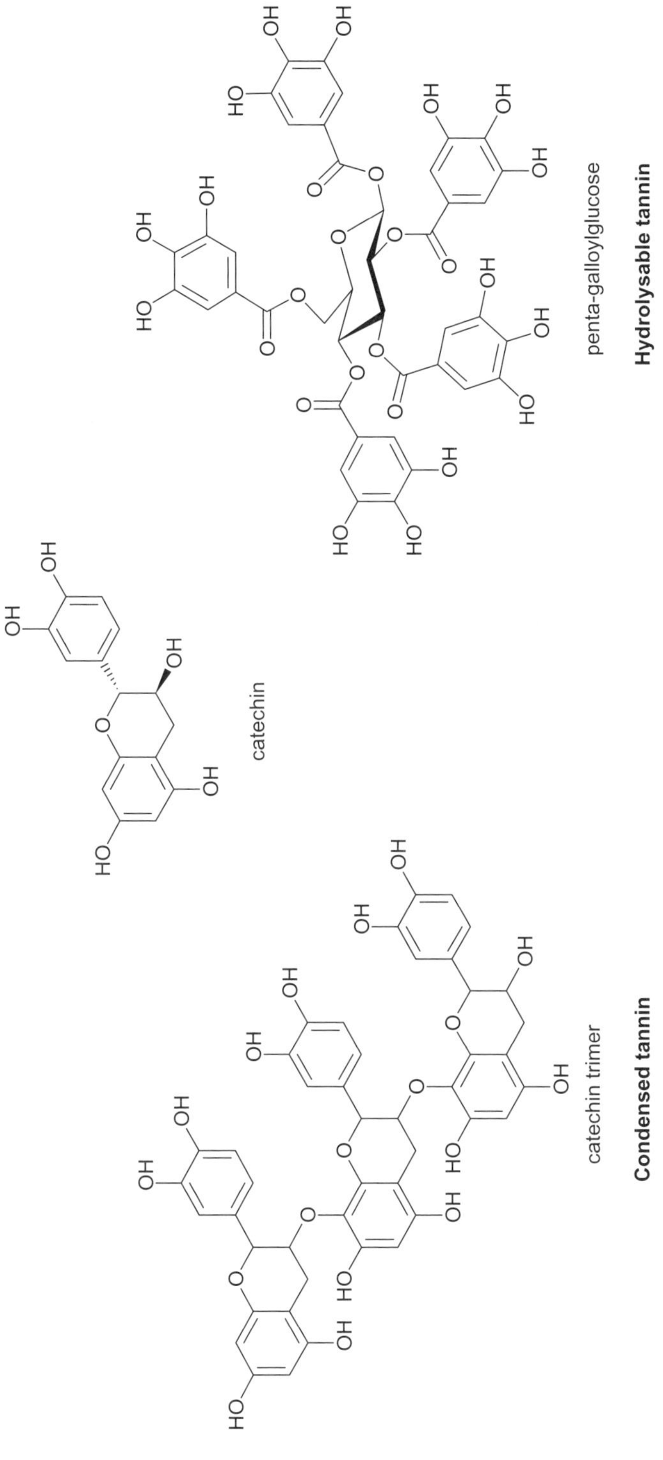

penta-galloylglucose

Hydrolysable tannin

catechin

catechin trimer

Condensed tannin

Figure 6.3 Structure of catechin and tannins.

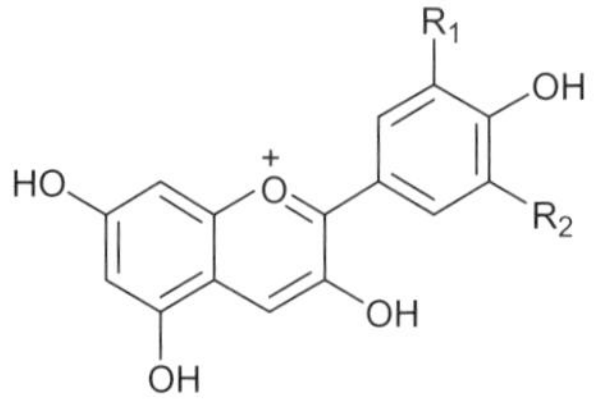

Anthocyanidins	R_1	R_2
Cyanidin	H	OH
Delphinidin	OH	OH
Malvidin	OMe	OMe
Pelargonidin	H	H
Petunidin	OMe	OH
Peonidin	OMe	H

Figure 6.4   Structure of the most common anthocyanins found in fruits.

6.2.1.9   *Anthocyanins*

Anthocyanins are widespread in nature, predominantly in fruits and flower tissues, in which they are responsible for the red, blue and purple colours. They are also found in leaves, stems, seeds and root tissue. In plants, they protect against excessive light by shading leaf mesophyll cells. Additionally, they play an important role in attracting pollinating insects.

Anthocyanins are *O*-glycoside and acylglycoside conjugates of anthocyanidins, which derive from the flavylium cation (2-phenylbenzopyrylium). The most common anthocyanidins (Figure 6.4) are pelargonidin, cyanidin, delphinidin, peonidin, petunidin and malvidin which are predominantly present in plants as sugar conjugates, with the most prevalent sugars being D-glucose, L-rhamnose, D-galactose, D-xylose and arabinose.[17]

6.2.2   Carotenoids

Carotenoids are a broad category of plant molecules, which includes the carotenes and xanthophylls, and are part of a larger class of molecules called terpenoids or isoprenoids. Carotenoids, as pigments, possess the ability to absorb visible light and appear coloured. While their nutraceutical role in humans is mostly related to molecular protection against free radical attack, carotenoids are produced by plants as photosynthetic pigments and as photoprotective entities.

The carotenoids are involved in photosynthesis in two ways. First, carotenoids can function as light-harvesting structures that then pass the energy to

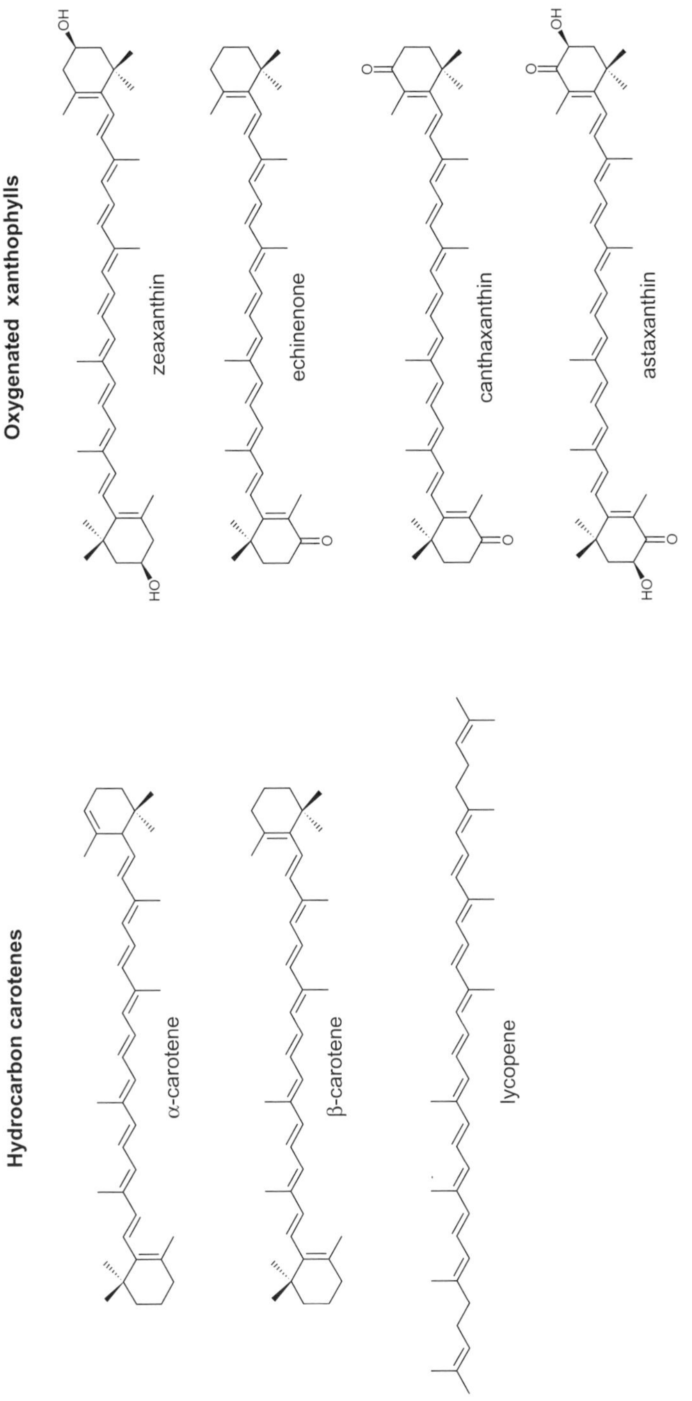

Figure 6.5 Structure of carotenoids.

chlorophyll. Carotene is commonly found as part of the light-harvesting apparatus of plants, though this activity may be more of a general role of the xanthophylls than the carotenes. Secondly, the greater function of the carotenes may be viewed as more protective in nature, as the carotenes appear to serve as a "sink" for triplet states created during the excitation of chlorophyll by light, thus protecting biological membranes and other molecules in plant tissue.

In fruits and vegetables, chloroplasts differentiate into organelles called chromoplasts. During this process they lose the ability to produce chlorophyll and synthesize a variety of yellow, red or orange carotenoid pigments. For example, as a tomato ripens and chloroplasts change into chromoplasts, less and less chlorophyll and photosynthetic enzymes are produced while more and more of the red carotenoid pigment, lycopene, is synthesized—the goal being to change the aesthetic characteristics of the seed-bearing fruit.[18]

In humans, the earliest role established for β-carotene was as a vitamin A precursor, a function it shares with several other provitamin A carotenoids. Carotenoids also play an important potential role in human health by acting as biological antioxidants, protecting cells and tissues from the damaging effects of free radicals and singlet oxygen. Lycopene is effective at quenching the destructive potential of singlet oxygen. Lutein and zeaxanthin are believed to function as protective antioxidants in the macular region of the human retina.[19] Other health benefits of carotenoids are related to the enhancement of immune system function, protection from sunburn, and inhibition of the development of certain types of cancers such as the protective role of lycopene against prostate cancer.[20]

Carotenoids have been classified into two groups—hydrocarbon carotenes and oxygenated xanthophylls (Figure 6.5).

6.2.3   Glucosinolates and Thiosulfinates

Glucosinolates are a class of nitrogen and sulfur containing secondary metabolites that are characteristic of the plant order of Capparales. They were first isolated from mustard seeds in the 1830s. More than 120 glucosinolates have now been described, mostly from the family Brassicaceae, but also in *ca.* 500 species, many of which are edible.[21] Glucosinolates are β-thioglycosyl *N*-hydroxysulfates comprising a sulfur-linked-β-D-glucopyranose moiety and an amino acid derived side chain. The highly variable structure of the side chain includes aliphatic (straight chain, branched chain, hydroxylated, sulfur-containing, ketoderivatives), alkenylic, aromatic, hydroxyalkyl benzoated or indolic moieties, and can be multiply glycosylated.

Spices and cruciferous vegetables such as cabbage, broccoli, cauliflower, cress, mustard and horseradish contain glucosinolates which are by themselves inert but can be converted to toxic metabolites when plant tissue is traumatized and the contents of different plant tissues mix. These substances are responsible for the smell associated with some cruciferous vegetables such as cabbage, broccoli and radishes, which may be a deterrent to animals. The derived

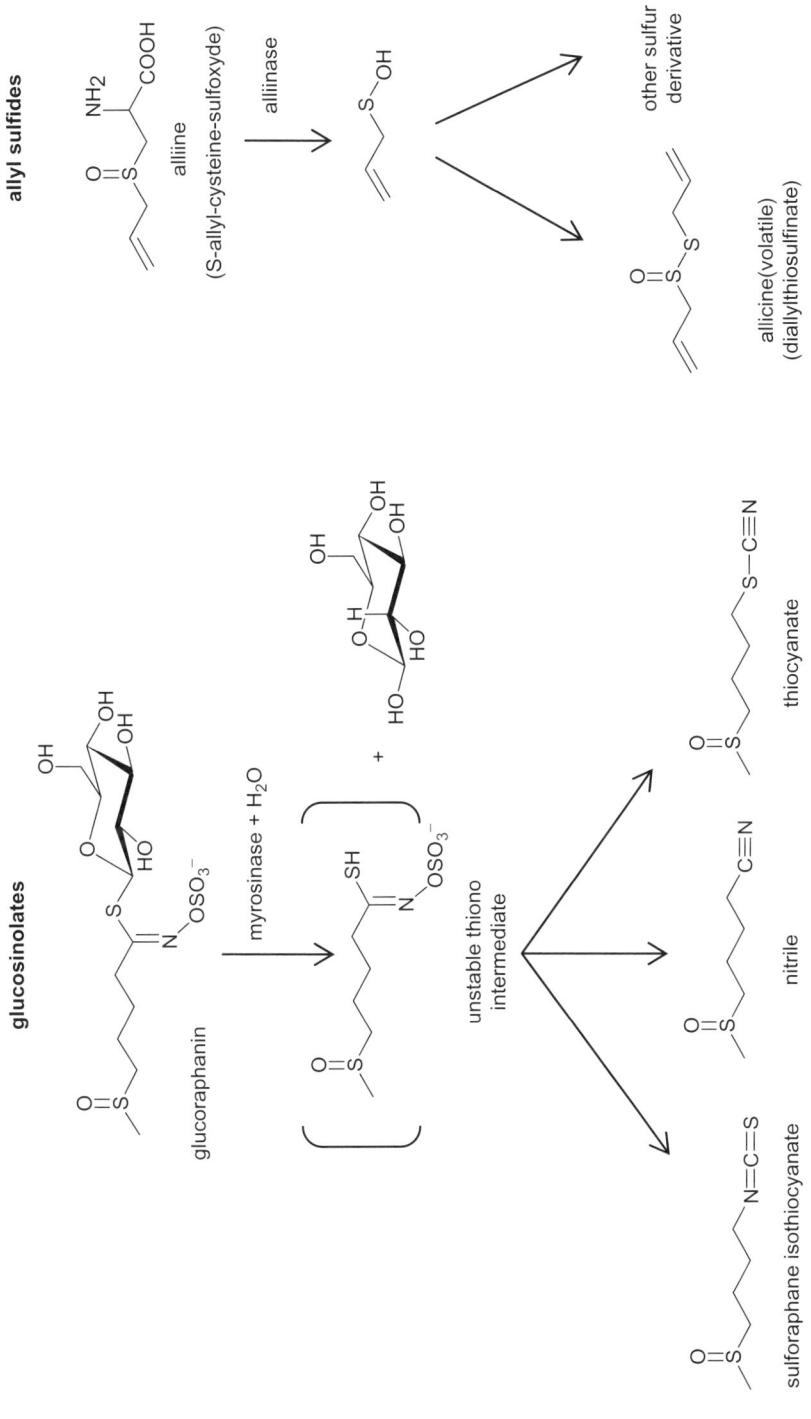

Figure 6.6 Structure of glucosinolates and thiosulfinates.

substances include isothiocyanates, thiocyanates and nitriles, and are created when plant tissue trauma allows glucosinolates from certain cells to mix biotics with enzymes in other cells (Figure 6.6).[18]

Intact glucosinolates and their breakdown products possess chemo-ecological functions and not only serve as defence mechanisms against herbivores and pathogens, but also as attractants to specialized toxin-tolerant insects.

Interest in the role of dietary glucosinolates has been stimulated by the observation of the cancer chemopreventive properties of certain breakdown products such as sulforaphane—derived from glucoraphanin (4-methylsulfinyl butyl glucosinolate) found in broccoli. Sulforaphane and other isothiocyanates prevent tumour growth in animal models by blocking cell cycle progression and the promotion of apoptosis. Glucosinolates are usually analyzed by reversed phase high performance liquid chromatography (HPLC) of their desulfo-derivatives. In these cases, the extracted glucosinolates are adsorbed onto a solid support, typically Sephadex A-25, followed by enzymatic desulfation and elution of the desulfoglucosinolates.[22] These types of constituents have recently been analyzed in their intact form by liquid chromatography-electrospray ionization-mass spectrometry (LC-ESI-MS).[23]

Other sulfur-containing compounds, such as thiosulfinates found in various *Allium* species, also exhibit various beneficial health properties. In this respect, onion (*Allium cepa*) and garlic (*Allium sativum*), which are among the oldest cultivated plants, are used both as a food and for medicinal applications. Thiosulfinates (*e.g.* allicin from garlic) result from the enzymatic cleavage of *S*-alk(en)yl-L-cystein *S*-oxides (*e.g.* alliin in garlic) that are present in intact plants (Figure 6.6). They are believed to be responsible for several health benefits attributed to *Allium* species, including the prevention of hypertension, type 2 diabetes, cancer and coronary heart disease. Due to their volatility, these compounds are also responsible for the pungency of these vegetables.[24] A comprehensive review of the analysis of these volatile constituents has been produced by Lanzotti.[25]

The structures and biotransformation of the main glucosinolates from Capparales and alkenyl cystein derivatives from *Allium* species are displayed in Figure 6.6.

6.3 Profiling and Analysis

Various analytical methods have been enlisted for the profiling and analysis of these natural products in either plant or vegetable material or in body fluids.[26] These methods are used either for targeted analysis (for quantification purposes for quality control or standardization) or for profiling of the secondary metabolite in complex matrices.

Profiling of vegetables or traditional ingredients has been performed mainly with chromatographic methods based on HPLC.[27] Various simple detectors such as UV or evaporative light scattering detection (ELSD) have been used, but MS is undoubtedly the detector of choice for the detection of nearly all natural products in complex matrices in either plants or body fluids.[28] Mass spectrometry has thus been deployed for the analysis of most of the classes of natural products mentioned in Section 2, either in hyphenation with HPLC (LC-MS or LC-MS/MS) for targeted and profiling analyzes or directly in matrices infused [flow injection analysis-mass spectrometry (FIA-MS)] for fingerprinting, mainly for metabolomics purposes.

6.3.1 Hyphenated Methods with MS (LC-MS and GC-MS)

LC-MS represents a key technique for the on-line identification of phytonutrients. As well as detection, a mass spectrometer offers the possibility of generating either nominal mass molecular ions or accurate mass measurements for the determination of empirical formulas.[29] Furthermore, the use of tandem or hybrid MS instruments[30] provides in-depth structural information through fragmentation of the molecular species by collision induced dissociation (CID) reactions. A number of reviews describe aspects related to on-line NP identification, either for screening and dereplication,[31–35] quality control, fingerprinting for authentication or standardization[30] or for the identification of biomarkers.[36]

For on-line identification purposes, the determination of the molecular weight is of great importance. This, however, necessitates the comparison of MS data obtained with different detection conditions in order to differentiate protonated or deprotonated molecules from adducts or fragments.[33,37] The use of high resolution instruments, such as a time-of-flight (ToF)[38,39] or a Fourier transform ion cyclotron resonance (FTICR)[40] mass spectrometer system, enables the direct determination of the molecular formula of crude mixtures. This strategic information allows for more precise targeting in the search of natural product libraries for dereplication purposes.[33]

Complementary structural information can be generated by CID in liquid chromatography tandem mass spectrometry (LC-MS/MS) or multi stage mass spectrometry (MS^n) experiments. The generated CID spectra are, however, not comparable to those recorded by electron ionization (EI), and this hampers a direct use of the standard EI-MS libraries for dereplication purposes. For automated dereplication procedures, specific MS/MS libraries have to be built based on standards available in a given laboratory, which consequently limits this approach. The development of hybrid instruments combined with high-resolution spectrometers such as ToF, FTICR or Orbitrap analyzers gives the possibility of recording MS/MS spectra with high mass accuracy, facilitating the interpretation of the obtained spectra.[41]

MS/MS spectra are mainly useful for the partial determination of sugar sequences of various glycosides,[42] the detection of characteristic losses such as those of prenylated compounds,[43] the classic fragmentation of flavonoids or related compounds for the determination of substituent positions on the A or B rings,[44] or the differentiation of isomers.[45] For other studies, the interpretation of the MS/MS spectra often requires the analysis of many related products[46] in order to extract structurally relevant information and to establish rules that can be used for structure prediction. For analysis of fully unknown constituents, this approach usually cannot provide enough information to ascertain the structure, so that the combination with other on-line information is mandatory. In this respect, NMR used on-line (LC-NMR) or at-line [liquid chromatography-solid phase extraction-nuclear magnetic resonance (LC-SPE-NMR), CapNMR[TM]] may contribute the missing structural information.[47,48] This method is, however, less sensitive and more time-consuming than MS, but is often mandatory to ascertain structure identification.

6.3.2 Direct Use of MS on Complex Mixtures

With the important development of multivariate data analysis, the spectroscopic methods used for HPLC detection (mainly NMR and MS) are more frequently applied to the analysis of crude extracts without prior HPLC separation for a rapid evaluation of the composition, especially in the quality control of herbal medical products and metabolomics.[49]

Numerous recent developments in MS ionization techniques accompanied by the utilization of accurate mass and tandem mass analyzers provide attractive techniques for metabolite fingerprinting. The high-throughput capacity of direct infusion mass fingerprinting of complex mixtures is similar to that of NMR fingerprinting. Direct MS analyzes of complex mixtures in conjunction with chemometric data analysis offer a viable solution when high-throughput screening is mandatory.[50] To date, however, only few investigations have involved the use of direct MS technology for fingerprinting studies.

6.3.3 Plant, Fruit or Vegetable Extract Profiling

Unlike herbal drugs or phytopharmaceuticals, the profiling of all constituents in food products is not required for the registration of a nutraceutical. However, for the understanding of the health benefits of such ingredients, many bioactive food components need to be analyzed for complete identification or quantification in plant, fruit or vegetable matrices. This is also important for safety and compliance, especially when herbal additives are used in food products.

For sample preparation of herbs and vegetables, crude extracts are usually obtained by maceration of the dried plant material with non-polar (*e.g.* CH_2Cl_2) or polar (*e.g.* CH_3OH) solvents. These extracts represent complex mixtures that may contain hundreds of constituents. If fresh vegetable matrices are analyzed, methods involving liquid–liquid partition might be employed. For some compound classes that are prone to degradation by enzymatic activity (*e.g.* glucosinolates), specific protocols involving extraction with hot solvents are needed.[23] In some cases, if sample enrichment is necessary, SPE pre-purification prior to HPLC can be used.

When the crude extracts are obtained, they are separated mainly on reversed phase columns using gradients of $CH_3CN : H_2O$ or $CH_3OH : H_2O$ with acidic or alkaline modifiers. Most of these analyzes are carried out on C18 columns, but for non-polar constituents such as carotenoids, C30 columns may provide better results.

Provided that the modifiers are volatile, these generic conditions are suitable for MS detection and represent a good starting point for profiling and quantitative analyzes. If specific markers need to be detected with high sensitivity, a specific optimization of the parameters on both the chromatographic and detector side is needed, as is the case for multiple reaction monitoring (MRM) experiments.

6.3.4 Complementary Analysis by LC-NMR and Related Methods

Due to the lack of efficient commercial MS/MS databases, the identification of both natural molecules in food products or metabolites in body fluids requires additional spectroscopic information for *de novo* structural identification. In this respect, LC-NMR provides a powerful tool that can yield important complementary information or, in some cases, a complete structural assignment of natural products on-line.[33,47,51,52] LC-NMR should ideally enable the complete structural characterization of any plant metabolite directly in an extract, provided that its corresponding LC peak is clearly resolved. In practice, however, this is not entirely true, as many factors hinder on-line structure determination.[47] These problems are mainly linked to the inherent low sensitivity of NMR for the detection of microgram or sub-microgram quantities of natural products separated by conventional HPLC and the need for solvent suppression.

Indeed, LC-NMR can be regarded as relatively insensitive due to the intrinsic properties of NMR detection. Consequently, the limits of detection (LODs) are several orders of magnitude higher than LC-UV or LC-MS. These limits also depend on the type of magnet used (400–900 MHz) and the type of flow probe employed (standard LC-NMR probes 40–120 μL, microcoils 20 nL–1 μL or cryo flow probes). The mode of operation also strongly affects sensitivity. Indeed, in the on-flow measurements, the time of acquisition is limited and only few transients are recorded when the LC peak crosses the flow cell, while in the stop-flow mode, an important number of transients can be acquired, improving the signal to noise (S/N) ratio. With a standard LC-NMR flow probe (60 μL on a 500 MHz), the detection limits are about 20 μg of NP injected on-column in the on-flow mode, but they can reach the ng range when LC-NMR is used at-line with sensitive microflow probes.[33]

LC-NMR, in the on-flow mode, can rapidly provide ^{1}H-NMR spectra of the main constituents of a crude extract without the need for complex automation. This type of information, in addition to other on-line spectroscopic data, *e.g.* photodiode array (PDA) and MS, may ascertain the structural determination and facilitate the dereplication process. Examples of on-flow LC-NMR analyzes for crude extract profiling are manifold (*e.g.* for flavonoids, alkaloids, terpenes, carotenoids),[33,48,51] although a relatively limited number of NPs (essentially only the very major ones) are detected in this way. The technique is also extremely useful for detecting compounds that are labile or might epimerize or interconvert as a result of their isolation.[53]

One of the limiting factors of on-line LC-NMR is the need for solvent suppression because HPLC separation is performed with CH_3CN-D_2O and CH_3OH-D_2O. In practice, the very large signals of CH_3CN or CH_3OH and the residual HOD signal are completely eliminated by powerful solvent suppression sequences. The main drawback of this procedure is that analyte signals localized under the solvent resonances will also be suppressed and thus some information can potentially be lost. Another problem occurring in LC-NMR is that the chemical shifts recorded in a typical reversed phase solvent will differ from those reported in standard deuterated NMR solvents. This can be a drawback if precise comparisons with literature data have to be performed.[47]

In order to circumvent these problems, new approaches in this field have favoured the use of LC-NMR at-line instead of on-line. In this case, trapping the HPLC peaks on SPE for preconcentration prior to NMR detection can be performed (SPE-NMR).[54,55] Another alternative is HPLC microfractionation of the extract, drying and re-injection of the concentrated LC peak in deuterated solvent and using microflow capillary LC-NMR probes (CapNMR™).[56–58] These at-line LC-NMR analyzers provide a good LC-peak preconcentration and high quality one-dimensional (1D) and two-dimensional (2D) NMR spectra in fully deuterated solvents without the need for solvent suppression. The work at-line has the disadvantage that more sample handling or automation is needed compared to the on-flow approach. However, these techniques are crucial for the complete *de novo* structure elucidation of given HPLC peaks since they can provide $^1H–^{13}C$ correlation experiments in the low μg range.[47]

6.3.5 Biofluid Analysis

As already mentioned, vegetable and fruit extracts represent complex mixtures. When ingested, the numerous natural products are metabolized or found in a non-modified form in the different biological fluids and tissues where they finally produce their effects. For a comprehensive view of the impact of these phytonutrients, a thorough monitoring of all of them in their intact and metabolized form would be required, but this is hardly feasible and often targeted analysis of given products is performed. This type of information is indeed important in assessing the health effects of a given phytonutrient since metabolism can significantly alter or enhance its biological activity. For example, ellagitannins from pomegranate juice have been found to be potent *in vitro* antioxidants, but they have been reported to be metabolized to poorly antioxidative hydroxy-6H-dibenzopyran-6-one derivatives by the colonic microflora of healthy humans.[59]

Thus, the *in vitro* antioxidant capacity, often used as a claim for many nutraceutical products, can be irrelevant in terms of *in vivo* antioxidant effects.

Some key issues such as bioavailability, metabolism, dose–response and toxicity of these bioactive food compounds or nutraceuticals themselves are not yet well-established and intense research efforts in this direction are needed.[60,61] Methods for the efficient analysis of these diverse phytonutrients prior to and after metabolism, and the assessment of the biological activity *in vivo*, are thus of utmost importance.

6.3.5.1 Sample Preparation

In most cases, biological samples such as body fluids cannot be assayed directly, but require pre-treatment to dispose of endogenous proteins, carbohydrates, salts and lipids. Although the sample pre-treatment for LC-MS/MS assays does not need to be as elaborated as other LC assays (especially those utilizing UV detection), it remains important to remove matrix components that might contaminate the system or cause ion suppression when high sensitivity is needed.[62] Protein precipitation, solid phase extraction and liquid–liquid extraction are the main sample preparation concepts combined with LC-MS/MS to analyze natural products in biofluids.[28]

6.3.5.2 Quantitative Targeted LC-MS and LC-MS/MS Methods

For the analysis of phytonutrients in body fluids, LC-MS represents a key technique which is mainly used for its high selectivity and sensitivity. Single stage MS is very scarcely used because of the complexity of the matrix and the low sensitivity when instruments are used in the full scan mode. Targeted tandem MS methods have mainly been deployed on various types of instruments, including hybrid systems.[28]

Among these instruments, the triple quadrupole (QqQ) systems are the most widely used. The versatility of tandem MS enables various selective screening strategies (*i.e.* full-scan, neutral-loss, precursor-ion and product-ion scan modes, and reaction-monitoring experiments). These experiments are especially useful for group-specific detection of metabolites. For example, phase II metabolites such as glucuronides and sulfates can be selectively detected by using positive ionisation (PI) ESI and a neutral loss scan of 176 and 80 Da, respectively.[63] In negative ionization (NI) ESI, sulfate conjugates produce abundant product ions at m/z 80 (SO_3^-) and m/z 97 (HSO_4^-), and glucuronides give ions at m/z 175 (deprotonated glucuronide moiety) and m/z 113 (fragment of glucuronide moiety), providing specific marker ions for the selective detection of sulfates and glucuronides in the precursor ion mode.[64] Multiple reaction monitoring provides the high sensitivity required in quantitative analysis in conjunction with the use of stable isotope standards.[65] The sensitivity of the full-scan mode is often not sufficient for metabolic studies. However, an instrument like a ToF-MS with a high selectivity of detection provides an interesting alternative through precise single monitoring MS experiments using very narrow mass ranges (a few mDA).[66]

6.3.5.3   *Profiling of Body Fluids with Metabolomics*

One goal of nutrition research is to determine the role of diet in metabolic regulation and to improve health. In this respect, with the advances in technology and bioinformatics, it is becoming possible to investigate the complex relationship between nutrition and metabolism from a comprehensive manner with "omics" technologies.[26] These new holistic analytical methods play an increasingly important role in nutrition, and novel sciences such as nutritranscriptomics, nutriproteomics and nutrimetabolomics have emerged recently.[67] The combination of various "omics" technologies in systems biology strategies can greatly facilitate the discovery of new biomarkers associated with specific nutrients or other dietary factors. In relation to phytonutrients and small natural molecules, mass spectrometry based metabolomics is more frequently applied[67] to nutrition issues, while NMR remains the most widely used tool.[68]

MS-based metabolomics provides great potential for high-throughput metabolite profiling or fingerprinting of low molecular mass metabolites in biological matrices. Such methods have been applied to both plant and vegetable matrices,[69] as well as biofluid analysis.[70] For these untargeted profiling studies, MS instruments in combination with some chromatographic methods such as gas chromatography (GC), liquid chromatography (LC) and capillary electrophoresis (CE) are widely used. MS on a very high resolution instrument can also be directly used for the analysis of such matrices without hyphenation (see Section 6.3.2). The recent introduction of ultra high performance liquid chromatography (UPLC) systems operating at very high pressures and using $<2\,\mu m$ packing columns yields a remarkable decrease in the analysis time, higher peak capacity, and increased sensitivity and reproducibility.[71] This technology has opened up the possibility of high-throughput LC-MS fingerprinting for metabolomics.[72] To date, however, only a few nutrimetabolomics investigations have utilized the MS technology.

Such a methodology has recently enabled the profiling of the chemical constituents of Pu-erh tea, black tea and green tea, as well as those of Pu-erh tea products of different ages. Differences in tea processing resulted in differences in the chemical constituents and the colour of tea infusions. Human biological responses to Pu-erh tea ingestion were also studied by UPLC-QqToF-MS in conjunction with multivariate statistical techniques. Metabolic alterations during and after Pu-erh tea ingestion were characterized by increased urinary excretion of 5-hydroxytryptophan, inositol and 4-methoxyphenylacetic acid, along with reduced excretion of 3-chlorotyrosine and creatinine. This study highlighted the potential of such a metabolomic approach to assess the effects nutritional interventions containing phytonutrients on human metabolism.[73]

Other studies are related to:

- cancer biomarker discovery;[74]
- the assessment of toxic effects, for example, for a comprehensive understanding of system responses to aristolochic acid intervention in rats;[75]
- for general toxicity induced by the use of *Tripterygium wilfordii*, a Chinese herb-derived remedy for rheumatoid arthritis.[76]

ESI-QqToF-MS has also been used for the differentiation of maturity and quality of bananas, grapes and strawberries.[77] Like the study of the changes in tea profiles and the effect in humans,[73] studies with NMR-based metabolomics have also been reported for the classification of chamomile flowers from different geographic locations[78] and for the detection of the metabolic effects of this herbal tea in urine samples.[79]

6.4   Analysis of Specific Phytonutrients

It is difficult to provide mass spectrometric rules common to all types of phytonutrients since they have all specific physicochemical properties. Based on the main type of natural products for which health benefit claims exist (see Section 2), a summary of the main MS analytical methods used for their detection, quantification and identification in both their original matrices and in biological fluids is given below.

6.4.1   Flavonoids, Isoflavones, Anthocyanins

As many polyphenols occur as glycosides in vegetables, the main features regarding the structure identification of this group of polyphenols are:

- the nature of the aglycone moiety;
- the type of carbohydrate or other functional group present;
- the stereochemical assignment of the terminal monosaccharide units;
- the sequence of the glycan part, interglycosidic linkages and the attachment of the functional groups to the aglycone.[44]

Most of these structural features can be solved by classical UV, MS and NMR techniques on isolated pure flavonoids and many papers have established rules for the identification of these polyphenols.[80] The direct analysis of flavonoids in plant extracts or biological matrices is, however, more complex and requires the development of dereplication protocols based on multi-hyphenated methods in which LC-PDA, LC-MS and LC-MS/MS and LC-MS[n] play a major role. In some cases, NMR data can be mandatory for a complete assignment and important information can be provided by LC-NMR.[47]

Flavonoids occur as rather complex mixtures with other constituents in crude plant extracts. They are preferably extracted by alcoholic solvents such as

CH₃OH, which enables the extraction of both aglycone and glycoside forms. The number of different flavonoids can be quite important. Their diversity is mainly related to the various glycosylation patterns on the relatively restricted number of aglycones.

6.4.1.1   LC-MS

In LC-MS, flavonoids can be detected in both PI and NI modes with either atmospheric pressure chemical ionization (APCI) or electrospray (ESI). $[M + H]^+$ and $[M - H]^-$ ions represent the most intense m/z ions. Because of the acidic nature of these polyphenols, the NI mode generally provides the highest sensitivity, but results in limited fragmentation. An LOD of 10 ng (S/N = 3) is obtainable in both total ion current (TIC) and UV chromatograms for these compounds, while the single ion monitoring single ion mass (SIM) mode allows the detection of compounds at less than 1 ng.[44] The abundance of these ions depends on the cone voltage applied and the nature of the eluent. Sometimes, dimers ($[2M + H]^+$ and $[2M - H]^-$) or adducts with solvents are observed.

6.4.1.2   Interpretation of Mass Spectra

Different papers have dealt with mass fragmentation pathways of the flavonoid aglycones under electron ionization.[81] Useful structural information is obtained by atmospheric pressure ionization (API) methods in conjunction with collision-induced MS/MS methods. The discussion here is limited to low energy CID spectra because this type of information can be recorded on most of the instruments used in analytical laboratories for dereplication purposes. The high energy CID mode provides more reproducible product ion spectra and offers interesting complementary information such as cross-ring cleavages of saccharides and homolytic glycoside cleavages; however, access to this type of information is restricted by the number of spectrometers available. Detailed and well-documented fragmentation studies of flavonoids in various CID regimes have been reported by Claeys and co-workers.[44]

Fragment ions provide important structural information for flavonoids and are used to establish the distribution of the substituents between the A- and B-rings. A careful study of fragmentation patterns can also be of particular value in the determination of the nature and site of attachment of the sugars in *O*- and *C*-glycosides. The fragments are annotated according to the nomenclature adopted by Domon and Costello[82] and Claeys and co-workers.[44,83] The main fragmentation schemes and nomenclature are illustrated in Figure 6.7.

6.4.1.2.1   Fragmentation of the Flavonoid Aglycones.   A vital piece of information is given by flavonoid aglycones fragmentation, in particular the one concerning the C-ring which results in $^{i,j}A^+$ and $^{i,j}B^+$ fragments and thus providing essential information on the nature and number of substituents of the A- and B-rings. In these fragments, i and j indicate the C-ring bonds that have been broken, and A and B refers to the intact A- and B-rings.

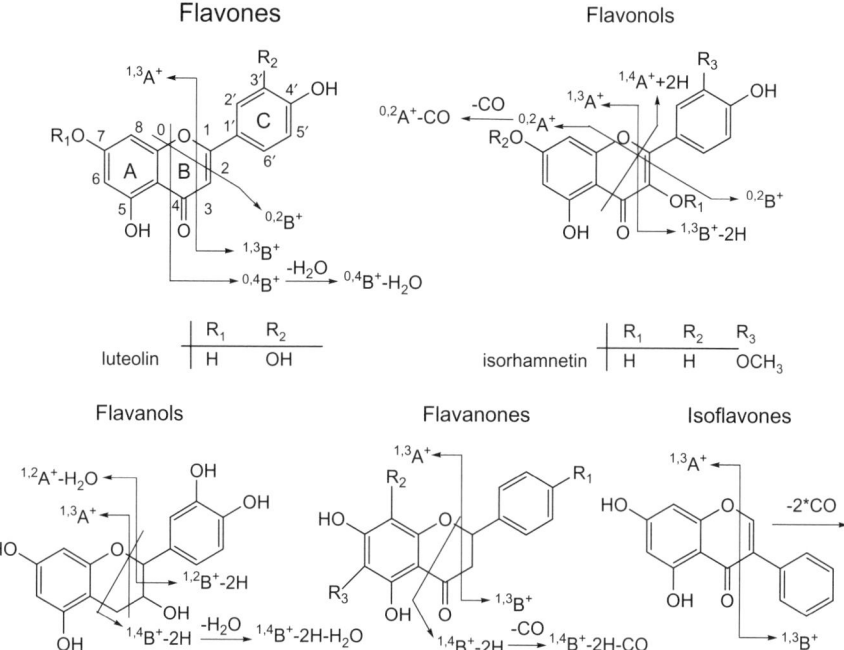

Figure 6.7 Fragmentation of the flavonoid aglycones.

For conjugated aglycones, an additional label 0 to the right of the letter differentiates these fragments from those of the glycosidic part (see Figure 6.9). For aglycones, the MS/MS spectra are mainly reported in the positive ion mode (PI). The negative ion mode, which is generally more sensitive, is more difficult to interpret and often generates mainly $^{1,3}A^-$ ions. Additionally, methoxylated flavonoids are characterized in NI by the loss of 15 Da corresponding to a [M-H-CH$_3$]$^{-\cdot}$ radical ion. Isoflavones will mainly provide the $^{0,3}B^-$ ion in this mode.

In the PI mode, although the $^{1,3}A^+$ ion is observed for all flavonoids, the fragmentation pathways depend largely on the substitution pattern and the type of aglycones considered. The main fragmentation pathways encountered in PI are summarized in Figure 6.7. For flavones, as illustrated in the MS/MS spectrum of luteolin in Figure 6.8A, the observed ions are $^{1,3}B^+$, $^{0,4}B^+$, $^{0,4}B^- - H_2O$ and $^{0,4}B^+$; $^{1,3}B^+$ and $^{0,4}B^+$ ions are characteristic of the presence of two hydroxyl groups on both the A- and B-rings. Flavanols will mainly exhibit $^{0,2}A^+$, $^{0,2}A^+ - CO$, $^{1,4}A^+ + 2H$ and $^{1,3}B^+ - 2H$; these fragments can be found, for example, in the MS/MS of isorhamnetin (Figure 6.8 B).

Differentiation between hydroxylated and/or methoxylated isomeric flavone/isoflavone aglycones can be performed in PI MS/MS. Differing to flavones, isoflavones will display a clearly distinguishable loss of two CO molecules leading to an overall neutral loss of 56 Da. This observation is well suited for the specific screening of plant constituents for isoflavones employing neutral

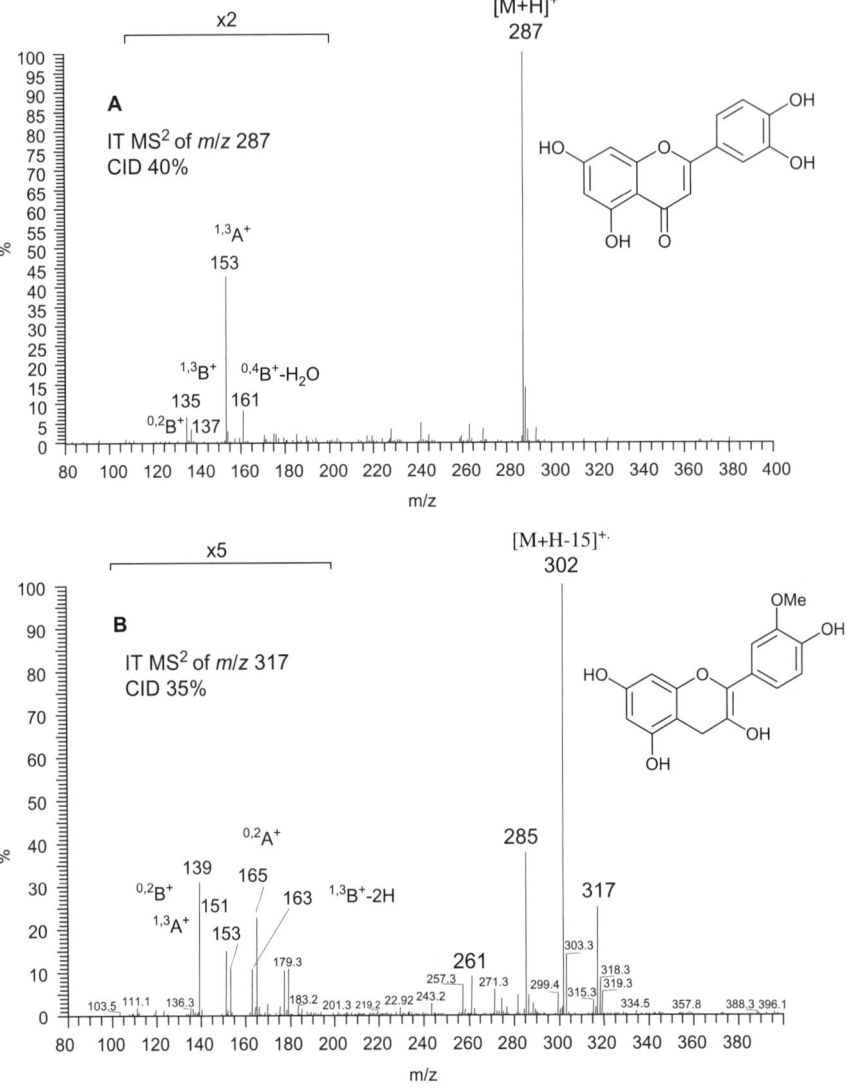

Figure 6.8   IT MS/MS spectra recorded for luteolin (A) and isorhamnetin (B).

loss MS/MS experiments and this has been applied to extracts of leaves of *Lupinus albus* and to soy flour.[46]

6.4.1.2.2   Fragmentation of the Flavonoid Glycosides.   LC-MS spectra of the glycosides enable differentiation between *O*-glycosides and *C*-glycosides. Product ions from glycoconjugates are denoted according to the nomenclature introduced by Domon and Costello.[82]   The spectra of *O*-glycosides are characterized by the presence of Y ions formed by a rearrangement reaction at the interglycosidic bond. For the different possible sugars, the following losses will be observed:

- hexose (-162 Da);
- deoxyhexose (-146 Da);
- pentose (-132 Da); and
- glucuronic acid (-176 Da).

In *C*-glycosides, on the contrary, the spectra are dominated by cross-ring cleavages of the saccharide moiety (X ions) such as $[M + H\text{-}90]^+$ ($^{0,3}X^+$), $[M + H\text{-}120]^+$ ($^{0,2}X^+$) and $[M + H\text{-}150]^+$ ($^{0,1}X^+$). In the case of *O*-diglycosides, the LC-MS or LC-MS/MS spectra provide information on the sequence of the carbohydrate and, in some cases, the glycosylation positions and interglycoside linkages can be deduced.

As an example, rutin (quercetin-3-*O*-rutinoside) $[M + H]^+$ 611 yields a Y_1^+ at m/z 445 corresponding to the loss of the terminal rhamnose unit (-146 Da) and a Y_0^+ at m/z 303 corresponding to the loss the glucose directly linked to the aglycone (Figure 6.9). These Y_n^+ ions are generated by a rearrangement reaction at the interglycosidic bond involving the hydroxyl hydrogen from the glycoside. A weak ion at m/z 449 could render the interpretation of the sugar sequence ambiguous and may lead to a false interpretation. This ion is referred to as Y* $[M + H\text{-}162]^+$ and corresponds to the loss of the inner sugar residue. The comparison with fragmentation obtained in the NI mode usually results in an unambiguous determination.[44]

The determination of the correct interglycosidic bond is possible in some cases by MS. It has been found that, for flavonoids substituted by the widespread rhamnosylglucose glycan part, a differentiation between the two most common interglycosidic linkage 1->2 (neohesperidoside) or 1->6 (rutinoside) can be made.[84] Alternative methods for differentiating flavonoid glycoside isomers are based on the fragmentation of their adducts with metal salts,[44] but these type measurements cannot be performed on-line and are not discussed here.

The position of sugars on the aglycones can be deduced if they are substituted by non-isomeric glycans. For example, in flavonol 3,7-*O*-diglycosides, the attachment positions can be determined from the intensity of the sugar losses because these compounds more readily lose the sugars at position 3 compared to 7. Sugars at position 5 on other flavones such as luteolin have also been revealed to be cleaved more easily than at position 7. However, these differences might be difficult to judge in the absence of known standards and, therefore, other techniques such as NMR or the use of UV shift reagents might be necessary.

6.4.1.2.3 Fragmentation of Catechins.

Catechins are the main polyphenols found in important food ingredients such as green tea. These compounds and their derivatives such as ($-$)-epigallocatechin-3-gallate (EGCG),

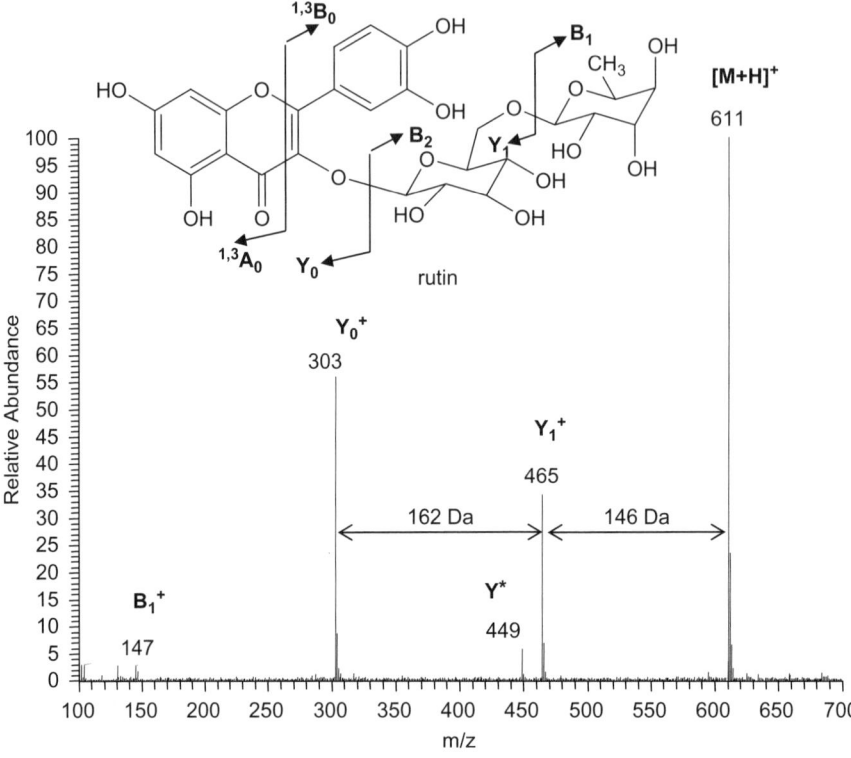

Figure 6.9   Fragmentation pattern of rutin.

(–)-epicatechin-3-gallate (ECG) have been analyzed in tea or cacao. Since these polyphenols have many hydroxyl groups, they are usually well analyzed in NI mode and several studies have been conducted for their sensitive detection in human body fluids.[85,86]

Studies concerning their fragmentation and identification are more scarce and, in these cases, the PI mode was found more informative.[87] Deprotonated molecules of such compounds are indeed rather stable and difficult to fragment.

As for flavones, a characteristic retro-Diels–Alder (RDA) fragmentation was observed for the catechin aglycone. Thus for catechin and epicatechin, the $[M+H]^+$ ions at m/z 291 lead to a diagnostic RDA fragment at m/z 139 (Figure 6.10A). Because epicatechin is an epimer of catechin, its mass spectrum is identical to that of catechin. The identities of epicatechin-3-*O*-gallate was ascertained on the basis of the $[M+H]^+$ ion at m/z 459 and [M + H-galloyl + H-H$_2$O]$^+$ at m/z 289 (Figure 6.10B).[87]

6.4.1.2.4   Fragmentation of Anthocyanidins.   Since these anthocyanins already have a positive charge, they can be readily analyzed in the ESI PI mode.

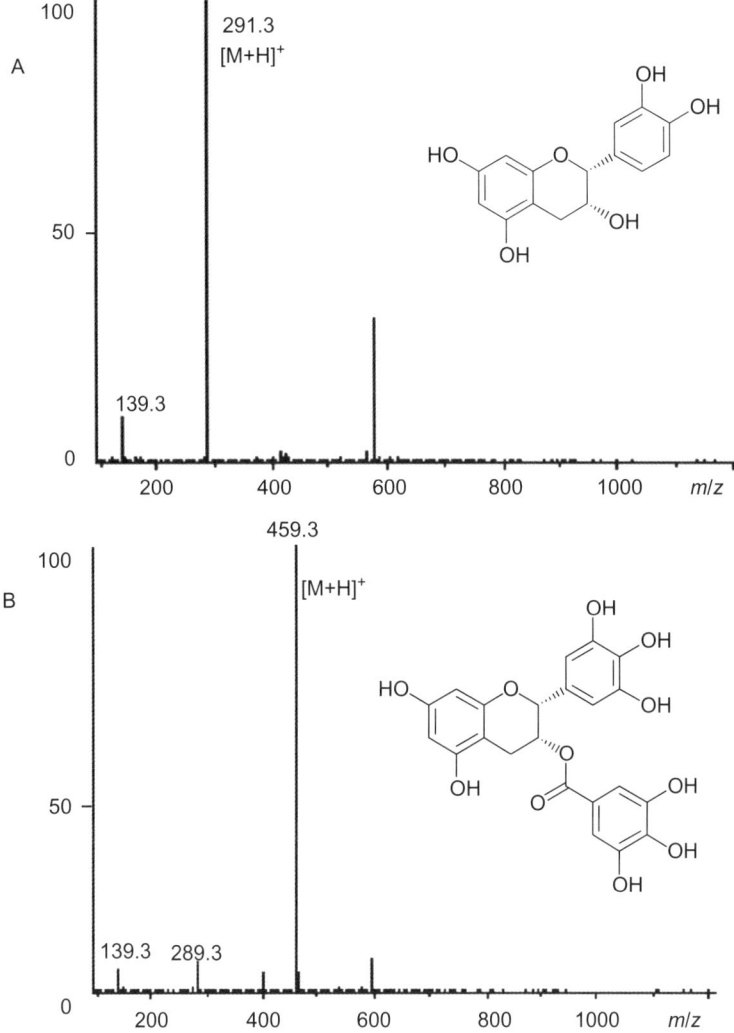

Figure 6.10 Representative PI ESI mass spectra obtained from the LC-MS TICs of catechu leaf extract. (A) Catechin and (B) epigallocatechin-3-*O*-gallate. Adapted from ref. 87.

A majority of the HPLC methods used today require high volumes of acids (1–15% v/v) as mobile phases to maintain a low pH and thus the stability of anthocyanins in solution in the form of the flavylium cation. In LC-ES-MS, the anthocyanidins and anthocyanins appear in the form of their molecular cation $[M^{+}]$.[88]

MS/MS fragmentation of anthocyanins has been less documented than for other flavonoids. Nevertheless, MS^{n} experiments performed on glucoside and glucuronide forms of peonidin have shown that, beside the characteristic loss of 162 or 176 Da for the conjugate part in the MS^{2} stage, the MS^{3} and MS^{4} stages

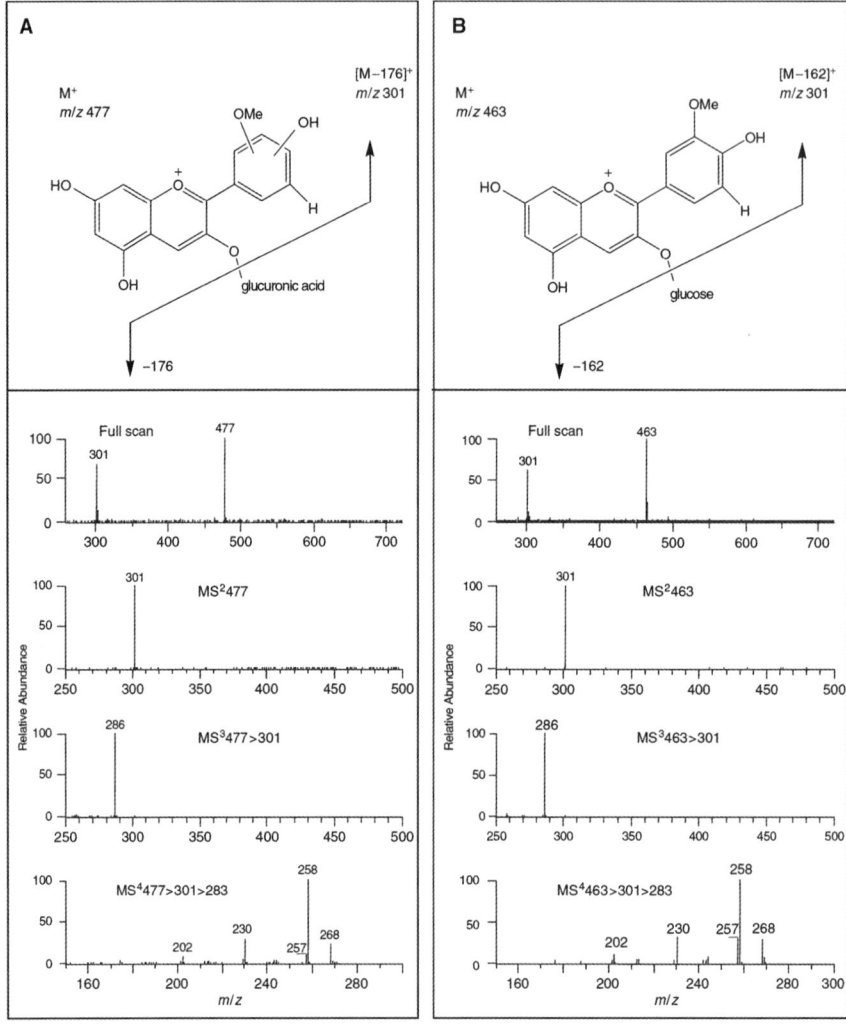

Figure 6.11 Comparison of the structure, full scan mass spectrum and MS^n spectrum of peonidin-*O*-monoglucuronide (A) and peonidin-*O*-glucoside (B) obtained by PI ESI from IT apparatus. Adapted from ref. 17.

performed on the aglycone part *m/z* 301 results in structurally informative fragments as shown in Figure 6.11.[17]

6.4.1.3   Resveratrol

Historically, GC-MS methods have been used to analyze resveratrol in wine using derivatization by bis-[trimethylsilyl]-trifluoroacetamide and the study of the fragmentation spectra of trimethylsilyl resveratrol derivatives, thereby reaching a detection limit of *ca.* $10 \mu g L^{-1}$.[89] The latest methods applied to resveratrol involve LC-ESI-MS in both PI and NI modes.[90] CID fragmentation of $[M - H]^-$ species of resveratrol has been recently studied.[91] The ESI-MS spectra in NI mode of *trans*-resveratrol showed the exclusive formation of $[M - H]^-$ at m/z 227, characteristic MS^2 fragments at m/z 185 and MS^3 fragments of the m/z 185 ion at m/z 183, 157 and 143. To explain these fragmentation patterns, both the phenol and the resorcinol moieties are considered as their pK_a values and mechanisms were proposed. LC-ESI-MS^n in the NI mode has been further applied to the determination of resveratrol in wine and plant extracts and also to resveratrol polymers in *Parthenocissus laetevirens* extracts, known to be an abundant source of stilbene derivatives.[92]

6.4.1.4 Analysis of Polyphenols in Biofluids

LC-MS/MS is the most widely used technique for flavonoid structure elucidation and a review of its application to the study of flavonoids in biofluids was published in 2004.[93] The key to polyphenol analysis in biofluids is to first understand how they are absorbed and metabolized, and then to characterize the relationship between these transformations and their bioactivities. A few papers recently published on this subject suggest that glucuronidation, sulfation and methylation are the main metabolic pathways to be considered.[96] Furthermore, oxidative metabolism also plays an important role and many oxidized forms of isoflavones from soy diet have been identified.[93]

These metabolic pathways have indeed been described for both flavonols and anthocyanins using the method of flavonoid glycoside determination described in Section 6.4.1.2.2. For example, Mullen *et al.* published a study on the determination of flavonol metabolites in human plasma and urine after the ingestion of red onion (*Allium cepa*), known to be a rich source of flavonols.[94] These metabolites were analyzed by LC-PDA-MS^2; using an ion trap mass spectrometer fitted with an electrospray interface. Methylation, sulfation and glucuronidation were observed with characteristic ions observed at $[M + 14]$, $[M + 176]$ and $[M + 80]$, respectively. Similarly, studies by Felgines *et al.* on the metabolism of anthocyanins described the same metabolic pathways for this class of flavonoids.[95,96] It is important to notice at this point that MS fragmentation does not allow determination of the regioselectivity of these metabolic transformations, and for this purpose, NMR is the technique of choice.

LC-MS is also used for the quantitative analysis of flavonoids and their metabolites. ESI and APCI interfaces have been used in both PI and NI modes; the data could be collected in single ion mass (SIM), SRM or MRM modes (see Table 6.2), allowing limits of quantification in the low $ng mL^{-1}$ range—even in complex matrices such as plasma or urine—which is sufficient for flavonoid study. It should be noted here that, because ESI is a very soft ionization technique, it is the most suitable for the study of metabolites, which often are labile conjugates.[97] Moreover, due to its specific and sensitive characteristics,

Table 6.2   Methods for the analysis of some polyphenols and their metabolites in biofluids. Adapted from ref. 28.

Polyphenols	Samples	Mass analyser[a]	Scan mode[b]
Anthocyanin[95,96]	Urine	QqQ	+; ESI; MRM
Anthocyanins[165]	Urine	QqQ	+; ESI; MS/MS
Anthocyanins[166]	Plasma	QqQ	+; ESI; MRM
Anthocyanins[167]	Plasma	Single-Q	+; ESI; SIM
Anthocyanins[168]	Tissue	QqQ	+; ESI; MRM
Cyanidin-3-glucuronide[169]	Plasma	Q-ToF	+; ESI; MS/MS
Delphinidin[170]	Plasma	ToF	+; ESI; Full scan
Hesperidin, naringin[171]	Serum	QqQ	−; APCI; SRM
Polyphenols[98]	Urine	QqQ	−; ESI; MRM
Phytoestrogens[172]	Urine	IT	−; ESI; SRM
Quercetin[173]	Plasma, urine	QqQ	−; ESI; MRM
Quercetin-4′-glucoside[174]	Plasma	IT	−; ESI; MS^n
Catechins[86]	Plasma, urine	IT	−; ESI; MS^n
Resveratrol[100]	Plasma, urine	QqQ	+; ESI; MRM

[a]QqQ = triple quadrupole, ToF = time of flight, Q = quadrupole, IT = ion trap.
[b]ESI = electrospray ionization, MRM = multiple reaction monitoring, MS/MS = tandem mass spectrometry, SIM = single ion mass, SRM = single reaction monitoring, APCI = atmospheric pressure chemical ionization, MS^n = multi stage mass spectrometry.

LC-MS/MS can be performed without complete separation of the different compounds, which makes it highly suitable for high-throughput analyzes. This has been illustrated by Ito *et al.* with the profiling of polyphenols in human urine.[98] In this study, polyphenol metabolites were deconjugated using β-glucuronidase and sulfatase, and the optimized LC-ESI-MS/MS method operating in MRM mode with NI allowed quantification in the low micromolar range of 15 different polyphenols in runs only six minutes long. Finally, most of metabolite analyzes are performed using triple quadrupole mass spectrometers, mainly because of their very good capability for quantitation in the MRM mode.[99]

An example of the type of detection that has been obtained by LC-ESI-MS for anthocyanins excreted in human urine following oral consumption of boysenberry extract and their relationship to the anthocyanins present in the original extract is shown in Figure 6.12.[17] This figure shows a comparison of the specific LC-UV traces (530 nm) of these compounds in the extract and in the urine sample. It demonstrates that, since their retention times differ, some of these anthocyanins have been metabolized. The partial structure determination of the corresponding glucuronide metabolites by different MS^n experiments is discussed in Figure 6.11. Compound (A) was the major metabolite and was identified as a peonidin monoglucuronide (m/z 477) resulting from methylation and glucuronidation of cyanidin. Other metabolites (labelled * on Figure 6.12) include cyanidin monoglucuronides (m/z 463) and pelargonidin monoglucuronide (m/z 447, resulting from cyanidin dehydroxylation).

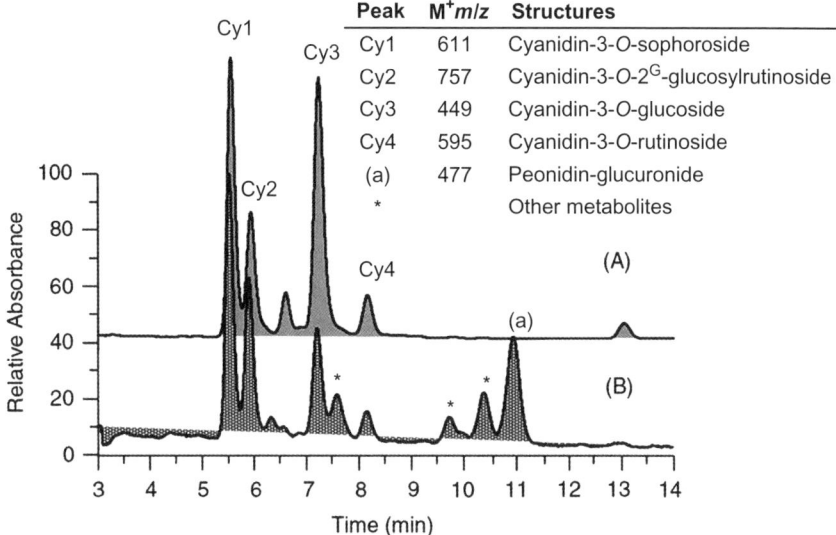

Peak	M⁺m/z	Structures
Cy1	611	Cyanidin-3-*O*-sophoroside
Cy2	757	Cyanidin-3-*O*-2^G-glucosylrutinoside
Cy3	449	Cyanidin-3-*O*-glucoside
Cy4	595	Cyanidin-3-*O*-rutinoside
(a)	477	Peonidin-glucuronide
*		Other metabolites

Figure 6.12 LC-MS profile of anthocyanins in a boysenberry extract (A) and in human urine (B). Inserted table presents the identified compounds using PI-ESI-IT. HPLC conditions: LiChroCart Superspher® 100 RP-18 end-capped column, 250 mm, 2 mm i.d., 35 °C, flow rate 0.25 mL min⁻¹, mobile phase: 5 : 95 formic acid : water v/v (solvent A) and 100% methanol (solvent B), using a gradient of 25–60% B over 40 min. Detection was performed at 530 nm. Adapted from ref. 17.

6.4.1.4.1 Catechins. LC-MS is also the preferred method for catechin analysis. It has been used for the quantification of catechins (catechin, epicatechin, epicatechin-3-*O*-gallate and epigallocatechin-3-*O*-gallate) in *Acacia catechu* extract using PI ESI-MS in the SIM mode.[87] Validation of this method showed an accuracy ranging from 1.06 to 11.76%; the precision (relative standard deviation) varied between 1.60 and 9.36% for these four analytes.

LC-MS has also been extended to the determination and quantification of catechin metabolites in plasma after tea polyphenol ingestion.[85] The identified metabolites using LC-ESI-MS in the NI mode included glucuronide and sulfate derivatives, which were further quantified in the MRM mode with limits of quantification between 3 and 8 nmol L⁻¹. Recently, a study on human urinary metabolites of catechins was published by Sang *et al.*[86] In this study, ring-fission metabolites as well as glucuronide, sulfate and methylated metabolites were detected. The typical A-ring fragmentation ion of catechins generated after RDA was also discussed. This showed that, although the exact regio-selectivity of metabolic reactions is difficult to assess, it is sometimes possible to determine at least on which ring it occurs.

Finally, LC-ESI-MS has been used in both PI and NI modes to identify and quantify resveratrol and its metabolites in complex matrices such as plasma and urine. As for other polyphenols, glucuronidation and sulfation have been observed.[100,101]

6.4.1.5   LC-NMR

For unknown flavonoids, dereplication protocols based on LC-MS or LC-MS/MS alone might not provide all the necessary information for unambiguous structural assignment and a multidimensional hyphenated approach involving LC-NMR on-line or at-line can be necessary (see description of these techniques in Section 6.3.4). For polyphenols, and in particular flavonoids, interesting information in LC-^{1}H-NMR spectra can be deduced from the chemical shifts such as the presence of methyl, methoxyl and methylene groups, as well as various types of aromatic protons. Aside from chemical shifts, ^{1}H-NMR coupling patterns, especially those of the aromatic protons of the polyphenolic nucleus, *meta* ($J = 1$–3 Hz) and *ortho* ($J = 7$–9 Hz), can easily be observed in LC-NMR. *Para* coupling ($J < 1$ Hz) is usually not resolved.

This type of information is very complementary to mass spectrometric information where A- or B-ring hydroxylation patterns cannot be precisely assessed. For a compound like apigenin, where only one hydroxyl is located on the B-ring, ^{1}H-NMR couplings will immediately solve the B-ring substitution determination since each of the three possibilities of localization of the OH group will give a unique splitting pattern. In the case of glycosides, the anomeric proton (H-1) alone can provide useful information regarding the saccharide moiety. For example H-1/H-2 coupling constants indicate which signal relates to which sugar in a polyglycoside polyphenol, and more frequently is also indicative of the α or β linkage of the glycosidic bond. Indeed, β-linked glucopyranosides that exhibit H-1/H-2 coupling constants of 7–8 Hz are readily distinguishable from the α-linked glucopyranosides with 3–4 Hz couplings. The chemical shift related to the glycosidic H-1 signal can also be of diagnostic value. For example, the H-1 signal of a sugar attached to another sugar can usually be distinguished from that of a sugar directly attached to the polyphenol aglycone in that it resonates upfield relative to the latter. Exceptions to this are known in the case where the primary glycoside is a C-glycoside (higher field H-1 signals compared to O-glycosides) or when the terminal sugar is apiofuranoside (relatively low field H-1 signals). Rules based on the chemical shift of the H-1 signal allow for a good prediction of their attachment position on the aglycone moiety.

Bidimensional experiments obtained in stop-flow LC-NMR or with at-line NMR methods, such as ^{1}H–^{1}H correlation experiments [correlation spectroscopy (COSY) or total correlation spectroscopy (TOCSY)], clarify the assignment of the coupled aromatic protons or the sugar protons in the case of glycosides. Indirect carbon information can be obtained through heteronuclear single quantum correlation (HSQC) or heteronuclear multiple bond correlation (HMBC), providing definitive information on interglycosidic bond and sugar

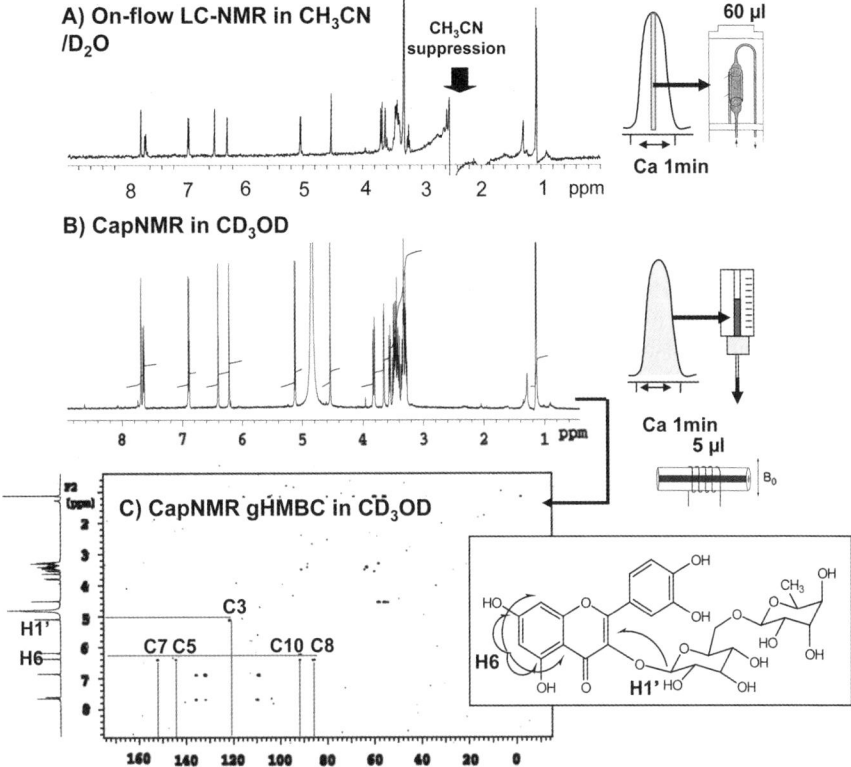

Figure 6.13 Analysis of flavonoids: complementary information from on-flow LC-NMR (A) and at-line CapNMR™ (B). The quality of the spectra obtained for the same amount of flavonoid glycoside is superior in CapNMR™ because the probe is more sensitive and a preconcentration of the whole LC peak is obtained prior analysis by microfractionation, drying and reinjection with a fully deuterated in a 5 μL volume that matches the volume of the CapNMR™ probe (B). In on-flow LC-NMR, only a slice of the corresponding LC peak is recorded while it is crossing the flow cell and solvent suppression is needed (A). (C) CapNMR™ gHMBC spectrum acquired overnight on 40 μg of the flavonoid glycoside on a 500 MHz instrument showing key correlation for *de novo* structure assignment.

sequences. An example of NMR data that have been acquired on a flavonoid glycoside such as rutin (20 μg) in both on-flow LC-NMR and at-line CapNMR™ is presented in Figure 6.13.[47] The most demanding long-range ^{1}H–^{13}C HMBC correlation experiment was obtained in 12 h, revealing the key correlations between H-6 and C-5, C-7 and C-10, H-8 and C-6, C-7, C-9 and C-1′, H-6′ and C-1′, C-2, C-4′ and C-5′, as well as those of the anomeric H-1′ of the glucose unit with C-3. For the same compound, HSQC was obtained in 3 h, and TOCSY and COSY in 10 min.[47]

Hyphenated NMR methods have been mainly used for the *de novo* characterization of flavonoid or related polyphenols in plant or vegetable matrices,[48,102] but these techniques can also be applied to the characterization of flavonoid metabolites in biofluids provided that enough of the matrices can be obtained.

6.4.2   Carotenoids

In recent years, there has been particular emphasis on obtaining more accurate data on the types and concentrations of carotenoids in foods for various health and nutrition reasons. The analysis of carotenoids is complicated because of the diversity and the presence of *cis,trans* isomeric forms of this group of compounds. Furthermore, due to their extended system of conjugated double bonds, carotenoids are unstable in the presence of light, heat or oxygen, and therefore their isolation, identification and quantitation is challenging. For this reason, several precautions are necessary when handling carotenoids. In addition, a wide variety of food products of vegetal and animal origin contain carotenoids, and a broad range of carotenoids can be found in these samples. In total, more than 700 carotenoids have been reported. In complex biological matrices such as human serum tissues and plant material, compounds can often interfere with their detection. Another problem associated with the analysis of carotenoids is the difficulty of obtaining standard compounds.[103]

6.4.2.1   LC-MS

The lability of carotenoids prohibits their analysis by GC-MS. Reversed-phase LC-UV and more conveniently LC-MS have been used for their detection. Furthermore, LC-MS analysis is particularly suited for the analysis of carotenoids from natural sources since these compounds are often found in trace quantities. Reversed phase chromatography is frequently used with a C18 stationary phase, usually with gradient elution. C30 columns are also often applied to enhance the separation of *cis* and *trans* isomers, and provide efficient separation of xanthophylls and isomers. A typical separation on C30 reversed phase is shown in Figure 6.14. In this case, more than 30 esterified and non-esterified carotenoids were analyzed from red pepper pods (*Capsicum annuum*) by LC-APCI-MS.[104]

One of the main differences between these solid phases is that lycopene will elute after β-carotene and α-carotene on C30, but before on C18 columns.[103] In contrast to other ionization techniques, xanthophylls and carotenes form both molecular ions and protonated molecules during PI APCI (see Figure 6.15) *versus* molecular ions and deprotonated molecules in NI mode. The relative abundance of these ions is strongly related to the solvent composition. APCI is an ideal method of ionization for low- to medium-polar compounds, which includes carotenoids and related compounds. Since their molecular mass does not exceed 2000 Da, even in the case of glycosides or esters with fatty acids, this

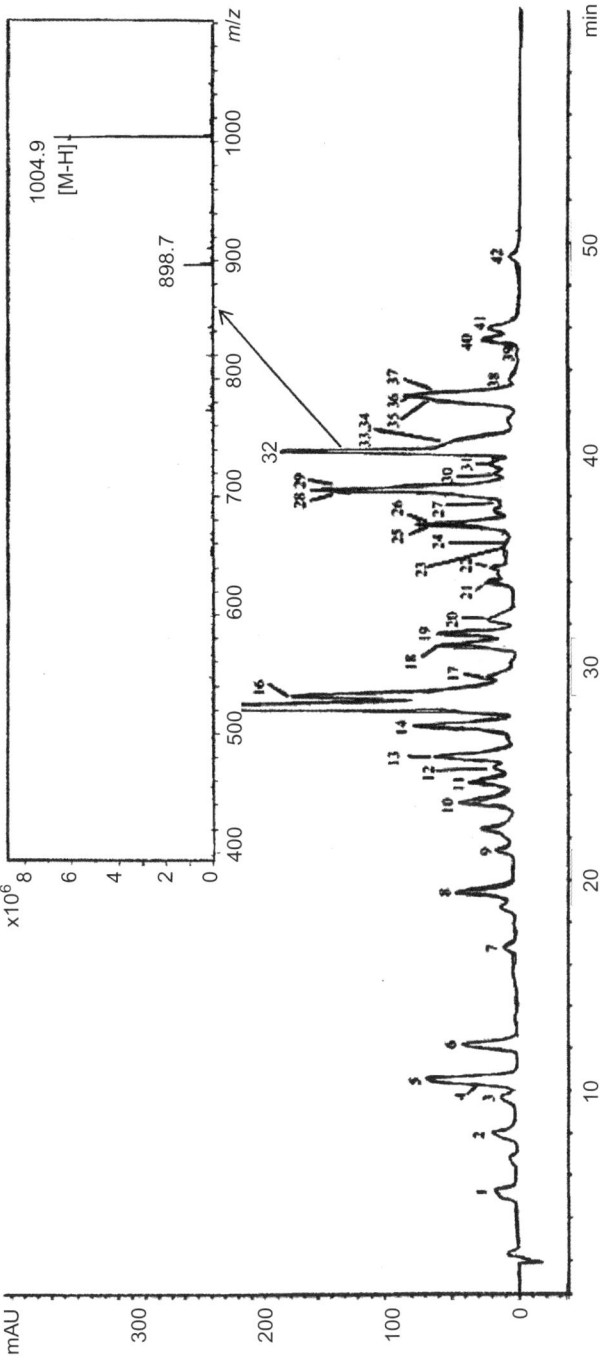

Figure 6.14 Separation of non-esterified carotenoids and carotenoid esters from red pepper pods by HPLC and MS[1] experiment in the negative ion mode of capsanthin-di-myristate identified in peak (32). HPLC conditions: YMC C30 column, 150 mm, 3.0 mm i.d., flow rate 0.42 mL min[−1], mobile phase: methanol/methyl *tert*-butyl ether (MTBE)/water (81 : 15 : 4, v/v/v; solvent A) and methanol/MTBE/water (4 : 92 : 4, v/v/v; solvent B) using a gradient program as follows: 0–30% B (22 min), 30–51.3% B (10 min), 51.3–62.7% B (23 min), 62.7–100% B (5 min), 100% B isocratic (5 min), 100–0% B (5 min). Total run time was 80 min. Detection was performed at 450 nm. Adapted from ref. 104.

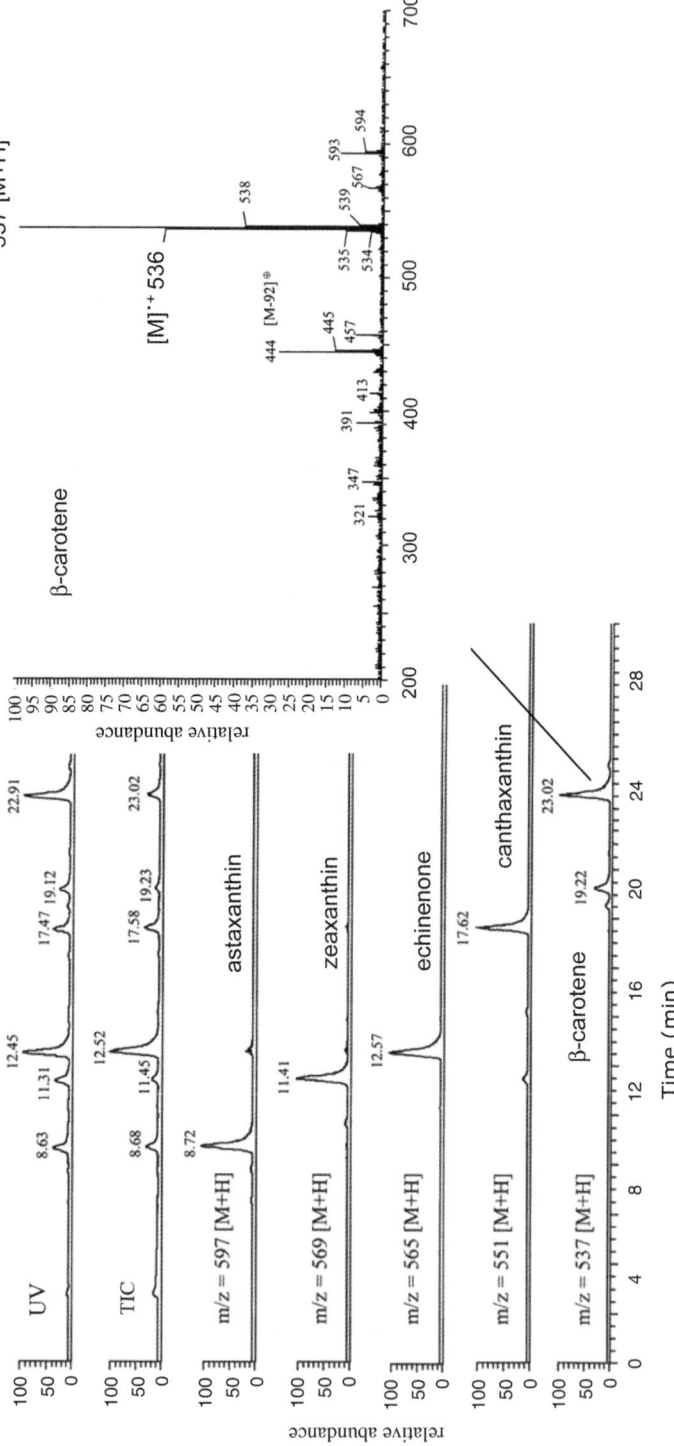

Figure 6.15 PI APCI-MS of β-carotene and C30 separation of a mixture of carotenoids. HPLC conditions: ABCR ProntoSil C30 column, 250 mm, 4.6 mm i.d., flow rate 1 mL min⁻¹, mobile phase: methanol/methyl *tert*-butyl ether (MTBE). Detection was performed at 450 nm. Adapted from ref. 107.

method is exceedingly suitable for their analysis.[105] LC-APCI-MS was found to produce protonated molecular ions as base peaks for most carotenoids.[106] All xanthophylls produced substantial $[M + H-H_2O]^+$ fragment ions, which were also found to be the base peaks for neoxanthin and lutein. In general, hydroxylated carotenoids were detected with a higher sensitivity than the non-hydroxylated ones.

As an example of the APCI-MS spectrum of a representative carotenoid, the PI LC-APCI-MS spectrum of β-carotene is shown in Figure 6.15. The base peak of the recorded mass spectrum of all-*trans* β-carotene with a molecular mass of 537 belongs to the protonated molecule $[M + H]^+$. As this is also the case with other easily ionized compounds, the molecular radical ion $[M]^+$ appeared at m/z 536 with high intensity. Another ion recorded at m/z 444, identified as $[M-92]^+$, is a typical fragment for β-carotene and is formed by free radical fragmentation from the radical cation $[M]^+ \cdot$. The type of separation obtained by C30 HPLC has also been demonstrated for a mixture of characteristic carotenoids.[107]

An advantage of APCI over ESI is its higher linear detector response over a concentration range of over three orders of magnitude for quantification purposes. However, APCI produces abundant fragmentation that tends to reduce the abundance of the molecular ions. ESI of carotenoids produces abundant molecular cations with little fragmentation and no molecular ions. Additionally, low detection limits are obtained with this technique. Since ESI fails to efficiently ionize non-polar compounds because of a lack of sites for protonation or deprotonation, the addition of silver was investigated to improve their ionization. The Ag^+-carotenoid and Ag^+-tocopherol adducts thus formed render these substances amenable to MS. The mass spectra revealed that all carotenoids and most tocopherols were partially oxidized to radical cations. Dichloromethane extracts of tomato, carrot and vegetable juices, a vitamin drink, and a commercial infant food product were thus analyzed by LC-MS after post-column argentation.[108]

In ESI, relatively few papers have dealt with only the single use of this technique for carotenoid identification. A balance between M^+ and $[M + H]^+$ was found to be dependent on the presence of a heteroatom.[109] The ESI detection limit was found to be 100-fold lower than the one of the PDA detector signal when carotenoids were separated on a C30 HPLC column.[110] The method was also found to be useful for the identification of oxidized carotenoids in biological samples with MS/MS.[111] In this case, the authors were able to identify two oxime derivatives based on the fragmentation pattern obtained.

Regarding identification, very few studies have dealt with the complete characterization of carotenoids. In this respect, a fast atom bombardment (FAB) method in combination with high energy collisionally activated dissociation (CAD) has been reported. Seventeen different carotenoids were differentiated by this method. In this case, M^+ ions were produced and the fragmentation pattern obtained revealed structural features indicative of the presence of hydroxyl groups, ring systems, ester and aldehydes groups, as well as the extent of the polyene conjugation.[112]

6.4.2.2   Oxidized Forms of Carotenoids

Under conditions of normal oxidative stress, carotenoids serve as protective antioxidants; however, when the oxidative stress exceeds the antioxidant capacity, carotenoids can be oxidized into numerous cleavage products. The determination and identification of oxidized carotenoids in biological samples remains a major challenge due to the small sample size and low stability of these compounds. In this respect, LC-MS/MS may be useful for the characterization of the oxidized forms. For example, the reaction of various zeaxanthin cleavage products with *O*-ethyl hydroxylamine to evaluate their levels in a biological sample was carried out in PI ESI mode. Protonated molecules $[M + H]^+$ of carotenoids upon collisionally induced dissociation produced a number of structurally characteristic product ions. A series of complicated clusters of product ions differing by 14 (CH_2) and 26 (C_2H_2) Da was characteristic of the polyene chain of intact carotenoids. All carotenoid ethyl oximes of zeaxanthin cleavage products were characterized by the loss of 60 and 61 Da in their MS/MS spectra, enabling the characterization of oxime derivatives in a human eye sample.[111]

6.4.2.3   Carotenoid Esters

Carotenoids in natural samples are predominantly esterified by fatty acids. In the xanthophyll group, the degree of esterification differs and is related to the number of hydroxyl groups present. Thus, mono- or diesters can be found. Most of the time, saponification is performed prior to the extraction, but important information is lost on the genuine form of the natural carotenoids.[105]

   The esterified carotenoid content in natural samples changes such as during the process of ripening in fruits. ESI and APCI are suitable for the analysis of the esterified forms in conjunction with C18 separation with a mobile phase containing predominantly methanol, acetone, 2-propanol and acetonitrile. Diesters have been analyzed by APCI with characteristic fragments corresponding to the loss of the fatty acid units. Direct analysis of diesters of carotenoids has been obtained recently with the separation of 24 mono- and diesters.[113] In *Capsicum annuum*, for example, more than 42 compounds were identified and capsanthin was found to be esterified with lauric, palmitic and myristic acids (see Figure 6.15). Comparison of the fragmentation of free carotenoids and their esterified forms enabled the extraction of a significant amount of useful structural information. The in-source fragmentation behaviour and CID of the $[M + H]^+$ ions of some capsanthin diesters allowed the differentiation of individual regioisomers based on their specific fragmentation patterns.[104]

   Carotenoids also exist in the form of glycosides. These types of compounds have been isolated from the snow alga, *Chlamydomonas nivalis*, and could be identified based on LC-APCI-MS. The employed method provided very informative fragmentation and an (all-*E*)-[di-(6-*O*-oleoyl-β-D-glucopyranosyloxy)]-astaxanthin was found to give a series of characteristic losses for the two fatty acid moieties and the glycoside part, as well as a combination of them (Figure 6.16).[114]

6.4.2.4 Analysis of Carotenoids in Biofluids

Some studies dealing with labelled carotenoids related to their administration to humans with subsequent monitoring by LC-APCI-MS have been described. The preparation of labelled carotenoids can be performed by photosynthesizing organisms cultivated with heavy water or with an atmosphere enriched in $^{13}CO_2$. The incorporation of deuterium has been measured in carrots or spinach with partial deuteration for both lutein and carotene.[115]

Deuterium-labelled carotenoids from intrinsically labelled spinach or collard greens were used to estimate the relative bioavailability of food-derived lutein in humans. Lutein molecules in the vegetables were partially deuterated with a highest abundance isotopomer at M-0 + 8 (unlabelled molecular mass, M-0 plus eight additional mass units from eight deuterium atoms in the molecules). This allowed labelled lutein to be distinguished from endogenous lutein in serum samples after consuming the labelled meal. The quantification of the labelled lutein was achieved by LC-APCI-MS and enabled the calculation of the enrichment for each time point after the dose. The method facilitated the study of lutein bioavailability from different foods of diverse carotenoid composition following various food preparation procedures.[116]

6.4.2.5 LC-NMR

One of the challenges in carotenoid identification is analysing the *cis,trans* isomerism. Other than all-*E*, natural material also contains mono- and di-*Z* geometrical isomers. In LC-PDA, this change is characterized by a hypso-chromic shift of 4–5 nm. This modification can be easily detected by ^{1}H-NMR and, in this respect, LC-NMR has been used to solve several problems related to the identification of carotenoids.[117] Several studies based on C30 separation and LC-APCI-MS and LC-NMR detections have been published.[118,119]

The analysis of lutein and zeaxanthin stereoisomers is of great importance, as they are the main constituents of the macula lutea (the central part of the human retina) and act as possible agents in the prevention and treatment of age-related macular degeneration (AMD). By combining a mild and quick extraction technique such as matrix solid-phase dispersion with HPLC, the extremely light and oxygen-sensitive lutein and zeaxanthin stereoisomers have been extracted, enriched and separated directly from the solid plant or tissue samples, excluding the preparation of artefacts. LC-APCI-MS on C30 enabled the lutein stereoisomers to be distinguished from the zeaxanthin stereoisomers within one chromatographic run in the upper picogram range, whereas LC-NMR coupling allowed the unequivocal identification of each stereoisomer at a concentration in the upper nanogram range (Figure 6.17).[118]

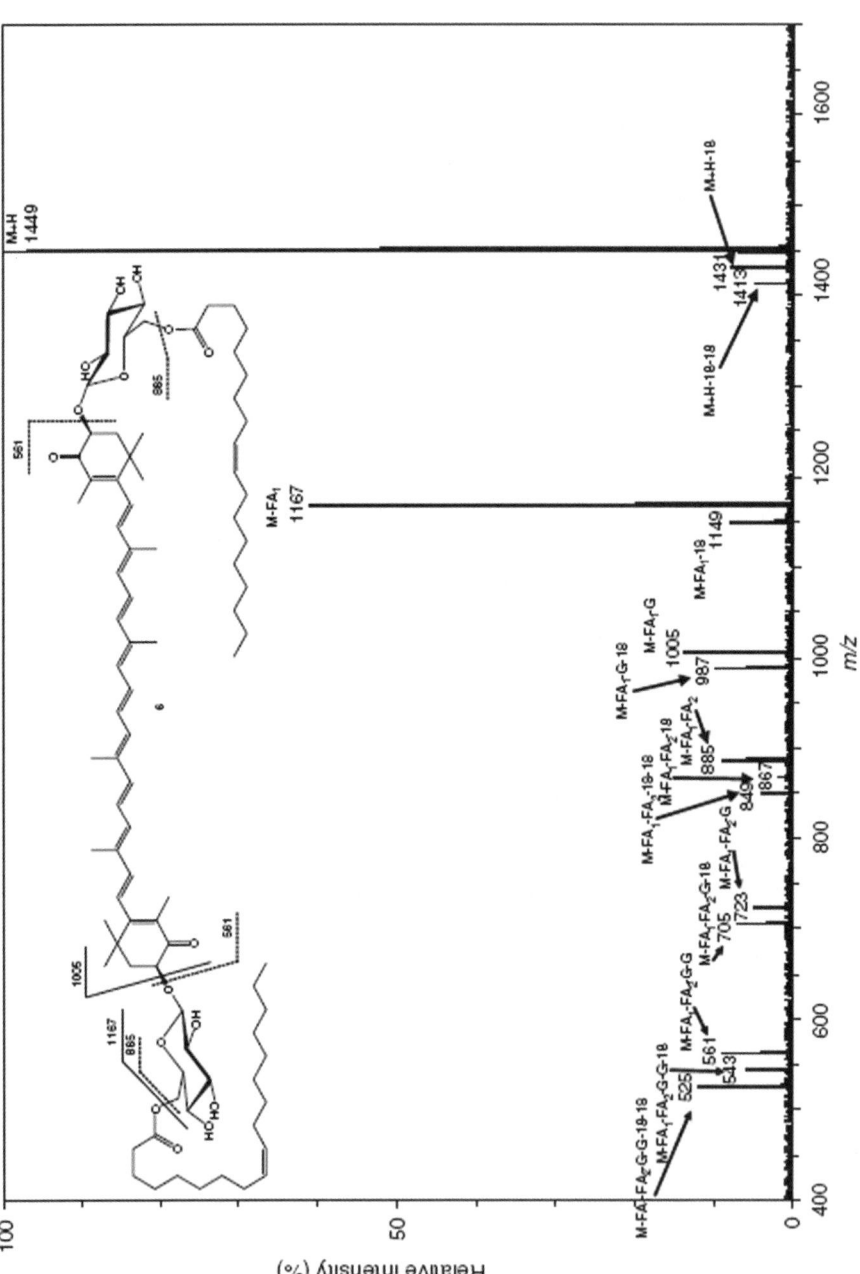

Figure 6.16 PI APCI spectrum of astaxanthin glucoside ester and major fragmentation pathway of protonated astaxanthin glucoside esters di-[(6-*O*-oleoyl-β-D-glucopyranosyl)oxy]-astaxanthin. Adapted from ref. 114.

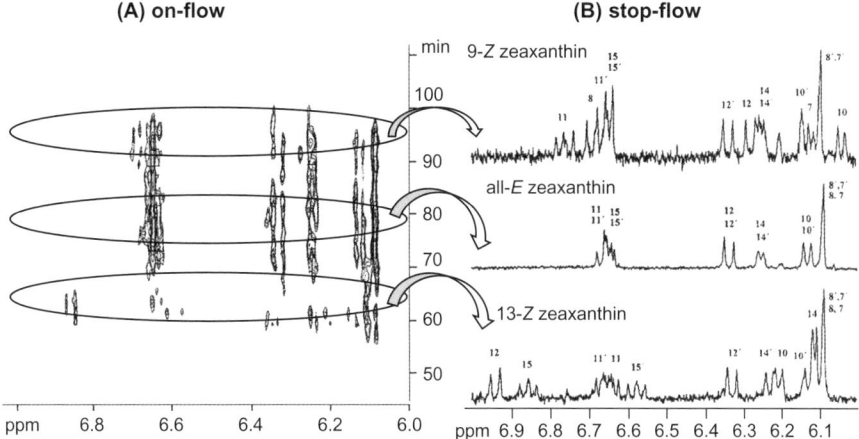

Figure 6.17 (A) On-flow LC-NMR: contour plot (600 MHz, olefinic region) of zeaxanthin stereoisomers. (B) Stop-flow ¹H NMR spectra (600 MHz, olefinic region) of (all-*E*)-zeaxanthin, (9-*Z*)-zeaxanthin and (13-*Z*)-zeaxanthin. Adapted from ref. 118.

The latest development in this LC-NMR approach involved capillary NMR. Such a miniaturized system has allowed the use of fully deuterated solvents for on-line LC-NMR coupling. The ¹H-NMR spectra of various carotenoids obtained in stop-flow mode gave a high signal-to-noise ratio with a sample amount in the low nanogram range. All necessary features for structure elucidation such as multiplet structure, coupling constants and integration values have been detected unambiguously.[119]

6.5 Nutritional Health Benefits

6.5.1 Bioavailability

The concept of bioavailability has been extensively investigated for several classes of micronutrients, and in particular phytochemicals. Indeed, in order for a bioactive food compound to exert any biological activity, it must reach the target organ at a minimal concentration that determines both biological effect and mechanism of action. The dietary intake of a bioactive compound does not necessarily reflect the dose reaching the target tissue and many intervention trials lacking validated measures of bioavailability have failed to demonstrate the anticipated results. Furthermore, depending on the individual predisposition (including genetics, medication and the age of the subjects), a bioavailable dose of an active compound may cause different magnitudes of effects in different people.[61]

In 2004, Holst and Williamson reviewed the methods for the investigation of the bioavailability of phytochemicals.[120] Following the classical pharmacological definition of bioavailability (absolute bioavailability being the exact amount of a compound that reaches the systemic circulation), they divided bioavailability into several linked and integrated steps: liberation, absorption, distribution, metabolism, and excretion (LADME). Holst and Williamson showed the diversity of processes and parameters responsible for the enormous variation in phytochemical bioavailability and, consequently, the need for accurate and validated biomarkers of exposure to determine the internal, target tissue and biologically effective dose.

The basis for validation of a biomarker of exposure is the understanding of major events describing the fate of the nutrients. A general scheme of these events is depicted in Figure 6.18. Parameters affecting excretion can be divided into exogenous factors such as:

- complexity of food matrix;
- chemical form of the compound;
- structures and amounts of co-ingested compounds; and
- endogenous factors (including mucosal mass, intestinal transit time, rate of gastric emptying, metabolism and extent of conjugation).

Based on the complexity of these factors, large inter-individual and intra-individual variations in bioavailability, sometimes ranging from 0% to 100% of the ingested dose, are observed.[120]

The absorption and metabolism of phytochemicals and drugs share similar principles and metabolic pathways (common transport systems). However, in the case of phytochemicals, it is important to consider the following differences.

First, the food matrix is much more complex than drug formulations. Therefore, the biomarker should be specific for the dietary component of interest, *i.e.* variation in its concentration should be exclusively due to changes in the intake of the dietary component of interest. In addition, the exact concentration of a bioactive component in a given food varies with the genotype of the plant or animal, growing conditions and processing methods.[121] Potential biomarkers of polyphenol intake derived from phase I and II metabolism of flavonoids and other polyphenols in the small intestine and liver have been reviewed. Metabolites such as glucuronide and sulfate conjugates (and related *O*-methylated forms) can be used to determine the extent of polyphenol absorption in human subjects and represent excellent biomarkers. In addition to these well-characterized metabolic forms, there is great potential for cellular metabolites of polyphenols, such as glutathionyl and cysteinyl conjugates,[122,123] as they are likely to be present in significant amounts in both blood and urine.

Second, several components have frequently interactive or synergistic effects. Furthermore, bioavailability can be very different between a purified component and the same components with its original food matrix.[61] Third, phytochemicals compete with other food constituents for common metabolic enzymes and transporters.

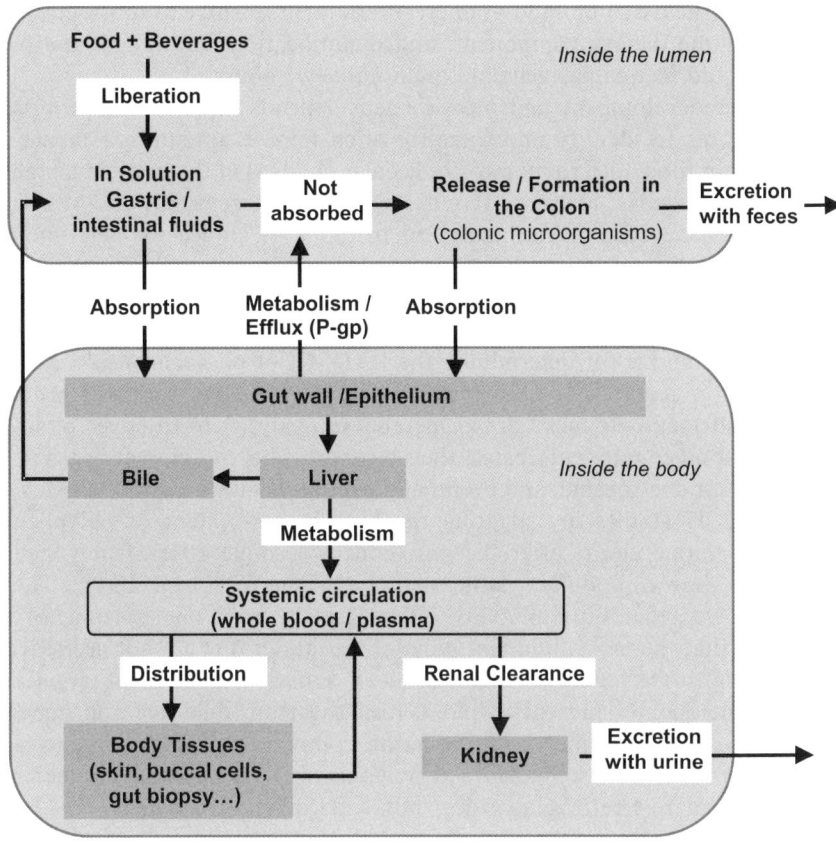

Figure 6.18 Basic events describing the fate of nutrients: (1) **Liberation**—the release and dissolution of a compound to become available for absorption (bioaccessibility); (2) **Absorption**—the movement of a compound from the site of administration to the blood circulation; (3) **Distribution**—the process by which a compound diffuses or is transferred from the intra-vascular space (blood) to the extra-vascular space (body tissues); (4) **Metabolism**—the biochemical conversion or transformation of a compound into a form that is easier to eliminate; and (5) **Excretion**—the elimination of unchanged compound or metabolites from the body, mainly via renal, biliary or pulmonary processes. The complex connection between the luminal content, lumen and body content results in a web of possible pathways for each compound. Reprinted from ref. 36.

Finally, food components are present at low doses and act through longer term (chronic) applications compared with higher (acute) doses of drugs. Indeed, tissue concentrations of phytonutrients compared with those of endogenous antioxidants suggest that many of the phytochemicals exert their activity indirectly as "low-dose toxins" by activating adaptive cellular response pathways to oxidative stress and environmental toxin exposure rather than acting as direct radical scavengers.[36] Therefore, concentrations of the biomarker in the tissue/biofluid of interest should be sensitive to subtle changes in intake of the dietary component, and quantification methods of the bio-markers should be qualitatively and quantitatively robust.

Biomarker development and measurement depends highly on appropriate analytical tools to identify and quantify often minute amounts of bioactive compounds in food, and to monitor molecular changes in the body in a highly specific and sensitive manner. Based on these requirements, LC-MS and LC-MS/MS have become the analytical methods of choice for determining the bioavailability of phytochemicals. Markers for bioavailability can be addressed in a hypothesis-driven, targeted fashion or holistically without any pre-assumption.[124] LC-MS and LC-MS/MS quickly became the analytical workhorse technique for determining the LADME of phytochemicals. More-over, nutrition research is taking advantage of proteomics (global protein analysis) and metabolomics (global metabolite analysis) to discover biologi-cally active food components, assess their bioavailability at different organ sites as well as their site-specific, and eventually overall, health benefits. Manach *et al.* reviewed 97 studies investigating the kinetics and extent of polyphenol absorption among adults after the ingestion of a single dose of polyphenol provided as pure compound, plant extract or whole food/beverage.[125] These authors showed that bioavailability differs greatly from one polyphenol to another, so that the most abundant polyphenols in our diet are not necessarily those leading to the highest concentrations of active metabolites in target tis-sues. The metabolites present in blood resulting from digestive and hepatic activity were found to differ from the native compounds. Gallic acid and iso-flavones were the most well-absorbed polyphenols,[126] followed by catechins, flavanones and quercetin glucosides, but with different kinetics. The least absorbed polyphenols are the proanthocyanidins, the galloylated tea catechins, and the anthocyanins.

Scholz and Williamson investigated the factors influencing the bioavailability of polyphenols—specifically flavanols, flavonols, flavanones and flavones.[127] Bioaccessibility, defined as the amount of compound reaching the enterocyte in a form suitable for absorption, is described as the most important factor determining absorption in the gut. Factors leading to an improved absorption of flavonols, notably quercetin and its metabolites, are primarily the nature of the attached sugar, and second, the solubility as modified by ethanol, fat and emulsifiers. In the case of green tea catechins, the absorption of flavanols is affected by epimerization reactions occurring during processing, by the pre-sence of lipids and carbohydrates, and is improved by the presence of piperine and tartaric acid. Flavanones, such as hesperidin, are strongly affected by the type of attached sugar. Proanthocyanidins, the galloylated tea catechins and the anthocyanins seem to be the least well-absorbed polyphenols in humans. Proanthocyanidins are not depolymerized in the stomach and reach the small intestine intact, where they are sparingly absorbed because of their high molecular weight.[128,129] The bioavailability of anthocyanins appears to be consistently very low across all studies with often less than 0.1% of the ingested dose appearing in the urine. Moreover, their bioabsorption occurs quickly following consumption.[130] Compared with other flavonoid groups such as flavonols, relatively little is known about the details and mechanisms of anthocyanin absorption and transport.

Since foodstuffs are a complex mixture of components, synergistic effects between compounds may take place, but interactions leading to a decrease of bioavailability may also occur. One area that is still highly controversial is the effect of milk proteins on polyphenol bioavailability. Milk contains various proteins, and interactions between polyphenols and proteins have been reported.[131] Several food products or beverages are traditionally consumed with or without mixing them with milk. Tea and chocolate are typical examples of products that are consumed through different ways among the populations of different countries. Tea is the most consumed drink in the world after water. In the UK, Ireland, Canada and India, tea is consumed with a substantial amount of milk.[132] In China and Japan, tea is consumed mostly without milk. In the United States, chocolate (mainly milk chocolate) makes a significant contribution to US per capita dietary antioxidants, with dietary chocolate being the third highest daily per capita antioxidant source.[133] In the UK and Germany, people consume mostly milk chocolate, whereas in France, dark chocolate consumption dominates. Given the benefits associated with the consumption of tea catechins and cocoa proanthocyanidins (see below), some research has been devoted to address polyphenol–milk protein interactions, but the results remain controversial to date.

Concerning tea, van Het Hof *et al.* claimed that catechins from green tea and black tea were rapidly absorbed, and that milk did not impair their bioavailability.[134] Similarly, Leenen *et al.* demonstrated that the consumption of a single dose of black or green tea induced a significant rise in plasma antioxidant activity *in vivo*; the addition of milk to tea did not abolish this increase.[135] However, more recently, Lorenz and co-workers showed that milk counteracts the favourable health effects of tea on vascular function. These results were obtained with 16 healthy female volunteers consuming either 500 mL of freshly brewed black tea, black tea with 10% skimmed milk, or boiled water as a control. Endothelial function determined through flow-mediated dilation (FMD) was measured by high-resolution vascular ultrasound before and 2 h after consumption. Black tea significantly improved FMD in humans compared with water, whereas the addition of milk completely blunted the effects of tea.[136]

Concerning chocolate, an article published in *Nature*[137] found that the bioavailability of epicatechin from milk chocolate was substantially reduced compared to dark chocolate, and even dark chocolate taken with a glass of milk. The hypothesis was that the milk proteins bind to polyphenols, making them less available. In agreement with these data, a bioefficacy study compared the effects of either dark or white chocolate bars on blood pressure and glucose and insulin responses to an oral glucose tolerance test in healthy subjects. After a seven-day cocoa-free run-in phase, 15 healthy subjects were randomly assigned to receive either 100 g dark chocolate bars, which contained 500 mg polyphenols, or 90 g white chocolate bars (negative control without polyphenols) for 15 days. Successively, subjects entered a further cocoa-free washout phase of seven days and were then crossed over to the other condition. Polyphenol bioefficacy was estimated through an oral glucose tolerance test performed at the end of each period to calculate the homeostasis model assessment of insulin resistance and the quantitative insulin sensitivity check index. Dark, but not white, chocolate was found to decrease blood pressure and improve insulin sensitivity in healthy persons.[138] However, in 2007, Roura *et al.* evaluated the possible interaction of milk on the absorption of (–)-epicatechin from cocoa powder in healthy humans. The study was carried out with 21 volunteers who received three interventions in a randomized crossover design with a one-week interval (250 mL of whole milk) (control), 40 g of cocoa powder dissolved in 250 mL of whole milk, and 40 g of cocoa powder dissolved with 250 mL of water. Quantification of (–)-epicatechin in plasma was determined by LC-MS/MS analysis after a solid-phase clean-up procedure. It was concluded that milk does not impair the bioavailability of polyphenols and thus their potential beneficial effect in chronic and degenerative disease prevention.[139]

In conclusion the ascertainment of the health-promoting effects of dietary polyphenols requires the determination of their bioavailability. However, despite extensive research activities in this area, it is not possible today to propose general principles with any predictive capacity.[127] Further studies will be required to investigate more deeply and systematically the various types of interactions affecting the bioavailability of different classes of phytonutrients.

6.5.2 Bioefficacy

During the last 15 years, research on the effects of dietary phytonutrients on human health has provided evidences supporting a role for certain classes of phytonutrients (*e.g.* polyphenols) in the prevention of degenerative diseases, particularly cardiovascular disease (CVD) and cancer. Polyphenols have strong antioxidant properties and can scavenge free radicals in the cell oxidation processes, but it has become clear that the mechanisms of action of polyphenols go beyond the modulation of oxidative stress.[12] The contribution of polyphenols to the antioxidant capacity of the human diet is much larger than that of any vitamins. Indeed, their total dietary intake could be as high as $1 \, g \, d^{-1}$, which is ten times higher than the intake of vitamin C and 100 times higher than the intakes of vitamin E and carotenoids.[13,14]

However, demonstrating the health beneficial properties (bioefficacy) of polyphenols and other phytonutrients is a more challenging task than measuring bioavailability in the blood or at the target tissue. Despite the significant number of human intervention studies showing biological effects, there is still a need for more investigations demonstrating health benefit of phytonutrients on a long-term basis. Even if the demonstrated health beneficial effects depend on the class of phytonutrients investigated, most of the trials were short term, and very few studies demonstrated a dose response relationship and convincing evidences. In addition, most studies, with the exception of those with isoflavones, administered food instead of pure compounds.[140]

Below we review a few food products and their efficacy on disease-risk reduction due to phytonutrients—mainly polyphenols, as this class of chemical is by far the most documented through clinical evidence.

6.5.2.1 Tea

Tea is the most consumed drink in the world after water. Green tea is a "non-fermented" tea, and contains more catechins than black tea or oolong tea. Since ancient times, green tea has been considered by traditional Chinese medicine as a healthful beverage. Recent human studies suggest that green tea may contribute to a reduction in the risk of cardiovascular disease and some forms of cancer (Table 6.3), as well as to the promotion of oral health and other physiological functions such as anti-hypertensive effect, body weight control, antibacterial and antivirasic activity, solar ultraviolet protection, bone mineral density increase, anti-fibrotic properties and neuroprotective power.[141] Indeed, tea polyphenols possess not only antioxidant properties but also antiviral, antibacterial, anti-inflammatory and anti-carcinogenic effects, as well as the ability to modulate certain signalling pathways such as nuclear factor-kB activation. Green tea polyphenols have been shown to have efficacy in various models of inflammatory bowel disease.[142] Increasing interest in green tea health benefits has led to its inclusion in the group of beverages with functional properties. Among all tea polyphenols, epigallocatechin-3-gallate has been shown to be responsible for much of the health promoting ability of green tea.[143] In addition, green tea contains certain minerals and vitamins which increase its antioxidant potential. However, in many cases, the concentrations of tea polyphenols required to observe biological effects *in vitro* exceed the concentrations achievable in blood and tissues by 10- to 100-fold. In addition, tea polyphenols are not stable and reactive oxygen species are generated under most cell culture conditions. It is not clear whether these reactions occur *in vivo*.[144] Finally, several studies conducted with human subjects report reduced body weight and body fat, as well as increased fat oxidation and thermogenesis, and thereby confirm findings in cell culture systems and animal models of obesity. These findings re-emphasize the need for well-designed and controlled clinical studies and since EGCG is regarded as the most active component of green tea, its specific effects on obesity should also be investigated in human trials.[145]

6.5.2.2 Coffee

With an average of six million tons produced per year, coffee is a widely consumed beverage and the major source of caffeoylquinic acids in the human diet. Coffee drinkers ingest up to 500 mg to 1 g cinnamates per day (completed with other dietary sources such as bran), whereas the daily intake of coffee abstainers is less than 100 mg.[148] When 25% milk was added to coffee, up to 40% of coffee chlorogenic acid is found to be bound to dairy proteins.

Table 6.3 Health beneficial effects of tea. Adapted from ref. 146 and ref 147.

Study reference	Subjects	Method	Outcomes
Regular consumption of green tea and the risk of breast cancer recurrence: follow-up study from the Hospital-based Epidemiologic Research Program at Aichi Cancer Center (HERPACC), Japan.[175]	1160 female invasive breast cancer; mean age = 51.5 years	Prospective study	A decrease in the risk of cancer recurrence is observed with a consumption of >3 cups per day of tea. This decrease is mainly in early stage cases.
Green tea and risk of breast cancer in Asian Americans.[176]	Asian-American women 501 breast cancer patients and 594 controls	Case-control study	There is a significant trend of decreasing risk of breast cancer with increasing amount of tea intake, respectively, in association with 0, 0–85 mL d^{-1} and >85.7 mL d^{-1}.
Green tea and the risk of breast cancer: pooled analysis of two prospective studies in Japan.[177]	35 004 women	Pooling of two prospective studies	Tea intake is not associated with a lower risk of breast cancer (relative risk for women drinking >5 cups per day compared with >1 cup per day.
Tea consumption and ovarian cancer risk: a case-control study in China.[178]	254 ovarian cancer patients and 652 controls	Case-control study, validated questionnaire	Increasing frequency and duration of tea drinking can reduce the risk of ovarian cancer.
Protective effect of green tea against prostate cancer: a case-control study in southeast China.[179]	130 prostate adenocarcinoma patients and 274 controls	Case-control study with face-to-face interview using a structured questionnaire	The prostate cancer risk declined with increasing frequency, duration and quantity of tea consumption for those consuming >1.5 kg of tea leaves annually.

Effect of life styles on the risk of subsite-specific gastric cancer in those with and without family history.[180]	887 gastric cancer and 28 619 control; age = 20–79 years	Case-control study	Consumption of more than 6 cups per day *vs.* never drinking decreased the risk of gastric cancer.
Green tea consumption and chronic atrophic gastritis: a cross-sectional study in a green tea production village.[181]	636 men and women; mean age: men = 59.2 years, women = 60.4 years	Cross-sectional study	Consumption of more than 10 cups per day reduces the risk of chronic atrophic gastritis.
Relationship between serologically diagnosed chronic atrophic gastritis, *Helicobacter pylori* and environmental factors in Japanese men.[182]	566 men; age = 50–55 years	Cross-sectional study	No relationship between tea consumption and risk of chronic atrophic gastritis.
Preventive effects of drinking green tea on cancer and cardiovascular disease: epidemiological evidence for multiple targeting prevention.[183]	8552 adults	Prospective cohort study	Consumption of more than 10 cups per day decreases the risk of stomach cancer.
A prospective study of green tea consumption and cancer incidence, Hiroshima and Nagasaki (Japan).[184]	14 873 men and 23 667 women	Prospective cohort study	No relation between tea consumption and reduced risk of stomach cancer (OR = 0.95, 95% CI = 0.76–1.2), colon cancer (OR = 1.0, 95% CI = 0.76–1.4) or rectum cancer (OR = 1.3, 95% CI = 0.77–2.1).[a,b]
Protective effect of green tea on the risk of chronic gastritis and stomach cancer.[185]	166 chronic atrophic gastritis, 133 gastric cancer and 433 controls	Case-control study	Significant inverse association between tea drinking and gastric cancer or chronic atrophic gastritis.
Green tea and the risk of gastric cancer in Japan.[186]	11 902 men and 14 409 women	Prospective cohort study	No association with risk of gastric cancer.

Table 6.3 Continued.

Study reference	Subjects	Method	Outcomes
A prospective study of stomach cancer death in relation to green tea consumption in Japan.[187]	30 370 men and 42 481 women	Prospective cohort study	No association between tea consumption and stomach cancer death.
Prospective study of education background and stomach cancer in Japan.[188]	18 746 men and 26 184 women	Prospective cohort study	No association between tea consumption and stomach cancer death; men.
Bladder cancer incidence in relation to vegetable and fruit consumption: a prospective study of atomic-bomb survivors.[189]	14 873 men and 23 667 women Hiroshima atomic bomb survivors	Prospective cohort study	Tea consumption is not related to risk of bladder cancer.
A population-based case-control study of lung cancer and green tea consumption among women living in Shanghai, China.[190]	649 lung cancer women and 675 controls women	Case-control study with face-to-face interviews	Among non-smoking women, consumption of green tea was associated with a reduced risk of lung cancer and the risks decreased with increasing consumption.
A prospective study of green tea consumption and cancer incidence, Hiroshima and Nagasaki (Japan).[184]	38 540 people; 14 873 men, mean age = 52.8 years and 23 667 women, mean age = 58.8 years	Prospective study	Tea consumption is unrelated to incidence of cancers under study.
Flavonoid intake and the risk of CVD in women.[191]	Women (38 445) free of CVD	Prospective study	Flavonoid intake not strongly associated with a reduced risk of CVD.
Inverse association of tea and flavonoid intakes with incident myocardial infarction: Rotterdam.[192]	Men (1836), women (2971)	Baseline semi-quantitative food frequency questionnaire	Relative risk of myocardial infarction lower in tea drinkers with a daily intake of 4375 mL than in non-tea drinkers

Study	Subjects	Method	Findings
Intake of flavonols and flavones and risk of coronary heart disease in male smokers.[193]	Male smokers 25 372; age = 50–69 years	Baseline assessment with validated questionnaire.	Intake of flavonols and flavones inversely associated with non-fatal myocardial infarction. Weaker inverse association with coronary death.
Catechin intake might explain the inverse relation between tea consumption and ischemic heart disease: Zutphen Elderly Study.[194]	Men 806; age = 65–84 years	Baseline diet history with dietician	Inverse association between tea intake and relative risk of mortality failed to reach significance.
Dietary catechins in relation to coronary heart disease death among postmenopausal women.[195]	Post-menopausal women (34 492)	Baseline food frequency questionnaire.	Tea catechins not associated with coronary heart disease death. Tea intake positively associated with healthy diet.
Coffee and tea consumption in the Scottish Heart Health Study follow up: conflicting relations with coronary risk factors, coronary disease, and all cause mortality.[196]	Men (5645), women (5800); age = 40–59 years	Food frequency questionnaire	Increasing tea consumption associated with coronary mortality and morbidity. Social class differences in tea consumption not controlled for.
Antioxidant flavonols and ischemic heart disease in a Welsh population of men: Caerphilly.[197]	Men (1900); age = 45–59 years	Semi quantitative food frequency questionnaire.	Tea consumption positively associated with coronary heart disease.
Tea consumption and the prevalence of coronary heart disease in Saudi adults: results from a Saudi national study.[198]	30–70 year old subjects (3430)	Case control study	Subjects drinking 4480 mL tea per day had lower prevalence of coronary heart disease versus non-tea drinkers even when risk factors accounted for. Dose–response between tea and lower coronary heart disease risk.

Table 6.3 Continued.

Study reference	Subjects	Method	Outcomes
Alcohol, smoking and coffee and risk of non-fatal acute myocardial infarction in Italy.[199]	507 cases with first episode of nonfatal myocardial infarction *versus* 478 controls admitted for acute diseases	Hospital based case control study	No significant association between tea and myocardial infarction. No details on numbers of tea consumers.
Coffee and tea intake and the risk of myocardial infarction.[200]	680 men; age = >76 years	Case control study	Tea drinking associated with lower risk of myocardial infarction.
Association between certain foods and risk of acute myocardial infarction in women.[201]	936 women; age = 21–69 years	Case control study	No significant association between tea and myocardial infarction.
Habitual tea consumption and risk of osteoporosis: a prospective study in the Women's Health Initiative Observational.[202]	91 women; age = 50–59 years	Structured lifestyle Q and history of fractures from medical records. Bone mineral density measured in sub-sample (4979).	Bone mineral density positively correlated with tea drinking. No significant association between tea drinking and risk of fractures at hip and forearm/wrist.
Tea drinking and bone mineral density in older women.[203]	1256 women; age = 65–76 years	Self-administered questionnaire. Sample classified into tea (90%) and non-tea drinkers (10%)	Tea drinkers had significantly greater (5%) mean bone mineral density measurements, adjusted for age, body mass index and smoking.

[a]odds ratio.
[b]confidence interval.

However, milk addition does not have a significant effect on beverage anti-oxidant power. Moreover, these interactions decrease during *in vitro* gastric and intestinal digestion, thus suggesting that interactions between chlorogenic acid and milk proteins in coffee and milk beverages may not have any significant effect on coffee antioxidant power before and after consumption.[149] Green coffee beans of Robusta coffee exhibited a two-fold higher antioxidant activity than Arabica coffee, but after roasting, this difference was no longer significant. In conclusion, these commonly consumed beverages have a significant antioxidant activity, the highest being soluble coffee on a cup-serving basis.[150]

6.5.2.3 Soy

Clinical studies show that ingestion of soy proteins reduces the risk factors for cardiovascular disease (Table 6.4). However, the claimed health benefits seem to be quite variable. The Nutrition Committee of the American Heart Association has assessed 22 randomized trials conducted since 1999 and found that isolated soy protein with isoflavones slightly decreased low density lipoprotein (LDL) cholesterol but had no effect on high density lipoprotein (HDL) cholesterol, triglycerides or blood pressure. Although the contributing factors to these discrepancies are not fully understood, the source of soybeans and processing procedures of the protein or isoflavones are believed to be important because of their effects on the content and integrity of certain bioactive protein subunits.[151]

There is a suggestion, but no conclusive evidence, that isoflavones from the sources studied so far have a beneficial effect on bone health. For menopausal symptoms, there is currently limited evidence that soybean protein isolates or soybean foods extracts are effective, though soybean isoflavone extracts are supposed to be effective in reducing hot flushes. Indeed, soy isoflavones interact with the estrogen receptor, which makes them weak or moderate phytoestrogens. The health benefits of soybean phytoestrogens in healthy postmenopausal women are subtle and even some well-designed studies do not show protective effects. Future studies should focus on high-risk postmenopausal women, especially in the areas of diabetes, cardiovascular disease, breast cancer and bone health.[152]

There are too few randomized clinical trials studies to reach conclusions on the effects of isoflavones on breast cancer, colon cancer, diabetes or cognitive function.

6.5.2.4 Fruits and Vegetables

Antioxidant compounds such as vitamin C, carotenoids and flavonoids found in fruits and vegetables may influence the risk of cardiovascular disease by preventing the oxidation of cholesterol in arteries. Oxidative stress is an important contributor to the risk of chronic diseases. Fruits and vegetables are

Table 6.4 Health beneficial effects of chocolate. Adapted from ref. 153.

Study	Subjects	Product tested	Outcomes
A dose–response effect from chocolate consumption on plasma epicatechin and oxidative damage.[204]	20 healthy adults; age = 20–56 years	Semi-sweet chocolate baking bits; one dose of 27, 53 and 80 g	Dose-dependent increase in plasma epicatechin. Non-significant trend for an increase in plasma antioxidant activity.
Cocoa inhibits platelet activation and function.[205]	30 healthy adults; age = 24–50 years; 10 per group	18.75 g procyanidin-rich cocoa powder in 330 mL water (one dose)	Suppression of platelet activation. Aspirin-like effect on primary hemostasis 6 h after consumption.
Epicatechin in human plasma: *in vivo* determination and effect of chocolate consumption on plasma oxidation status.[206]	10 healthy adults (age = 26–49 years) + 3 healthy adults (age = 28–36 years) consuming control	105 g (of which 80 g chocolate) semi-sweet baking bits (one dose)	12-fold increase in plasma epicatechin 2 h later, increase in plasma total antioxidant activity and decrease in TBARS.[a]
Daily cocoa intake reduces the susceptibility of LDL to oxidation as demonstrated in healthy human volunteers.[207]	15 healthy men, 9 in active group; age = 32.5 ± 6.4 years	12 g cocoa powder, three times per day for two weeks	Increase in LDL oxidation lag time, no change in plasma lipids or antioxidants. Higher excretion of epicatechin/metabolites in urine.
Effects of cocoa powder and dark chocolate on LDL oxidative susceptibility and prostaglandin concentrations in humans.[208]	23 healthy adults; age = 21–62 years	22 g cocoa powder and 16 g dark chocolate per day for four weeks	Increase in LDL oxidation lag time, increase in serum antioxidant capacity, increase in HDL cholesterol.
The effects of flavanol-rich cocoa and aspirin on *ex vivo* platelet function.[209]	16 healthy adults; age = 22–49 years	18.75 g cocoa powder in 300 mL water with sugar, with and without aspirin (one dose)	After 6 h, cocoa inhibited epinephrine stimulated platelet activation and function.
Cocoa products decrease low density lipoprotein oxidative susceptibility but do not affect biomarkers of inflammation in humans.[210]	25 healthy adults; age = 20–60 years	36.9 g dark chocolate and 30.95 g cocoa powder in a drink per day for six weeks	LDL oxidizability was lower, but no effect on inflammation markers, or plasma antioxidant capacity.

Chocolate consumption and platelet function.[211]	18 healthy adults	25 g semi-sweet chocolate chips (one dose)	Increase in plasma epicatechin after 2h with concurrent increase in prostacyclin–leukotriene ratio. Reduction in platelet-related haemostasis.
Chocolate and blood pressure in elderly individuals with isolated systolic hypertension.[212]	13 elderly adults with mild hypertension; age = 55–64 years	100 g dark chocolate per day for 14 d	Lower systolic and diastolic blood pressure.
Flavanol-rich cocoa induces nitric oxide (NO) dependent vasodilation in healthy humans.[213]	27 healthy adults; age = 18–72 years	High polyphenol cocoa drink 4×230 mL d^{-1} for 4 d	Improved peripheral vasodilation after 4 d, large acute response after 90 min.
Vascular effects of cocoa rich in flavan-3-ols.[214]	20 adults (all with 1 CHD risk factor); age = 41 years ± 14; 77% were smokers	100 mL high cocoa polyphenol drink (one dose)	NO bioactivity and arterial flow-mediated dilation increased.
Plasma antioxidants from chocolate—dark chocolate may offer its consumers health benefits the milk variety cannot match.[215]	12 healthy adults; age = 25–35 years	100 g dark chocolate (with and without 200 mL milk) (one dose)	Dark chocolate increased plasma antioxidant capacity and epicatechin. Consuming milk with it reduced these effects. Milk chocolate had less effect than both these treatments.
Dark chocolate consumption increases HDL cholesterol concentration and chocolate fatty acids may inhibit lipid peroxidation in healthy humans.[216]	45 healthy adults; age = 19–49 years	75 g dark chocolate or high phenolic dark chocolate for three weeks	Both dark chocolates increased HDL cholesterol and lipid peroxidation decreased (but also with white chocolate control). No change in plasma antioxidant capacity.
Flavonoid-rich dark chocolate improves endothelial function and increases plasma epicatechin concentrations in healthy adults.[217]	21 healthy adults; age = 21–55 years	46 g d^{-1} high phenolic dark chocolate for 14 d	Improved endothelium-dependent flow-mediated dilation, no change in blood pressure, oxidative markers or blood lipids. Higher plasma epicatechin.

Table 6.4 Continued.

Study	Subjects	Product tested	Outcomes
Flavanol-rich cocoa drink lowers plasma F-2-isoprostane concentrations in humans.[218]	20 healthy males; age = 20–40 years	High polyphenol cocoa drink, 100 mL (one dose)	F2 isoprostanes improved 2 and 4 h after exercise.
Short-term administration of dark chocolate is followed by a significant increase in insulin sensitivity and a decrease in blood pressure in healthy persons.[138]	15 healthy adults; age = 34 ± 7.6 years	Dark chocolate, 100 g (one dose)	Insulin sensitivity higher and insulin resistance lower. Systolic blood pressure lower.
Regular consumption of a flavanol-rich chocolate can improve oxidant stress in young soccer players.[219]	28 healthy males under exercise stress; age = 18–20 years	105 g d^{-1} milk chocolate for 14 d	Decrease in diastolic and mean blood pressure, plasma cholesterol, LDL, MDA,[b] urate and lactate dehydrogenase activity, increase in vitamin E–cholesterol ratio. No change in plasma epicatechin, but samples were fasting.
Effect of dark chocolate on arterial function in healthy individuals.[220]	17 healthy adults; age = 24–32 years	100 g dark chocolate (one dose)	Increase in resting and hyperaemic brachial artery diameter. Increase in flow-mediated dilation at 60 min. Aortic augmentation index decreased. No significant change in malondialdehyde, and total antioxidant capacity and pulse wave velocity.

Study	Subjects	Dose	Results
Cocoa reduces blood pressure and insulin resistance and improves endothelium-dependent vasodilation in hypertensives.[221]	11 adult smokers; age = average 31 years	High polyphenol cocoa drink, 100 mL (one dose)	Increased circulating NO, flow-mediated dilation, both correlated to increases in flavanol metabolites. Effects were reversed with NG-monomethyl-L-arginine to prove link to NO.
(−)-Epicatechin mediates beneficial effects of flavanol-rich cocoa on vascular function in humans.[222]	16 healthy males; age = 25–32 years	300 mL high polyphenol cocoa drink (one dose)	Acute elevations in levels of circulating NO species, an enhanced flow-mediated dilation response of conduit arteries, and an augmented microcirculation.
Dark chocolate improves endothelial and platelet function.[223]	20 male smokers; age = age not given	40 g dark chocolate (one dose)	Improved flow-mediated dilation after 2h lasting for 8h. Reduction in platelet function. Increased plasma total antioxidant status.
Aging and vascular responses to flavanol-rich cocoa.[224]	15 young (age = <50 years) and 19 older (age = >50 years)	High polyphenol cocoa drink 4×230 mL d^{-1} for 4–6 d	NO synthesis after cocoa was suppressed in older volunteers. Flow-mediated dilation was enhanced in both groups but more in older group. Pulse wave amplitude enhanced in both groups, with acute rises with cocoa ingestion, more robustly in older subjects. No change in blood pressure.

Table 6.4 Continued.

Study	Subjects	Product tested	Outcomes
Cardioprotective effects of chocolate and almond consumption in healthy women.[225]	49 women with cholesterol 4.1–$7.8\,mmol\,L^{-1}$; age = 22–65 years	$41\,g\,d^{-1}$ of high polyphenol dark chocolate either with or without almonds $60\,g\,d^{-1}$ for six weeks plus dietary advice	Dark chocolate decreased TAG[c] by 21%, 19% when eaten with almonds, 13% with almonds alone and 11% with no intervention. Circulating intercellular adhesion molecule with dark chocolate alone.
Dark chocolate improves coronary vasomotion and reduces platelet reactivity.[226]	22 heart transplant recipients (18 men, 4 women)	40 g of flavonoid-rich dark (70% cocoa) chocolate	Dark chocolate induces coronary vasodilation, improves coronary vascular function, and decreases platelet adhesion 2 h after consumption.

[a]Thiobarbituric acid reactive substances.
[b]Malondialdehyde
[c]Triacylglycerols.

good sources of antioxidant phytochemicals that mitigate the damaging effect of oxidative stress.

Carotenoids are recognized as playing an important role in the prevention of human diseases and maintaining good health. In addition to being potent antioxidants, some carotenoids also contribute to dietary vitamin A. Recent interest in carotenoids has focused on the role of lycopene in human health. Unlike some other carotenoids, lycopene does not have provitamin A properties. Because of the unsaturated nature of lycopene, it is considered to be a potent antioxidant and a singlet oxygen quencher.[154] Lycopene from tomato products has been shown to lower biomarkers of oxidative stress and carcinogenesis in healthy and type II diabetic patients, and prostate cancer patients, respectively. Processed tomato products such as tomato juice, tomato paste, tomato puree, tomato ketchup and tomato oleoresin have been shown to provide bioavailable sources of lycopene, with consequent increases in plasma lycopene levels *versus* baseline. Dietary fats enhance this process and should be consumed together with food sources of lycopene. However, there are limited *in vivo* data on the health benefits of lycopene alone. Most of the clinical trials with tomato products suggest a synergistic action of lycopene with other nutrients in lowering biomarkers of oxidative stress and carcinogenesis. Consumption of processed tomato products containing lycopene is therefore of significant health benefit and can be attributed to a combination of naturally occurring nutrients in tomatoes.[155]

Despite the fact that a higher intake of fruit and vegetables can help prevent the morbidity and mortality associated with heart disease, more association studies are needed to ascertain the link between the intake of single carotenoids and the risk of cardiovascular disease.[156] Similarly, the impact of pure fruit and vegetable juices on cancer risk is weakly positive, although contradictory findings hamper conclusions.[157]

6.5.2.5 Chocolate

Chocolate makes a significant contribution per capita to dietary antioxidants in the United States, and by inference in the European Union. In the US diet, chocolate is the third highest daily per capita antioxidant source. Several health benefit effects of chocolate have been reported (Table 6.5). Epicatechin, a major polyphenol in chocolate and chocolate extracts, is a powerful inhibitor of plasma lipid oxidation. After consumption of dark chocolate and cocoa powder, the lower density lipoproteins isolated from plasma were protected from oxidation compared with the lipoproteins isolated after cocoa butter consumption, which were put under oxidative stress. These results may suggest that cocoa is more beneficial to health than teas and red wine in terms of its higher antioxidant capacity.[158]

Chocolate has been shown to have potentially beneficial effects with respect to heart disease.[133] In an animal model of atherosclerosis, cocoa powder at a human dose equivalent of two dark chocolate bars per day significantly

Table 6.5 Health beneficial effects of soy. Adapted from ref. 152.

Study	Subjects	Product tested	Outcomes
Effect of soy protein isolate on bone metabolism.[227]	50 subjects; age = 40–62 years	Soyabean protein isolate, daily dose (2, 32, 60 mg aglycone equivalents), 9 months	No effect on bone mineral density.
Soy protein and isoflavones: their effects on blood lipids and bone density in post-menopausal women.[228]	66 subjects; age = 49–83 years	Soyabean protein isolate, daily dose (35, 56 mg aglycone equivalents), 6 months	With 56 mg, 2.2% increase for bone mineral density for lumbar spine.
Effect of soy protein containing isoflavones on cognitive function, bone mineral density and plasma lipids in postmenopausal women: a randomized controlled trial.[229]	175 subjects; age = 60–75 years	Soyabean protein isolate, daily dose (99 mg aglycone equivalents), 12 months	No effect on bone mineral density.
Soy isoflavones have a favourable effect on bone loss in Chinese postmenopausal women with lower bone mass: a double-blind, randomized, controlled trial.[230]	175 subjects; age = 48–62 years	Soyabean germ extract, daily dose (25, 50 mg aglycone equivalents), 12 months	50 mg: no effect on bone mineral density; 0.5% increase for bone mineral content for total hip and trochanter.
Effect of isoflavones extracted from red clover (Rimostil) on lipid and bone metabolism.[231]	46 subjects; age = 57 years	Red clover extract (*Trifolium pratense* L), daily dose (7, 35, 53 mg aglycone equivalents), 6 months	17 mg: no effect. 35 mg: 4.1% increase bone mineral density for proximal radius and ulna. 53 mg: 3% increase bone mineral density for proximal radius and ulna.
Effects of genistein and hormone-replacement therapy on bone loss in early post-menopausal women: a randomized double-blind placebo-controlled study.[232]	90 subjects; age = 47–57 years	Genistein, 54 mg aglycone equivalents, 12 months	3% increase bone mineral density for hip and spine.

inhibited atherosclerosis, lowered cholesterol, LDL and triglycerides, raised HDL, and protected the lower density lipoproteins from oxidation. Dark chocolate rich in cocoa leads to an acute improvement in arterial and endothelial function.[159] In overweight adults, the acute ingestion of both solid dark chocolate and liquid cocoa improved endothelial function and lowered blood pressure.[160] This provides further insights into the favourable effects of chocolate to the cardiovascular system, but the antioxidant and biological effects of chocolate may be explained not solely by the established absorption of catechin monomers but also by the absorption of microbial phenolic acid metabolites.[128]

6.5.2.6   *Wine*

Epidemiological studies suggest that consumption of wine, grape products and other foods containing polyphenols is associated with reduced risk for cardiovascular disease. The benefits of wine consumption appear to be greater than other alcoholic beverages.[161] Red wine is highly consumed in France (approximately four times greater than that in the UK). This high red wine consumption is thought to be the reason for the "French paradox", where despite saturated fat intakes and cholesterol levels similar to the UK and USA, France has been shown to have a coronary mortality rate close to that of China or Japan.[161,163]

Over the past couple of decades, many studies have shown promising health benefits associated with wine consumption. Moderate consumption of wine leads to an improved lipid metabolism, an increased antioxidant activity and an improved anticoagulant status. These effects are thought to be due to some flavonoids, in particular quercetin. Flavonoids are known to prevent endothelial dysfunction and to reduce blood pressure and oxidative stress.[162] Some clinical studies have shown that flavonoids improve endothelial function in patients with hypertension and ischemic heart disease. Evidence suggests that alcohol has a positive synergistic effect with wine polyphenols on some atherosclerotic risk factors.[163]

Moreover, data suggest that red wine is more cardioprotective than white wine, possibly due to the increased content of a flavanoid antioxidant found in red wine, *i.e.* resveratrol, a unique component of grapes. Many studies have provided evidence that resveratrol possesses antioxidant and anti-apoptotic effects apart from activation of longevity proteins (*e.g.* SIRT-1). The angiogenic, anti-hypercholesterolemic and anti-diabetic effects of resveratrol and the mechanisms involved in reduced ventricular remodelling and increased cardiac functions have recently been reported, with different strategic target molecules involved in resveratrol-mediated cardioprotection, leading to its beneficial effects during health and disease.[164]

6.6   Conclusions

Phytonutrients play an increasingly important role as essential constituents of dietary supplements or functional foods, and in various health issues related to nutrition. Based on many epidemiological and clinical studies, an increasing number of claims not only on the beneficial roles of some of these NPs (*e.g.* flavonoids or carotenoids) but also on their adverse effects are reported. The comprehension of the mode of action of each of these natural products is extremely complex and requires a thorough analysis of these constituents in their original matrices (fruit, vegetables, herbs or enriched extracts), as well as an assessment of their bioavailability by sensitive detection methods. For this, mass spectrometry plays an increasingly important role because of its versatility, sensitivity and selectivity, as well as its capability to provide key structural information for identification purposes. As discussed, MS is well adapted to the profiling of many of these phytonutrients and the quantitative analysis of some targeted biomarkers.

With the recent use of systems biology approaches for the study of the health benefits of functional foods and food ingredients, new and exciting research opportunities in nutrition have appeared. In this respect, emerging sciences such as nutrimetabolomics enable the holistic study of the human metabolism of food products. Such approaches rely strongly on state-of-the-art mass spectrometric methods for profiling complex biological matrices and can potentially highlight key biomarkers for health benefit assessment.

With the tremendous developments in instrumentation and data treatment over the last few years, it is a safe bet to assume that mass spectrometry will continue to accelerate the pace at which significant discoveries are made on the role played by phytonutrients in nutrition and health. However, this progress should also rely on a good comprehension of the biological role of different phytonutrients *in vivo* on all possible biological targets with which they might interact. This represents a worthy challenge for nutrition chemists to undertake toward a comprehensive understanding the meaning of what is "healthy food".

References

1. American Dietetic Association, *J. Am. Diet, Assoc.*, 2004, **104**, 814–826.
2. E. H. Jeffery, S. M. Kundrat and A. S. Keck, in: *Biochemical Physiological, Molecular Aspects of Human Nutrition,* ed. M. H. Stipanuk, Saunder, Elsevier, St Louis, 2006, pp. 13–45.
3. R. E. C. Wildman and M. Kelley, in: *Handbook of Nutraceuticals and Functional Foods,* ed. R. E. C. Wildman, CRC Press, Boca Raton, FL, 2006, pp. 1–21.
4. J. Larsson, J. Gottfries, S. Muresan and A. Backlund, *J. Nat. Prod.*, 2007, **70**, 789–794.
5. A. L. Harvey, *Curr. Opin. Chem. Biol.*, 2007, **11**, 480–484.
6. S. S. Percival and E. Turner, in: *Handbook of Nutraceuticals and Functional Foods,* ed. R. E. C. Wildman, CRC Press, Boca Raton, 2006, pp. 269–284.
7. E. M. Williamson, *Drug Saf.*, 2003, **26**, 1075–1092.
8. C. M. Hasler and J. B. Blumberg, *J. Nutr.*, 1999, **129**, 756S–757S.

9. A. King and G. Young, *J. Am. Diet Assoc.*, 1999, **99**, 213–218.
10. L. Bravo, *Nut. Rev.*, 1998, **56**, 317–333.
11. J. B. Harborne, *The Flavonoids: Advances in Research since 1980,* Chapman and Hall, New York, 1988.
12. A. Scalbert, I. T. Johnson and M. Saltmarsh, *Am. J. Clin. Nutr.*, 2005, **81**, 215S–217S.
13. C. Manach, A. Scalbert, C. Morand, C. Remesy and L. Jimenez, *Am. J. Clin. Nutr.*, 2004, **79**, 727–747.
14. A. Scalbert and G. Williamson, *J. Nutr.*, 2000, **130**, 2073S–2085.
15. M. Iriti and F. Faoro, *Nat. Prod. Commun.*, 2009, **4**, 611–634.
16. A. Crozier, J. Burns, A. A. Aziz, A. J. Stewart, H. S. Rabiasz, G. I. Jenkins, C. A. Edwards and M. E. Lean, *Biol. Res.*, 2000, **33**, 79–88.
17. J. M. Cooney, D. J. Jensen and T. K. McGhie, *J. Sci. Food Agric.*, 2004, **84**, 237–245.
18. R. E. C. Wildman, *Handbook of Nutraceuticals and Functional Foods,* CRC Press, Boca Raton, FL, 2000.
19. A. El-Agamey, G. M. Lowe, D. J. McGarvey, A. Mortensen, D. M. Phillip, T. G. Truscott and A. J. Young, *Arch. Biochem. Biophys.*, 2004, **430**, 37–48.
20. A. V. Rao and S. Agarwal, *Nut. Res.*, 1999, **19**, 305–323.
21. J. W. Fahey, A. T. Zalcmann and P. Talalay, *Phytochemistry*, 2001, **56**, 5–51.
22. I. Minchinton, J. Sang, D. Burke and R. J. W. Truscott, *J. Chromatogr.*, 1982, **247**, 141–148.
23. T. Mohn, B. Cutting, B. Ernst and M. Hamburger, *J. Chromatogr. A*, 2007, **1166**, 142–151.
24. E. Block, *Angew. Chem. Int. Ed. Engl.*, 1992, **31**, 1135–1178.
25. V. Lanzotti, *J. Chromatogr. A*, 2006, **1112**, 3–22.
26. M. Kussmann, S. Rezzi and H. Daniel, *Curr. Opin. Biotechnol.*, 2008, **19**, 83–99.
27. J.-L. Wolfender, *Planta Med.*, 2009, **75**, 719–734.
28. J. Xing, C. F. Xie and H. X. Lou, *J. Pharm. Biomed. Anal.*, 2007, **44**, 368–378.
29. W. A. Korfmacher, *Drug Discov. Today*, 2005, **10**, 1357–1367.
30. K. W. Cheng, F. Cheng and M. Wang, in: *Bioactive Natural Products: Detection, Isolation, and Structural Determination,* ed. S. M. Colegate and R. J. Molyneux, CRC Press, London, 2008, pp. 245–266.
31. M. Careri, A. Mangia and M. Musci, *J. Chromatogr. A*, 1998, **794**, 263–297.
32. X. G. He, *J. Chromatogr. A*, 2000, **880**, 203–232.
33. J. L. Wolfender, E. F. Queiroz and K. Hostettmann, *Exp. Opin. Drug Discov.*, 2006, **1**, 237–260.
34. K. V. Sashidhara and J. N. Rosaiah, *Nat. Prod. Commun.*, 2007, **2**, 193–202.
35. M. Yang, J. Sun, Z. Lu, G. Chen, S. Guan, X. Liu, B. Jiang, M. Ye and D.-A. Guo, *J. Chromatogr. A*, 2008, **1216**, 2045–2062.

36. M. Kussmann, M. Affolter, K. Nagy, B. Holst and L. B. Fay, *Mass Spectrom. Rev.*, 2007, **26**, 727–750.
37. J. L. Wolfender, S. Rodriguez and K. Hostettmann, *J. Chromatogr. A*, 1998, **794**, 299–316.
38. G. X. Xie, R. Plumb, M. M. Su, Z. H. Xu, A. H. Zhao, M. F. Qiu, X. B. Long, Z. Liu and W. Jia, *J. Sep. Sci.*, 2008, **31**, 1015–1026.
39. E. Grata, J. Boccard, D. Guillarme, G. Glauser, P. A. Carrupt, E. Farmer, J. L. Wolfender and S. Rudaz, *J. Chromatogr. B*, 2008, **871**, 261–270.
40. H. Suzuki, R. Sasaki, Y. Ogata, Y. Nakamura, N. Sakurai, M. Kitajima, H. Takayama, S. Kanaya, K. Aoki, D. Shibata and K. Saito, *Phytochemistry*, 2008, **69**, 99–111.
41. G. Hopfgartner, in: *Encyclopedia of Mass Spectrometry Hyphenated Methods,* **Vol. 8**, ed. W. M. A. Niessen, Elsevier Science, Oxford, 2006, pp. 465–474.
42. F. Sun, Q. He, P. Y. Shi, P. G. Xiao and Y. Y. Cheng, *Rapid Commun. Mass Spectrom.*, 2007, **21**, 3743–3750.
43. N. Fuzzati, N. Gabetta, I. Strepponi and F. Villa, *J. Chromatogr. A*, 2001, **926**, 187–198.
44. F. Cuyckens and M. Claeys, *J. Mass Spectrom.*, 2004, **39**, 1–15.
45. P. Waridel, J. L. Wolfender, K. Ndjoko, K. R. Hobby, H. J. Major and K. Hostettmann, *J. Chromatogr. A*, 2001, **926**, 29–41.
46. F. Kuhn, M. Oehme, F. Romero, E. Abou-Mansour and R. Tabacchi, *Rapid Commun. Mass Spectrom.*, 2003, **17**, 1941–1949.
47. J. L. Wolfender, E. F. Queiroz and K. Hostettmann, *Magn. Reson. Chem.*, 2005, **43**, 697–709.
48. J. L. Wolfender, E. F. Queiroz and K. Hostettmann, in: *Bioactive Natural Products; Detection, Isolation and Structural Determination,* ed. S. M. Colegate and R. J. Molyneux, CRC Press, London, 2008, pp. 143–190.
49. C. Seger and S. Sturm, *J. Proteome Res.*, 2007, **6**, 480–497.
50. M. Bedair and L. W. Sumner, *TrAC, Trends Anal. Chem.*, 2008, **27**, 238–250.
51. J. W. Jaroszewski, *Planta Med.*, 2005, **71**, 691–700.
52. Z. Yang, *Pharm. Biomed. Anal.*, 2006, **40**, 516–527.
53. A. L. Cogne, E. F. Queiroz, A. Marston, J. L. Wolfender, S. Mavi and K. Hostettmann, *Phytochem. Anal.*, 2005, **16**, 429–439.
54. V. Exarchou, M. Krucker, T. A. van Beek, J. Vervoort, I. P. Gerothanassis and K. Albert, *Magn. Reson. Chem.*, 2005, **43**, 681–687.
55. J. W. Jaroszewski, *Planta Med.*, 2005, **71**, 795–802.
56. D. L. Olson, J. A. Norcross, M. O'Neil-Johnson, P. F. Molitor, D. J. Detlefsen, A. G. Wilson and T. L. Peck, *Anal. Chem.*, 2004, **76**, 2966–2974.
57. J. F. Hu, E. Garo, H. D. Yoo, P. A. Cremin, L. Zeng, M. G. Goering, M. O'Neil-Johnson and G. R. Eldridge, *Phytochem. Anal.*, 2005, **16**, 127–133.
58. G. Glauser, D. Guillarme, E. Grata, J. Boccard, A. Thiocone, P. A. Carrupt, J. L. Veuthey, S. Rudaz and J. L. Wolfender, *J. Chromatogr. A*, 2008, **1180**, 90–98.

59. B. Cerda, J. C. Espin, S. Parra, P. Martinez and F. A. Tomas-Barberan, *Eur. J. Nutr.*, 2004, **43**, 205–220.
60. J. C. Espin, M. T. Garcia-Conesa and F. A. Tomas-Barberan, *Phytochemistry*, 2007, **68**, 2986–3008.
61. B. Holst and G. Williamson, *Curr. Opin. Biotechnol.*, 2008, **19**, 73–82.
62. H. Guo, A. H. Liu, M. Ye, M. Yang and D. A. Guo, *Rapid Commun. Mass Spectrom.*, 2007, **21**, 715–729.
63. M. Liu, X. Q. Li, C. Weber, C. Y. Lee, J. Brown and R. H. Liu, *J. Agric. Food Chem.*, 2002, **50**, 2926–2930.
64. P. Jacob, M. Wilson, L. Yu, J. Mendelson and R. T. Jones, *Anal. Chem.*, 2002, **74**, 5290–5296.
65. N. R. Kitteringham, R. E. Jenkins, C. S. Lane, V. L. Elliott and B. K. Park, *J. Chromatogr. B*, 2009, **877**, 1229–1239.
66. K. L. Simpson, A. D. Whetton and C. Dive, *J. Chromatogr. B*, 2009, **877**, 1240–1249.
67. X. W. Zhang, Y. L. Yap, D. Wei, G. Chen and F. Chen, *Biotechnol. Adv.*, 2008, **26**, 169–176.
68. K. S. Solanky, N. J. Bailey, B. M. Beckwith-Hall, S. Bingham, A. Davis, E. Holmes, J. K. Nicholson and A. Cassidy, *J. Nutr. Biochem.*, 2005, **16**, 236–244.
69. J. L. Wolfender, G. Glauser, J. Boccard and S. Rudaz, *Nat. Prod. Commun.*, 2009, **4**, 1417–1430.
70. G. Theodoridis, H. G. Gika and I. D. Wilson, *Trends Anal. Chem.*, 2008, **27**, 251–260.
71. D. T. T. Nguyen, D. Guillarme, S. Rudaz and J. L. Veuthey, *J. Sep. Sci.*, 2006, **29**, 1836–1848.
72. I. D. Wilson, J. K. Nicholson, J. Castro-Perez, J. H. Granger, K. A. Johnson, B. W. Smith and R. S. Plumb, *J. Proteome Res.*, 2005, **4**, 591–598.
73. G. Xie, M. Ye, Y. Wang, Y. Ni, M. Su, H. Huang, M. Qiu, A. Zhao, X. Zheng, T. Chen and W. Jia, *J. Agric. Food Chem.*, 2009, **57**, 3046–3054.
74. X. Zhang, D. Wei, Y. Yap, L. Li, S. Guo and F. Chen, *Mass Spectrom. Rev.*, 2007, **26**, 403–431.
75. Y. Ni, M. Su, Y. Qiu, M. Chen, Y. Liu, A. Zhao and W. Jia, *FEBS Lett.*, 2007, **581**, 707–711.
76. M. Chen, Y. Ni, H. Duan, Y. Qiu, C. Guo, Y. Jiao, H. Shi, M. Su and W. Jia, *Chem. Res. Toxicol.*, 2008, **21**, 288–294.
77. H. Chen, Y. Sun, A. Wortmann, H. Gu and R. Zenobi, *Anal. Chem.*, 2007, **79**, 1447–1455.
78. Y. L. Wang, H. R. Tang, J. K. Nicholson, P. J. Hylands, J. Sampson, I. Whitcombe, C. G. Stewart, S. Caiger, I. Oru and E. Holmes, *Planta Med.*, 2004, **70**, 250–255.
79. Y. L. Wang, H. R. Tang, J. K. Nicholson, P. J. Hylands, J. Sampson and E. Holmes, *J. Agric. Food Chem.*, 2005, **53**, 191–196.

80. K. R. Markham, *Techniques of Flavonoid Identification,* Academic Press, London, 1982.
81. P. A. Hedin and V. A. Phillips, *J. Agric. Food Chem.*, 1992, **40**, 607–611.
82. B. Domon and C. Costello, *Glycoconj. J.*, 1988, **5**, 397–409.
83. Y. L. Ma, Q. M. Li, H. Van den Heuvel and M. Claeys, *Rapid Commun. Mass Spectrom.*, 1997, **11**, 1357–1364.
84. F. Cuyckens, R. Rozenberg, E. De Hoffmann and M. Claeys, *J. Mass Spectrom.*, 2001, **36**, 1203–1210.
85. M. D. L. Mata-Bilbao, C. Andres-Lacueva, E. Roura, O. Jauregui, C. Torre and R. M. Lamuela-Raventos, *J. Agric. Food Chem.*, 2007, **55**, 8857–8863.
86. S. M. Sang, M. J. Lee, I. Yang, B. Buckley and C. S. Yang, *Rapid Commun. Mass Spectrom.*, 2008, **22**, 1567–1578.
87. D. Shen, Q. Wu, M. Wang, Y. Yang, E. J. Lavoie and J. E. Simon, *J. Agric. Food Chem.*, 2006, **54**, 3219–3224.
88. A. Chandra, J. Rana and Y. Q. Li, *J. Agric. Food Chem.*, 2001, **49**, 3515–3521.
89. G. J. Soleas, D. M. Goldberg, E. P. Diamandis, A. Karumanchiri, J. Yan and E. Ng, *Am. J. Enol. Vitic.*, 1995, **46**, 346–352.
90. R. Flamini, *Mass Spectrom. Rev.*, 2003, **22**, 218–250.
91. L. Stella, M. D. Rosso, A. Panighel, A. D. Vedova, R. Flamini and P. Traldi, *Rapid Commun. Mass Spectrom.*, 2008, **22**, 3867–3872.
92. J. Chen, S. He, H. Mao, C. Sun and Y. Pan, *Rapid Commun. Mass Spectrom.*, 2009, **23**, 737–744.
93. J. K. Prasain, C. C. Wang and S. Barnes, *Free Radic. Biol. Med.*, 2004, **37**, 1324–1350.
94. W. Mullen, A. Boitier, A. J. Stewart and A. Crozier, *J. Chromatogr. A*, 2004, **1058**, 163–168.
95. C. Felgines, S. Talavera, M.-P. Gonthier, O. Texier, A. Scalbert, J.-L. Lamaison and C. Remesy, *J. Nutr.*, 2003, **133**, 1296–1301.
96. C. Felgines, S. Talavera, O. Texier, A. Gil-Izquierdo, J.-L. Lamaison and C. Remesy, *J. Agric. Food Chem.*, 2005, **53**, 7721–7727.
97. H. Keski-Hynnila, M. Kurkela, E. Elovaara, L. Antonio, J. Magdalou, L. Luukkanen, J. Taskinen and R. Kostiainen, *Anal. Chem.*, 2002, **74**, 3449–3457.
98. H. Ito, M. P. Gonthiera, C. Manach, C. Morand, L. Mennen, C. Remesy and A. Scalbert, *BioFactors*, 2004, **22**, 241–243.
99. R. Kostiainen, T. Kotiaho, T. Kuuranne and S. Auriola, *J. Mass Spectrom.*, 2003, **38**, 357–372.
100. A. Burkon and V. Somoza, *Mol. Nutr. Food Res.*, 2008, **52**, 549–557.
101. D. J. Boocock, K. R. Patel, G. E. S. Faust, D. P. Normolle, T. H. Marczylo, J. A. Crowell, D. E. Brenner, T. D. Booth, A. Gescher and W. P. Steward, *J. Chromatogr. B*, 2007, **848**, 182–187.
102. J. L. Wolfender, G. Marti and F. Queiroz Emerson, *Curr. Org. Chem.*, 2010, in press.
103. J. Oliver and A. Palou, *J. Chromatogr. A*, 2000, **881**, 543–555.

104. U. Schweiggert, D. R. Kammerer, R. Carle and A. Schieber, *Rapid Commun. Mass Spectrom.*, 2005, **19**, 2617–2628.

105. T. Rezanka, J. Olsovska, M. Sobotka and K. Sigler, *Curr. Anal. Chem.*, 2009, **5**, 1–25.

106. P. A. Clarke, K. A. Barnes, J. R. Startin, F. I. Ibe and M. J. Shepherd, *Rapid Commun. Mass Spectrom.*, 1996, **10**, 1781–1785.

107. T. Lacker, S. Strohschein and K. Albert, *J. Chromatogr. A*, 1999, **854**, 37–44.

108. C. Rentel, S. Strohschein, K. Albert and E. Bayer, *Anal. Chem.*, 1998, **70**, 4394–4400.

109. T. Guaratini, R. Vessecchi, E. Pinto, P. Colepicolo and N. P. Lopes, *J. Mass Spectrom.*, 2005, **40**, 963–968.

110. R. B. Vanbreemen, *Anal. Chem.*, 1995, **67**, 2004–2009.

111. J. K. Prasain, R. Moore, J. S. Hurst, S. Barnes and F. van Kuijk, *J. Mass Spectrom.*, 2005, **40**, 916–923.

112. R. B. Vanbreemen, H. H. Schmitz and S. J. Schwartz, *J. Agric. Food Chem.*, 1995, **43**, 384–389.

113. J. C. Young, E. S. M. Abdel-Aal, I. Rabalski and B. A. Blackwell, *J. Agric. Food Chem.*, 2007, **55**, 4965–4972.

114. T. Rezanka, L. Nedbalova, K. Sigler and V. Cepak, *Phytochemistry*, 2008, **69**, 479–490.

115. K. Putzbach, M. Krucker, K. Albert, M. A. Grusak, G. W. Tang and G. G. Dolnikowski, *J. Agric. Food Chem.*, 2005, **53**, 671–677.

116. A. Lienau, T. Glaser, G. W. Tang, G. G. Dolnikowski, M. A. Grusak and K. Albert, *J. Nutr. Biochem.*, 2003, **14**, 663–670.

117. K. Albert, *On-line LC-NMR and Related Techniques,* John Wiley & Sons, Chichester, 2002.

118. M. Dachtler, T. Glaser, K. Kohler and K. Albert, *Anal. Chem.*, 2001, **73**, 667–674.

119. K. Putzbach, M. Krucker, M. D. Grynbaum, P. Hentschel, A. G. Webb and K. Albert, *J. Pharm. Biomed. Anal.*, 2005, **38**, 910–917.

120. B. Holst and G. Williamson, in: *Phytochemicals in Health and Disease*, ed. Y. Bao and R. Fenwick, Marcel Dekker, New York, 2004, **vol. 2**, pp. 25–56.

121. J. P. E. Spencer, M. M. A. Mohsen, A. M. Minihane and J. C. Mathers, *Br. J. Nutr.*, 2008, **99**, 12–22.

122. J. P. E. Spencer, M. M. A. El Mohsen and C. Rice-Evans, *Arch. Biochem. Biophys.*, 2004, **423**, 148–161.

123. J. P. E. Spencer, G. G. C. Kuhnle, R. J. Williams and C. Rice-Evans, *Biochem. J.*, 2003, **372**, 173–181.

124. A. Fardet, R. Llorach, A. Orsoni, J.-F. Martin, E. Pujos-Guillot, C. Lapierre and A. Scalbert, *J. Nutr.*, 2008, **138**, 1282–1287.

125. C. Manach, G. Williamson, C. Morand, A. Scalbert and C. Remesy, *Am. J. Clin. Nutr.*, 2005, **81**, 230S–242S.

126. A. Cassidy, *J. AOAC Int.*, 2006, **89**, 1182–1188.

127. S. Scholz and G. Williamson, *Int. J. Vitam. Nutr. Res.*, 2007, **77**, 224–235.

128. L. Y. Rios, M. P. Gonthier, C. Remesy, I. Mila, C. Lapierre, S. A. Lazarus, G. Williamson and A. Scalbert, *Am. J. Clin. Nutr.*, 2003, **77**, 912–918.
129. R. R. Holt, S. A. Lazarus, M. C. Sullards, Q. Y. Zhu, D. D. Schramm, J. F. Hammerstone, C. G. Fraga, H. H. Schmitz and C. L. Keen, *Am. J. Clin. Nutr.*, 2002, **76**, 798–804.
130. T. K. McGhie and M. C. Walton, *Mol. Nutr. Food Res.*, 2007, **51**, 702–713.
131. M. J. T. J. Arts, G. R. M. M. Haenen, H. P. Voss and A. Bast, *Food Chem. Toxicol.*, 2001, **39**, 787–791.
132. J. H. Weisburger, *Cancer Lett.*, 1997, **114**, 315–317.
133. J. A. Vinson, J. Proch, P. Bose, S. Muchler, P. Taffera, D. Shuta, N. Samman and G. A. Agbor, *J. Agric. Food Chem.*, 2006, **54**, 8071–8076.
134. K. H. Van Het Hof, G. A. A. Kivits, J. A. Weststrate and L. B. M. Tijburg, *Eur. J. Clin. Nutr.*, 1998, **52**, 356–359.
135. R. Leenen, A. J. C. Roodenburg, L. B. M. Tijburg and S. A. Wiseman, *Eur. J. Clin. Nutr.*, 2000, **54**, 87–92.
136. M. Lorenz, N. Jochmann, A. von Krosigk, P. Martus, G. Baumann, K. Stangl and V. Stangl, *Eur. Heart J.*, 2007, **28**, 219–223.
137. M. Serafini and A. Crozier, *Nature (London)*, 2003, **426**, 788.
138. D. Grassi, C. Lippi, S. Necozione, G. Desideri and C. Ferri, *Am. J. Clin. Nutr.*, 2005, **81**, 611–614.
139. E. Roura, C. Andres-Lacueva, R. Estruch, M. L. Mata-Bilbao, M. Izquierdo-Pulido, A. L. Waterhouse and R. M. Lamuela-Raventos, *Ann. Nutr. Metab.*, 2007, **51**, 493–498.
140. G. Williamson and C. Manach, *Am. J. Clin. Nutr.*, 2005, **81**, Suppl–255S.
141. C. Cabrera, R. Artacho and R. Gimenez, *J. Am. Coll. Nutr.*, 2006, **25**, 79–99.
142. G. W. Dryden, M. Song and C. McClain, *Curr. Opin. Gastroenterol.*, 2006, **22**, 165–170.
143. N. Khan and H. Mukhtar, *Life Sci.*, 2007, **81**, 519–533.
144. S. Sang, J. D. Lambert and C. S. Yang, *J. Sci. Food Agric.*, 2006, **86**, 2256–2265.
145. S. Wolfram, Y. Wang and F. Thielecke, *Mol. Nutr. Food Res.*, 2006, **50**, 176–187.
146. C. Cabrera, R. Artacho and R. Gimenez, *J. Am. Coll. Nutr.*, 2006, **25**, 79–99.
147. E. J. Gardner, C. H. S. Ruxton and A. R. Leeds, *Eur. J. Clin. Nutr.*, 2006, **61**, 3–18.
148. M. N. Clifford, *J. Sci. Food Agric.*, 2000, **80**, 1033–1043.
149. C. J. Dupas, A. C. Marsset-Baglieri, C. S. Ordonaud, F. M. G. Ducept and M. N. Maillard, *J. Food Sci.*, 2006, **71**, S253–S258.
150. M. Richelle, I. Tavazzi and E. Offord, *J. Agric. Food Chem.*, 2001, **49**, 3438–3442.
151. C. W. Xiao, *J. Nutr.*, 2008, **138**, 1244S–1249.

152. A. Cassidy, P. Albertazzi, I. L. Nielsen, W. Hall, G. Williamson, I. Tetens, S. Atkins, H. Cross, Y. Manios, A. Wolk, C. Steiner and F. Branca, *Proc. Nutr. Soc.*, 2006, **65**, 76–92.

153. K. A. Cooper, J. L. Donovan, A. L. Waterhouse and G. Williamson, *Br. J. Nutr.*, 2008, **99**, 1–11.

154. A. V. Rao and L. G. Rao, *Pharmacol. Res.*, 2007, **55**, 207–216.

155. A. Basu and V. Imrhan, *Eur. J. Clin. Nutr.*, 2006, **61**, 295–303.

156. S. Voutilainen, T. Nurmi, J. Mursu and T. H. Rissanen, *Am. J. Clin. Nutr.*, 2006, **83**, 1265–1271.

157. H. S. Ruxton Carrie, J. Gardner Elaine and D. Walker, *Int. J. Food Sci. Nutr.*, 2006, **57**, 249–272.

158. K. W. Lee, Y. J. Kim, H. J. Lee and C. Y. Lee, *J. Agric. Food Chem.*, 2003, **51**, 7292–7295.

159. K. Aznaouridis, C. Vlachopoulos, N. Alexopoulos, N. Ioakeimidis, D. Tsekoura, C. Pitsavos and C. Stefanadis, *Am. J. Hypertens.*, 2004, **17**, 137A–137A.

160. Z. Faridi, V. Y. Njike, S. Dutta, A. Ali and D. L. Katz, *Am. J. Clin. Nutr.*, 2008, **88**, 58–63.

161. M. M. Dohadwala and J. A. Vita, *J. Nutr.*, 2009, **139**, 1788S–1793.

162. F. Perez-Vizcaino, J. Duarte and R. Andriantsitohaina, *Free Radic. Res.*, 2006, **40**, 1054–1065.

163. K. A. Cooper, M. Chopra and D. I. Thurnham, *Nutr. Res. Rev.*, 2004, **17**, 111–129.

164. S. V. Penumathsa and N. Maulik, *Can. J. Physiol. Pharmacol.*, 2009, **87**, 275–286.

165. Q. Tian, M. M. Giusti, G. D. Stoner and S. J. Schwartz, *J. Agric. Food Chem.*, 2006, **54**, 1467–1472.

166. H. Matsumoto, Y. Nakamura, H. Iida, K. Ito and H. Ohguro, *Exp. Eye Res.*, 2006, **83**, 348–356.

167. S. Passamonti, U. Vrhovsek, A. Vanzo and F. Mattivi, *FEBS Lett.*, 2003, **544**, 210–213.

168. S. Talavera, C. Felgines, O. Texier, C. Besson, A. Gil-Izquierdo, J.-L. Lamaison and C. Remesy, *J. Agric. Food Chem.*, 2005, **53**, 3902–3908.

169. T. Ichiyanagi, Y. Shida, M. M. Rahman, Y. Hatano and T. Konishi, *J. Agric. Food Chem.*, 2005, **53**, 7312–7319.

170. T. Ichiyanagi, M. M. Rahman, Y. Kashiwada, Y. Ikeshiro, Y. Shida, Y. Hatano, H. Matsumoto, M. Hirayama, T. Tsuda and T. Konishi, *Free Radic. Biol. Med.*, 2004, **36**, 930–937.

171. X. Li, H. Xiao, X. Liang, D. Shi and J. Liu, *J. Pharm. Biomed. Anal.*, 2004, **34**, 159–166.

172. A. A. Franke, L. J. Custer, L. R. Wilkens, L. Le Marchand, A. M. Y. Nomura, M. T. Goodman and L. N. Kolonel, *J. Chromatogr. B*, 2002, **777**, 45–59.

173. Y.-L. Ma, I. Vedernikova, H. Van den Heuvel and M. Claeys, *J. Am. Soc. Mass Spectrom.*, 2000, **11**, 136–144.

174. B. A. Graf, W. Mullen, S. T. Caldwell, R. C. Hartley, G. G. Duthie, M. E. J. Lean, A. Crozier and C. A. Edwards, *Drug Metab. Dispos.*, 2005, **33**, 1036–1043.

175. M. Inoue, K. Tajima, M. Mizutani, H. Iwata, T. Iwase, S. Miura, K. Hirose, N. Hamajima and S. Tominaga, *Cancer Lett.*, 2001, **167**, 175–182.

176. A. H. Wu, M. C. Yu, C.-C. Tseng, J. Hankin and M. C. Pike, *Int. J. Cancer*, 2003, **106**, 574–579.

177. Y. Suzuki, Y. Tsubono, N. Nakaya, Y. Koizumi and I. Tsuji, *Br. J. Cancer*, 2004, **90**, 1361–1363.

178. M. Zhang, W. Binns Colin and H. Lee Andy, *Cancer Epidemiol. Biomarkers Prev.*, 2002, **11**, 713–718.

179. L. Jian, L. P. Xie, A. H. Lee and C. W. Binns, *Int. J. Cancer*, 2004, **108**, 130–135.

180. X. Huang, K. Tajima, N. Hamajima, M. Inoue, T. Takezaki, T. Kuroishi, K. Hirose, S. Tominaga, J. Xiang and S. Tokudome, *J. Epidemiol.*, 1999, **9**, 40–45.

181. K. Shibata, M. Moriyama, T. Fukushima, A. Kaetsu, M. Miyazaki and H. Une, *J. Epidemiol.*, 2000, **10**, 310–316.

182. Y. Kuwahara, S. Kono, H. Eguchi, H. Hamada, K. Shinchi and K. Imanishi, *Scand. J. Gastroenterol.*, 2000, **35**, 476–481.

183. K. Nakachi, S. Matsuyama, S. Miyake, M. Suganuma and K. Imai, *BioFactors*, 2000, **13**, 49–54.

184. J. Nagano, S. Kono, D. Preston and K. Mabuchi, *Cancer Causes Control*, 2001, **12**, 501–508.

185. V. W. Setiawan, Z.-F. Zhang, G.-P. Yu, Q.-Y. Lu, Y.-L. Li, M.-L. Lu, M.-R. Wang, C. H. Guo, S.-Z. Yu, R. C. Kurtz and C.-C. Hsieh, *Int. J. Cancer*, 2001, **92**, 600–604.

186. Y. Tsubono, Y. Nishino, S. Komatsu, C.-C. Hsieh, S. Kanemura, I. Tsuji, H. Nakatsuka, A. Fukao, H. Satoh and S. Hisamichi, *N. Engl. J. Med.*, 2001, **344**, 632–636.

187. Y. Hoshiyama, T. Kawaguchi, Y. Miura, T. Mizoue, N. Tokui, H. Yatsuya, K. Sakata, T. Kondo, S. Kikuchi, H. Toyoshima, N. Hayakawa, A. Tamakoshi, Y. Ohno and T. Yoshimura, *Br. J. Cancer*, 0000, **87**, 309–313.

188. Y. Fujino, A. Tamakoshi, Y. Ohno, T. Mizoue, N. Tokui and T. Yoshimura, *Prev. Med. (Baltimore)*, 2002, **35**, 121–127.

189. J. Nagano, S. Kono, D. L. Preston, H. Moriwaki, G. B. Sharp, K. Koyama and K. Mabuchi, *Int. J. Cancer*, 2000, **86**, 132–138.

190. L. Zhong, M. S. Goldberg, Y. T. Gao, J. A. Hanley, M. E. Parent and F. Jin, *Epidemiology*, 2001, **12**, 695–700.

191. H. D. Sesso, J. M. Gaziano, S. Liu and J. E. Buring, *Am. J. Clin. Nutr.*, 2003, **77**, 1400–1408.

192. J. M. Geleijnse, L. J. Launer, D. A. van der Kuip, A. Hofman and J. C. Witteman, *Am. J. Clin. Nutr.*, 2002, **75**, 880–886.

193. T. Hirvonen, P. Pietinen, M. Virtanen, M. L. Ovaskainen, S. Hakkinen, D. Albanes and J. Virtamo, *Epidemiology*, 2001, **12**, 62–67.

194. I. C. Arts, P. C. Hollman, E. J. Feskens, H. B. Bueno de Mesquita and D. Kromhout, *Am. J. Clin. Nutr.*, 2001, **74**, 227–232.

195. I. C. Arts, D. R. Jacobs Jr, L. J. Harnack, M. Gross and A. R. Folsom, *Epidemiology*, 2001, **12**, 668–675.

196. M. Woodward and H. Tunstall-Pedoe, *J. Epidemiol. Community Health*, 1999, **53**, 481–487.

197. M. G. L. Hertog, P. M. Sweetnam, A. M. Fehily, P. C. Elwood and D. Kromhout, *Am. J. Clin. Nutr.*, 1997, **65**, 1489–1494.

198. I. A. Hakim, M. A. Alsaif, M. Alduwaihy, K. Al-Rubeaan, A. R. Al-Nuaim and O. S. Al-Attas, *Prev. Med. (Baltimore)*, 2003, **36**, 64–70.

199. A. Tavani, M. Bertuzzi, E. Negri, L. Sorbara and C. La Vecchia, *Eur. J. Epidemiol.*, 2001, **17**, 1131–1137.

200. H. D. Sesso, J. M. Gaziano, J. E. Buring and C. H. Hennekens, *Am. J. Epidemiol.*, 1999, **149**, 162–167.

201. A. Gramenzi, A. Gentile, M. Fasoli, E. Negri, F. Parazzini and C. La Vecchia, *BMJ*, 1990, **300**, 771–773.

202. Z. Chen, M. B. Pettinger, C. Ritenbaugh, A. Z. LaCroix, J. Robbins, B. J. Caan, D. H. Barad and I. A. Hakim, *Am. J. Epidemiol.*, 2003, **158**, 772–781.

203. V. M. Hegarty, H. M. May and K.-T. Khaw, *Am. J. Clin. Nutr.*, 2000, **71**, 1003–1007.

204. J. F. Wang, D. D. Schramm, R. R. Holt, J. L. Ensunsa, C. G. Fraga, H. H. Schmitz and C. L. Keen, *J. Nutr.*, 2000, **130**, 2115S–2119S.

205. D. Rein, T. G. Paglieroni, T. Wun, D. A. Pearson, H. H. Schmitz, R. Gosselin and C. L. Keen, *Am. J. Clin. Nutr.*, 2000, **72**, 30–35.

206. D. Rein, S. Lotito, R. R. Holt, C. L. Keen, H. H. Schmitz and C. G. Fraga, *J. Nutr.*, 2000, **130**, 2109S–2114S.

207. N. Osakabe, S. Baba, A. Yasuda, T. Iwamoto, M. Kamiyama, T. Takizawa, H. Itakura and K. Kondo, *Free Radic. Res.*, 2001, **34**, 93–99.

208. Y. Wan, J. A. Vinson, T. D. Etherton, J. Proch, S. A. Lazarus and P. M. Kris-Etherton, *Am. J. Clin. Nutr.*, 2001, **74**, 596–602.

209. D. A. Pearson, T. G. Paglieroni, D. Rein, T. Wun, D. D. Schramm, J. F. Wang, R. R. Holt, R. Gosselin, H. H. Schmitz and C. L. Keen, *Thromb. Res.*, 2002, **106**, 191–197.

210. S. Mathur, S. Devaraj, S. M. Grundy and I. Jialal, *J. Nutr.*, 2002, **132**, 3663–3667.

211. R. R. Holt, D. D. Schramm, C. L. Keen, S. A. Lazarus and H. H. Schmitz, *JAMA*, 2002, **287**, 2212–2213.

212. D. Taubert, R. Berkels, R. Roesen and W. Klaus, *JAMA*, 2003, **290**, 1029–1030.

213. N. D. Fisher, M. Hughes, M. Gerhard-Herman and N. K. Hollenberg, *J. Hypertens.*, 2003, **21**, 2281–2286.

214. C. Heiss, A. Dejam, P. Kleinbongard, T. Schewe, H. Sies and M. Kelm, *JAMA*, 2003, **290**, 1030–1031.

215. M. Serafini, R. Bugianesi, G. Maiani, S. Valtuena, S. De Santis and A. Crozier, *Nature (London)*, 2003, **424**, 1013.

216. J. Mursu, S. Voutilainen, T. Nurmi, T. H. Rissanen, J. K. Virtanen, J. Kaikkonen, K. Nyyssonen and J. T. Salonen, *Free Radic. Biol. Med.*, 2004, **37**, 1351–1359.

217. M. B. Engler, M. M. Engler, C. Y. Chen, M. J. Malloy, A. Browne, E. Y. Chiu, H.-K. Kwak, P. Milbury, S. M. Paul, J. Blumberg and M. L. Mietus-Snyder, *J. Am. Coll. Nutr.*, 2004, **23**, 197–204.

218. I. Wiswedel, D. Hirsch, S. Kropf, M. Gruening, E. Pfister, T. Schewe and H. Sies, *Free Radic. Biol. Med.*, 2004, **37**, 411–421.

219. C. G. Fraga, L. Actis-Goretta, J. I. Ottaviani, F. Carrasquedo, S. B. Lotito, S. Lazarus, H. H. Schmitz and C. L. Keen, *Clin. Dev. Immunol.*, 2005, **12**, 11–17.

220. C. Vlachopoulos, K. Aznaouridis, N. Alexopoulos, E. Economou, I. Andreadou and C. Stefanadis, *Am. J. Hypertens.*, 2005, **18**, 785–791.

221. D. Grassi, S. Necozione, C. Lippi, G. Croce, L. Valeri, P. Pasqualetti, G. Desideri, J. B. Blumberg and C. Ferri, *Hypertension*, 2005, **46**, 398–405.

222. H. Schroeter, C. Heiss, J. Balzer, P. Kleinbongard, C. L. Keen, N. K. Hollenberg, H. Sies, C. Kwik-Uribe, H. H. Schmitz and M. Kelm, *Proc. Natl. Acad. Sci. U. S. A.*, 2006, **103**, 1024–1029.

223. F. Hermann, L. E. Spieker, F. Ruschitzka, I. Sudano, M. Hermann, C. Binggeli, T. F. Lüscher, W. Riesen, G. Noll and R. Corti, *Heart*, 2006, **92**, 119–120.

224. N. D. Fisher and N. K. Hollenberg, *J. Hypertens.*, 2006, **24**, 1575–1580.

225. S. B. Kurlandsky and K. S. Stote, *Nutr. Res.*, 2006, **26**, 509–516.

226. J. Flammer Andreas, F. Hermann, I. Sudano, L. Spieker, M. Hermann, A. Cooper Karen, M. Serafini, F. Luscher Thomas, F. Ruschitzka, G. Noll and R. Corti, *Circulation*, 2007, **116**, 2376–2382.

227. J. C. Gallagher, R. Satpathy, K. Rafferty and V. Haynatzka, *Menopause*, 2004, **11**, 290–298.

228. S. M. Potter, J. A. Baum, H. Teng, R. J. Stillman, N. F. Shay and J. W. Erdman Jr, *Am. J. Clin. Nutr.*, 1998, **68**, 1375S–1379S.

229. S. Kreijkamp-Kaspers, L. Kok, D. E. Grobbee, E. H. F. de Haan, A. Aleman, J. W. Lampe and Y. T. van der Schouw, *JAMA*, 2004, **292**, 65–74.

230. Y.-M. Chen, S. C. Ho, S. S. H. Lam, S. S. S. Ho and J. L. F. Woo, *J. Clin. Endocrinol. Metab.*, 2003, **88**, 4740–4747.

231. P. B. Clifton-Bligh, R. J. Baber, G. R. Fulcher, M. L. Nery and T. Moreton, *Menopause*, 2001, **8**, 259–265.

232. N. Morabito, A. Crisafulli, C. Vergara, A. Gaudio, A. Lasco, N. Frisina, R. D'Anna, F. Corrado, M. A. Pizzoleo, M. Cincotta, D. Altavilla, R. Ientile and F. Squadrito, *J. Bone Miner. Res.*, 2002, **17**, 1904–1912.

Section 3
Addressing the Health Aspects of Nutrition

CHAPTER 7

Addressing the Health Beneficial Aspects of Nutrition—The Example of the Obesity Epidemic

MARIA LANKINEN[a,b] AND MATEJ OREŠIČ[a]

[a] VTT Technical Research Centre of Finland, Espoo, Finland; [b] Department of Clinical Nutrition, School of Public Health and Clinical Nutrition, University of Kuopio, Kuopio, Finland

We have known for more than 2000 years that health and nutrition are intimately linked: Hippocrates recommended *ca.* 400 BC to "leave your drugs in the chemist's pot if you can heal the patient with food". While modern nutrition focuses rather on prevention than cure (the latter being the classical pharmaceutical domain of care), Hippocrates' concept still holds very much true and it particularly applies to the interplay of nutrition with energy balance, cardiovascular health, immunity and protein turnover, which are the four themes addressed in this "nutrition and health" section of the book. Introducing the impact of mass spectrometry on nutritionally actionable health aspects, we focus first on the global health issue of obesity.

No issue so vividly illustrates the power of diet to alter health as the prevalence of obesity, defined as body mass index (BMI) ≥ 30 (weight in kilograms divided by height in square metres), which is increasing throughout the world and presents a growing global health problem. A total of 1.6 billion

RSC Food Analysis Monographs No. 9
Mass Spectrometry and Nutrition Research
Edited by Laurent B. Fay and Martin Kussmann

people worldwide are overweight, with at least 400 million being obese and with the numbers for 2015 projected to exceed 700 million. Worldwide, 20 million children under five years of age are estimated to be overweight (WHO fact sheet no. 311, www.who.int/mediacentre/factsheets/fs311/en/). Given the current prevalence of childhood obesity and its geographic distribution throughout the world, the term "pandemic" is appropriate for describing the current status of childhood obesity. For these reasons, the present introduction to addressing health aspects of nutrition by mass spectrometry emphasizes research on energy imbalance in the body.

The aetiology of obesity is complex. Multiple studies have demonstrated that genes play an important role in the development of obesity. However, changes in genes and biology alone can not explain this rapid rise in obesity. It is thus evident that lifestyle factors such as lack of physical activity and an unbalanced diet also have an important role in the obesity epidemic.[1,2] Excess weight gain is a consequence of long-term imbalance between energy intake and expenditure. However, the basic equation "energy intake = energy expenditure" is more complex when considering the multitude of genetic, biological, psychological, socio-cultural and environmental factors which affect both sides of the equation.[1,2] Many dietary determinants such as energy density, content of macronutrients, fibre and water, glycemic index, palatability and variety within food groups are supposed to influence energy intake.[3,4] The central nervous system with its complex hormonal and neuropeptide signalling pathways also has an important role in food intake control.[5]

In obese individuals, the mass of adipose tissue (a major endocrine organ) is elevated. Therefore it is not surprising that obesity is the main risk factor for a number of metabolic abnormalities such as dyslipidemia, insulin resistance, (chronic) inflammation and hypertension. Obesity co-morbidities also include coronary heart disease, stroke, certain types of cancer, type 2 diabetes, gall bladder disease, osteoarthritis and gout, and pulmonary diseases, including sleep apnea. Due to these numerous co-morbidities, obesity reduces quality of life and leads to enormous expenditure on healthcare expenditure.[1,6]

One of the most common complications of obesity is metabolic syndrome, which is characterized by abnormalities of insulin, glucose and lipid metabolism, hypertension and abdominal obesity; it may progress over years to diseases such as type 2 diabetes mellitus and atherosclerotic vascular disease.[7–10] The pathogenesis of metabolic syndrome is heterogeneous, but many lifestyle and genetic factors clearly interact to produce it.[8] Although multiple influences contribute to the metabolic syndrome, the syndrome appears to be relatively uncommon in the absence of some excess of body fat.

Even with the aetiology of the metabolic syndrome remaining partly elusive, current evidence suggests that relatively modest lifestyle changes—including weight loss, dietary changes towards current recommendations and increased physical activity—can greatly reduce the risk of diabetes in persons with impaired glucose tolerance.[11,12] At an epidemiological level, as obesity increases so does the prevalence of the metabolic syndrome.[13] However, at the individual level, the association may not be so clear cut.[14] For example, lipodystrophic individuals, who have inherent failure in adipose tissue development and function, are extremely lean but develop severe insulin resistance.[14] Conversely, some morbid obese individuals do not develop metabolic syndrome. Several hypotheses exist on how the excess adiposity antagonizes insulin action in peripheral tissues.[14,15]

When an individual reaches a point when their adipocytes exceed their storage capacity (*i.e.* usually when they are already obese), fat begins to accumulate in tissues not suited for lipid storage.[14] This ectopic deposition of excess lipids in non-adipose organs such as liver, muscle and pancreatic beta-cells is believed to cause adverse affects on cellular metabolism, most notably insulin action, *via* a process called lipotoxicity. One clinically manifest complication caused by the deposition of excess lipids in liver is non-alcoholic fatty liver disease (NAFLD).[16] This is a chronic liver condition characterized by insulin resistance and hepatic fat accumulation. Approximately 70% of those with type 2 diabetes mellitus also have NAFLD.[16] Saturated free fatty acids such as derived from *de novo* fatty acid synthesis or *via* adipose tissue lipid overspill may especially harm the liver because they are poorly incorporated into triglycerides and, thereby, cause endoplasmic reticulum stress and apoptosis. Notably, lipidomic analyses revealed depletion of possibly beneficial long-chain polyunsaturated fatty acids in the human non-alcoholic fatty liver, accompanied by an increase of saturated fatty acids.[17] Similarly in muscle, the excess lipid accumulation may impair insulin action.[18] When fatty acids are taken up by muscle, they either undergo β-oxidation in the mitochondria or they can be stored as triglycerides. Due to rate-limiting steps in the β-oxidation and the limited storage capacity of muscle, the increased fatty acid flux seen in obesity causes the accumulation of fatty acid intermediates in the muscle. Fatty acid intermediates include diacylglycerols, ceramides and acylcarnitines, which ultimately trigger activation of a number of serine kinases including JNK1, IKKβ and PKC-θ, that in turn represent a convergence point of inflammation and insulin resistance. Ceramides can also attenuate insulin signalling by inhibiting the Akt/PKB pathway and, indeed, lipid-induced insulin resistance in rats is prevented by pharmacological inhibition of ceramide production.[19]

Another suggested mechanism of how increased adiposity may affect insulin action is *via* induction of a chronic inflammatory state.[15,20] Adipokines, which are released from either adipocytes or from macrophages infiltrating adipose tissue, induce a low-grade chronic inflammatory state which may play a central role in insulin resistance.[21] Deregulation of inflammatory pathways in adipose tissue has also been found among genetically identical twins, who were discordant for obesity.[22] Obese twins, who also had more subcutaneous, intra-abdominal and liver fat, activated inflammatory pathways in the subcutaneous abdominal adipose tissue more strongly and were less insulin sensitive than non-obese co-twins.

In addition, some of the specific lipid molecules mentioned above (*e.g.* fatty acid intermediates) are associated with inflammation and may play a role in obesity-related insulin resistance.[20] Ceramides, diacylglycerols and pro-inflammatory lysophosphatidylcholines have all been identified as potential mediators of lipid-induced insulin resistance.[23–28] This may provide a link between the lipotoxicity and inflammation hypothesis.

Humans are not alone in their body. In fact, human cells are a eukaryotic minority in a prokaryotically dominated intestinal ecosystem. The adult intestine contains trillions of microbes, which are able to perform metabolic functions including the ability to harvest otherwise inaccessible nutrients from our diet. Therefore, the potential role of gut microbiota in obesity onset and development is of strong scientific and medical/nutritional interest. Mass spectrometry-based metabolomics studies show that the gut microbiota has large effects on mammalian blood metabolites.[29] Comparisons of mice raised in a germ-free environment with those that acquired microbiota since birth have revealed that colonized, conventionally raised mice gain more total body fat than their germ-free counterparts when fed the same diet, even though they consume less chow per day.[30,31] Even the colonization of adult germ-free mice produces a dramatic increase in body weight, despite an associated decrease in food consumption.[32] Furthermore, colonization with "obese donor microbiota" results in a significantly greater increase in total body weight than colonization with "lean donor microbiota".[32]

Mass spectrometry plays a key role understanding the physiological importance of different regimes on the metabolic status of obese people during intervention studies. In one such study, Lankinen and colleagues applied lipidomic and metabolomic platforms to study carbohydrate modification on serum metabolic profiles in subjects with metabolic syndrome and combined such information with the adipose tissue gene expression data using multivariate statistical modelling.[33] The study found that dietary carbohydrate modification alters concentrations of pro-inflammatory lysophospholipids, and may, thus, contribute to pro-inflammatory processes which in turn promote adverse changes in insulin and glucose metabolism. In another study, the effect of lean fish and fatty fish on serum lipidomic profiles was investigated in subjects with coronary heart disease in an eight-week intervention study.[34] Multiple bioactive lipid species, including potentially lipotoxic ceramides, lysophosphatidylcholines and diacylglycerols, decreased significantly in the fatty fish group. The observed changes may thus be related to the protective effects of fatty fish on the progression of atherosclerotic vascular diseases or insulin resistance.

Two more recent studies have shown that diet-induced weight loss leads to marked changes in serum lipidomic profiles.[35,36] Interestingly, while the total serum triglycerides did not change significantly when comparing the concentration changes within the control and weight reduction groups, the mass spectrometry based lipidomic analysis revealed that changes were highly significant particularly among the triglyceride molecular species containing saturated fatty acids.[35] These triglycerides were also found to mainly associate with the very low density lipoprotein (VLDL) fraction, as well as correlate with waist circumference in another study where lipidomic analysis was performed on each lipoprotein fraction separately.[37]

These examples taken from modern obesity research show that mass spectrometry has developed into a tool to assess the health and safety aspects of food. Mass spectrometry has therefore developed into a key platform for both holistic and targeted specific analysis of nutritional health effects. Mass spectrometry is today used to demonstrate the impact of nutrition on almost all health aspects of consumers (*e.g.* growth and development, energy balance, immunity and physical performance). The next chapters are dedicated to cardiovascular health, immune modulation and protein turnover.

First, Bruce German outlines how mass spectrometry has emerged as the central analytical platform in life science research at the same time as cardiovascular disease has become one of the greatest life science challenges. He shows that the analytical power of mass spectrometry has provided much of the insights into the mechanisms underlying cardiovascular disease, the diagnostics being developed to identify those suffering from the disease (even predicting those at risk of suffering) and the therapeutics being discovered to resolve it. On the other hand, cardiovascular disease is a complex human health issue that has triggered continuous innovation of instrumentation platforms. German addresses those aspects of mass spectrometry that have been instrumental in explaining how diet relates to risk for, and protection from, heart disease. He furthermore recognizes that science is far from solving cardiovascular disease, that this disorder remains the number one death cause, and that further developments in mass spectrometry will be necessary to finally solving it.

The next chapter by Martin Kussmann addresses the manifold interactions between nutritional status and immune health, and how mass spectrometry and proteomics can elucidate this interplay, for example through revelation of immune biomarkers. Immunological conditions like inflammation and allergy are discussed and the particular susceptibility of the immune system to environmental, such as nutritional, influences early in life is reflected. Special sections are dedicated to human gut ecology, its immunological impact as well as to the immune bioactives and health effects of human milk.

Finally, Michael Affolter addresses the dynamics of protein turnover and degradation, and how mass spectrometry has contributed and will further contribute to protein turnover. In particular the advancements of classical isotope ratio mass spectrometry and MS-based proteomic platforms now enable protein turnover to be addressed in a holistic fashion but at "individual protein resolution", meaning that such analyses are no longer limited to rather coarse tissue-level considerations. Affolter discusses then the impact of nutrition on protein and proteome turnover. He elaborates on muscle protein metabolism and how this is affected by either exercise or sarcopenia and nutritional intervention.

References

1. World Health Organization, *World Health Organ. Tech. Rep. Ser.*, 2000, **894**, 1.

2. M. S. Bray, *Obesity (Silver Spring)*, 2008, **16**, S72.
3. M. A. McCrory, P. J. Fuss, E. Saltzman and S. B. Roberts, *J. Nutr.*, 2000, **130**, 276S.
4. M. A. McCrory, P. J. Fuss, J. E. McCallum, M. Yao, A. G. Vinken, N. P. Hays and S. B. Roberts, *Am. J. Clin. Nutr.*, 1999, **69**, 440.
5. M. W. Schwartz, S. C. Woods, D. Porte Jr, R. J. Seeley and D. G. Baskin, *Nature*, 2000, **404**, 661.
6. D. L. Thompson, *AAOHN. J.*, 2007, **55**, 265.
7. S. M. Grundy, *J. Clin. Endocrinol. Metab.*, 2007, **92**, 399.
8. D. E. Laaksonen, L. Niskanen, H. M. Lakka, T. A. Lakka and M. Uusitupa, *Ann. Med.*, 2004, **36**, 332.
9. G. M. Reaven, *Diabetes*, 1988, **37**, 1595.
10. P. Mathieu, P. Pibarot and J. P. Després, *Vasc. Health. Risk. Manag.*, 2006, **2**, 285.
11. J. Tuomilehto, J. Lindström, J. G. Eriksson, T. T. Valle, H. Hämäläinen, P. Ilanne-Parikka, S. Keinänen-Kiukaanniemi, M. Laakso, A. Louheranta, M. Rastas, V. Salminen and M. Uusitupa, *N. Engl. J. Med.*, 2001, **344**, 1343.
12. W. C. Knowler, E. Barrett-Connor, S. E. Fowler, R. F. Hamman, J. M. Lachin, E. A. Walker and D. M. Nathan, *N. Engl. J. Med.*, 2002, **346**, 393.
13. Y. W. Park, S. Zhu, L. Palaniappan, S. Heshka, M. R. Carnethon and S. B. Heymsfield, *Arch. Intern. Med.*, 2003, **163**, 427.
14. S. Virtue and A. Vidal-Puig, *PloS Biol.*, 2008, **6**, e237.
15. S. A. Summers, *Proc. Lipid. Res.*, 2006, **45**, 42.
16. K. Cusi, *Curr. Opin. Endocrinol. Diabetes Obes.*, 2009, **16**, 141.
17. A. Kotronen, T. Seppänen-Laakso, J. Westerbacka, T. Kiviluoto, J. Arola, A.-L. Ruskeepää, M. Orešič and H. Yki-Järvinen, *Diabetes*, 2009, **58**, 203.
18. S. Schenk, M. Saberi and J. M. Olefsky, *J. Clin. Invest.*, 2008, **118**, 2992.
19. W. L. Holland, J. T. Brozinick, L. P. Wang, E. D. Hawkins, K. M. Sargent, Y. Liu, K. Narra, K. L. Hoehn, T. A. Knotts, A. Siesky, D. H. Nelson, S. K. Karathanasis, G. K. Fontenot, M. J. Birnbaum and S. A. Summers, *Cell Metab.*, 2007, **5**, 167.
20. S. E. Shoelson, L. Herrero and A. Naaz, *Gastroenterology*, 2007, **132**, 2169.
21. B. Antuna-Puente, B. Feve, S. Fellahi and J.-P. Bastard, *Diabetes Metab.*, 2008, **34**, 2.
22. K. H. Pietiläinen, J. Naukkarinen, A. Rissanen, J. Saharinen, P. Ellonen, H. Keränen, A. Suomalainen, A. Götz, T. Suortti, H. Yki-Järvinen, M. Orešič, J. Kaprio and L. Peltonen, *PLoS ONE*, 2008, **5**, e51.
23. M. P. Wymann and R. Scneiter, *Nat. Rev. Mol. Cell Biol.*, 2008, **9**, 162.
24. S. M. Turpin, G. I. Lancaster, I. Darby, M. A. Febbraio and M. J. Watt, *Am. J. Physiol. Endocrinol. Metab.*, 2006, **291**, E1341.

25. W. L. Holland, T. A. Knotts, J. A. Chavez, L.-P. Wang, K. L. Hoehn and S. A. Summers, *Nutr. Rev.*, 2007, **65**, S39.
26. S. Timmers, P. Schrauwen and J. de Vogel, *Physiol. Behav.*, 2008, **94**, 242.
27. E. D. Motley, S. M. Kabir, C. D. Gardner, K. Eguchi, G. D Frank, T. Kuroki, M. Ohba, T. Yamakawa and S. Eguchi, *Hypertension*, 2002, **39**, 508.
28. V. D. F. de Mello, M. Lankinen, U. Schwab, M. Kolehmainen, S. Lehto, T. Seppänen-Laakso, M. Orešič, L. Pulkkinen, M. Uusitupa and A. T. Erkkilä, *Diabetologia*, 2009, epub ahead of print.
29. W. R. Wikoff, A. T. Anfora, J. Liu, P. G. Schultz, S. A. Lesley, E. C. Peters and G. Siuzdak, *PNAS.*, 2008, **106**, 3698.
30. F. Bäckhed, R. E. Ley, J. L. Sonnenburg, D. A. Peterson and J. I. Gordon, *Science*, 2005, **307**, 1915.
31. F. Bäckhed, J. K. Manchester, C. F. Semenkovich and J. I. Gordon, *PNAS*, 2007, **104**, 979.
32. P. J. Turnbaugh, R. E. Ley, M. A. Mahowald, V. Magrini, E. R. Mardis, J. I. Gordon, *Nature*, **444**, 1027.
33. M. Lankinen, U. Schwab, P. V. Gopalacharyulu, T. Seppänen-Laakso, L. Yetukuri, M. Sysi-Aho, P. Kallio, T. Suortti, D. E. Laaksonen, H. Gylling, K. Poutanen, M. Kolehmainen and M. Orešič, *Nutr. Metab. Cardiovasc. Dis.*, 2009. doi:10.1016/j.numecd.2009.04.009.
34. M. Lankinen, U. Schwab, A. Erkkilä, T. Seppänen-Laakso, M.-L. Hannila, H. Mussalo, S. Lehto, M. Uusitupa, H. Gylling and M. Orešiè, *PLoS ONE*, 2009, **4**, e5258.
35. U. Schwab, T. Seppänen-Laakso, L. Yetukuri, J. Ågren, M. Kolehmainen, D. E. Laaksonen, A.-L. Ruskeepää, H. Gylling, M. Uusitupa and M. Orešič, *PLoS ONE*, 2008, **3**, e2630.
36. J. T. Smilowitz, M. M. Wiest, S. M. Watkins, D. Teegarden, M. B. Zemel, J. B. German and M. D. Van Loan, *J. Nutr.*, 2008, **139**, 222.
37. A. Kotronen, V. R. Velagapudi, L. Yetukuri, J. Westerbacka, R. Bergholm, K. Ekroos, J. Makkonen, M.-R. Taskinen, M. Orešiè and H. Yki-Järvinen, *Diabetologia*, 2009, **52**, 684.

CHAPTER 8

Mass Spectrometry, Diet and Cardiovascular Disease: What will They Mean for Food?

J. BRUCE GERMAN[a,b,*]

[a] Department of Food Science & Technology, University of California, Davis, CA 95616 USA; [b] Nestlé Research Centre, Vers-chez-les-Blanc, PO Box 44, CH-1000 Lausanne 26 Switzerland

8.1 Introduction

The importance of the heart to health and its penchant for failure as coronary heart disease is not new to modern human health. In fact, this little organ, primarily muscle, is a most remarkable feat of physiology. Working ceaselessly at a beat per second for our entire lives, with a greater than three-fold dynamic range in output and metabolic demand, is it any wonder it is, for most humans, the first system to fail. In truth, the massive demands we make on this tough little organ are at the core of the problem. Cardiovascular disease is not the result of a single electrochemical failure, the immediate consequence of a single pathogenic organism or the acute collapse of a sustaining structure. Instead cardiovascular disease in the majority of cases is caused by the gradual deterioration of the integrity and functions of the arterial vessels that supply the

*Funding received from: University of California Discovery Program (05GEB01NHB); National Institute of Environmental Health Sciences (P42ES004699); California Dairy Research Foundation; and CHARGE study (P01 ES11269).

RSC Food Analysis Monographs No. 9
Mass Spectrometry and Nutrition Research
Edited by Laurent B. Fay and Martin Kussmann
© The Royal Society of Chemistry 2010
Published by the Royal Society of Chemistry, www.rsc.org

muscle with fuel and transport away its metabolic products. It is thus not the cardiac heart muscle *per se* that fails, but the network of smooth muscle supported and endothelial lined plumbing that fails to support the heart.

Building a detailed understanding of how the vascular network works and ultimately fails has been the focused goal of a substantial fraction of life science over the past 100 years. It was first recognized over a century ago, in examining individuals who died because their hearts inexplicably stopped pumping, that the linings of the major vessels were diseased. The first descriptions were of vessel walls thickened and engorged with plaque—a whitish waxy substance. Diligent chemical analyses ultimately recognized this substance to be largely cholesterol. Ever since clinicians pulled apart these vividly diseased tissues, the study of cholesterol has occupied the careers of scientists from every aspect of the life sciences. As has been true for most of life science research, innovations in technology and particularly analytical chemistry have enabled breakthroughs in understanding the biological processes that lead to artery failure. Perhaps the greatest enabling technologies in the past 50 years—certainly in understanding the chemical nature of the cells, tissues and debris—have been in the area of mass spectrometry. This chapter discusses the central role of mass spectrometry to solving the problems of heart disease from the focus of the risk factors that have emerged to guide individuals and populations away from the disease.

The establishment first of the concept of measureable risk factors and then the use of risk factors as targets of intervention has changed the scientific and public health strategies as the means to prevent heart disease. Various aspects of lifestyle have been considered to alter risk of disease including diet. As different dietary habits were associated with risk, the concept that specific foods could both increase and decrease risk emerged both as a scientific target and a commercial opportunity. More recently, this science has led to the emergence of functional foods and the principles of regulated marketing claims in support of higher value of functional foods. The future research and ultimate prevention of this entire class of disease will certainly require even more of a role for mass spectrometry.

It speaks to the past success of mass spectrometry in heart disease research that, although not available yet, most clinicians await impatiently the arrival of a hand-held fast-throughput mass spectrometer to provide comprehensive risk analysis of many aspects of health to routine practice. More provocatively, it can be proposed now that, because heart disease is so multi-factorial, solving this disease will have the net effect of providing the means for science to execute on a wide range of health benefits.

Heart disease research was the first to champion the concept of measuring metabolites in humans (cholesterol) as a means to identifying not existing disease but of the varying individual risk of future disease. Success in acting on such varying risk of cardiovascular disease (CVD) is now leading research and development to many other risks and in so doing achieving a view to personalizing health in general. Providing the tools to measure health empowers individuals and the industries that serve them. Consumers will demand to know more about their personal health and life science industries will be seeking to provide them personal solutions. In such a future, measuring our personal "chemistry" will be of ever increasing importance to our health and performance.

8.2 Cardiovascular Disease: A Multi-faceted Deterioration of Arterial Structure and Function

Cardiovascular disease is the quintessential chronic, degenerative disease (Figure 8.1). It requires decades to develop. It damages a range of tissues and functions. It is largely irreversible. It involves multiple interacting causal mechanisms. It displays many risk factors and predisposing traits. It responds to various targets of intervention. In sum, cardiovascular disease is a syndrome of overlapping destructive processes that have the common property of degrading the integrity, metabolism and function of the arteries that supply the heart muscle. In most people, if such deterioration of artery structure and function is not stopped, it ultimately leads to the heart's failure. The major cellular events and tissue stages associated with cardiovascular disease have been described in detail elsewhere.[1] Interestingly, few articles examining and summarizing scientific research into the causes and solutions to heart disease have taken the perspective of the enabling technologies that propelled the

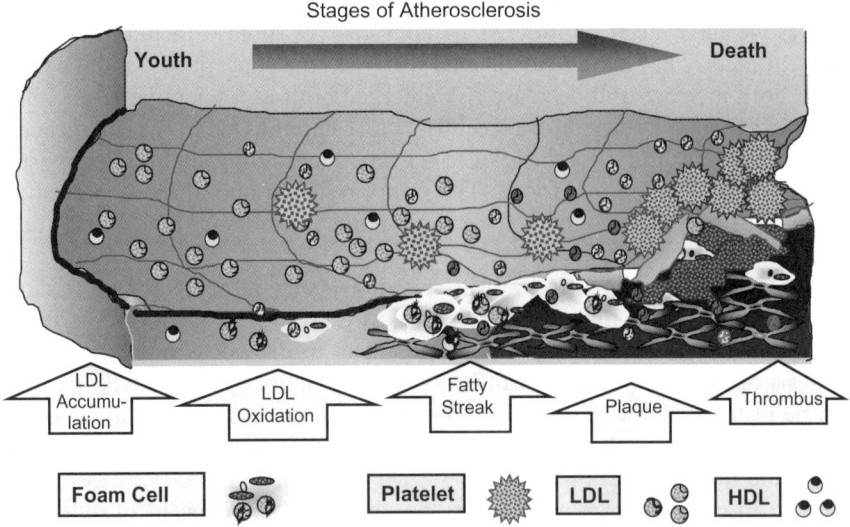

Figure 8.1 Stages of atherosclerosis. The deleterious processes begin with elevated lipoproteins, continue with damage to the endothelial layer and progress to accumulation of modified LDL in subendothelial macrophages. Then comes smooth muscle proliferation in the subendothelium before a mature plaque partially occludes the artery, awaiting a catastrophic rupture of the plaque and a blockage—a heart attack.

research. This was a critical omission. The discoveries of the causes and solutions to all diseases both propel and in turn follow the invention and application of the scientific tools that enabled them. As the tools of microbiology have been central to the discovery and eradication of pathogenic diseases, the tools of chemistry have been central to the study of non-pathogenic diseases. Mass spectrometry has been the most ubiquitous and powerful tool in the chemistry kit for almost a century.

8.2.1 Cholesterol: the Cadeau Empoisonné

Cholesterol is the most studied biomolecule of life science, yet remains poorly understood. Eukaryotes appear to have acquired the ability to synthesize sterols from acetate early in evolution through gene transfer from the Archaebacteria. Unfortunately for human health, eukaryotes did not acquire at the same time the genes to break cholesterol back down to acetate. This failure is at the heart of the problem. The various processes that lead to artery disease were recognized in the 19th century to be driven by this single molecule, cholesterol. Arteries dissected from cadavers who died from heart disease showed such an accumulation of thick waxy build-up that early chemical analyses easily solved the problem of what was there—cholesterol. This realization led to a concerted, worldwide research effort through the first half of the 20th century. The accumulation of this single molecule within artery walls led researchers to pursue a detailed understanding of its biosynthesis. In parallel, analytical chemists sought ways to measure its concentration within the blood of people. During the second half of the century, these twin research areas were brought to practice in measuring cholesterol levels in populations and in developing targets for small molecule inhibitors of cholesterol synthesis. Both were successful.

8.2.2 Cholesterol as a Plasma Diagnostic

During the 1950s, several very large population studies were conducted to establish whether there was genuinely a link between cholesterol levels in an individual human's plasma and their risk of dying of heart disease.[2] The results were convincing (Figure 8.2).

The Multiple Risk Factor Intervention Trial (MRFIT) study demonstrated convincingly that those individuals whose blood cholesterol levels were higher than normal ($200 \, \text{mg} \, \text{dl}^{-1}$) were at increasingly elevated risk.[3] In response to this compelling predictor of disease, the National Institutes of Health in the United States launched a nationwide, multi-million dollar strategy to alert the population, individually. The Cholesterol Education Program (www.nhlbi.nih.gov/about/ncep/index.htm), which started in 1986, was both bold and ambitious. The entire American adult population was encouraged to have their blood cholesterol measured. The Surgeon General made recommendations for diet and lifestyle changes to those whose cholesterol levels put them at risk. There was also a

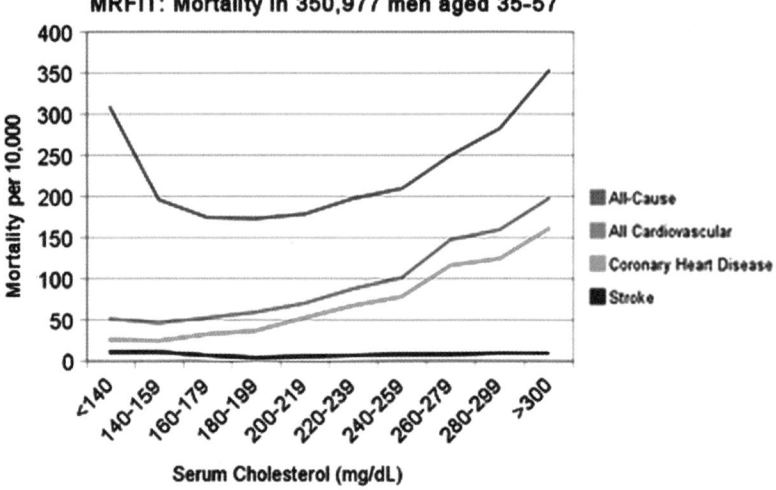

Figure 8.2   Relationship between measured serum cholesterol and coronary artery disease in the very large MRFIT trial conducted in the United States.

major push from funding agencies to develop therapeutic solutions, not solely for heart disease, but for cholesterol as a risk of heart disease. And, thus there was a massive investment in research and developments of potential solutions for individuals with elevated cholesterol. From that investment emerged an unprecedented understanding of lipid metabolism and various targets involved in the regulation of total cholesterol levels in blood.

8.2.3   Cholesterol Biosynthesis

Revealing the biosynthetic pathways of cholesterol was one of the crowning achievements of biochemistry. Enzyme after complex enzyme was identified, isolated and characterized in this quest. Once in place, the pathways were interrogated for their flux-limiting steps. HMG-CoA reductase (3-hydroxy-3-methyl-glutaryl-CoA reductase) proved to be the enzymatic step that was driving the net formation of cholesterol. With this information in hand, the race was on to screen for small molecules whose chemistry was to inhibit this enzyme and whose consumption could be harnessed to decrease cholesterol levels in human blood. That search culminated in a rather odd place, fungi! Researchers found a group of compounds in complex yeasts that acted to inhibit the HMG-CoA Reductase enzyme.[4] It appears that the yeast evolved a secondary pathway using these compounds as defences against other eukaryotes. The compounds were rushed into development as a new class of pharmaceutical: risk reduction drugs. Safety was assured painstakingly, as this is after all a class of compound that acts to inhibit the synthesis of one of the most ubiquitous compounds in human cellular biology. It was necessary to establish that all the various compounds downstream of HMG-CoA from isoprenes to steroid hormones were not adversely affected by inhibiting the reductase early in their synthesis.[4] Mass spectrometry was indispensible in this safety evaluation. All of the myriad metabolites of the isoprene pathway needed to be characterized in the presence and absence of the statin compounds. Since the approval of lovastatin by the US Food and Drug Administration (FDA), a variety of structural analogs has been developed, tested and launched as successful therapeutic drugs.

The identification of all the enzymes and intermediates in cholesterol biosynthesis in the first half of the 20th century was one of the great feats of biochemistry research. Yet, a detailed understanding of the quantitative flux through these pathways *in vivo* could not be accomplished by simple measures of pathway intermediates. This understanding required measures of rates. First generation studies were able to take advantage of the ability of mass spectrometry to discriminate masses and the availability of the stable, heavy isotope of carbon ^{13}C as the means to trace cholesterol. These studies used ^{13}C in cholesterol both as an intact molecule for absorption,[5] then as a precursor for cholesterol synthesis, then combining multiple isotopic enrichments to measure both synthesis and absorption simultaneously.[6] These mass spectrometry enabled studies were not only instrumental in documenting that humans could be hypercholesterolemic due to both hyperabsorption and hypersynthesis, but that dietary components could lower total plasma cholesterol by either absorption blocking or synthesis inhibition.

The next progressive analytical breakthrough in measuring rates of cholesterol synthesis and metabolism came in the form of an inspiring combination of chemistry, mass spectrometry and mathematics—mass isotopomer analysis.[7] Chemistry was necessary to synthesize small molecule precursors of metabolic pathways that were enriched with known specific activities of a stable isotope atom in a particular site in the molecule. Once in place as substrates, these stable isotopically enriched precursors could be administered to whole animals including humans as a simple tracer against a background of any experimental design. As a function of time, biofluids are sampled and the entire range of metabolic products of the original tracer substrate are measured for their quantitative mass abundances by highly accurate mass spectrometry.[8] Fragmentation of molecular ions provides isotopic ratios within the various metabolic products, yielding the relative abundances of different mass isotopomers in the multiple metabolic products of the original tracer substrate. Innovative mathematical analyses are then able to compare the statistical distributions predicted from the binomial or multinomial expansion to the pattern of the various isotopomer frequencies observed in the metabolic products. These computational tools are then able to calculate the quantitative enrichment of the biosynthetic products by the isotopically labelled precursor substrates.[9] The mathematical solution of the isotope content in the final products and substrate molecules that were incorporated biosynthetically into each product yields true quantitative determinations of the flux through the entire pathway.[8]

Not surprisingly, considering its health importance, cholesterol and fat biosynthesis were immediate targets of this new, mass spectrometry based analytical strategy.[10,11] This basic toolset has become the definitive platform for studying the relationships between drugs, diet and genetics in regulating cholesterol biosynthesis.[12–14] Interestingly, in spite of detailed understanding of the basic biosynthetic pathways of cholesterol biosynthesis, the absolute amounts of cholesterol in cells and humans are regulated primarily not by the molecules that make it, but the molecules that secrete it and by the molecules that must transport it.

8.2.4   Cholesterol Metabolism to Bile Salts

Cholesterol as a biological molecule in animals presents an interesting problem. We cannot break it down. As a result, cholesterol must be metabolized to a more polar and soluble product and excreted as a waste product. Because this is the major metabolic product of cholesterol, bile acid biosynthesis and excretion represents a major contributor to whole body cholesterol status. The oxidation of cholesterol follows the bile acid pathway. The rate determining step of bile acid synthesis was determined to be the 7-α-hydroxlase enzyme. The quantitative description of this complex pathway of cholesterol metabolism was made possible by isotope dilution mass spectrometry.[15] Once identified as the rate determining step in bile acid synthesis, scientists began to search for exogenous agents that act upon it. Pharmacologists have screened chemicals as therapeutic agents for biliary diseases and to lower cholesterol. From this search, various drugs have been developed that target the 7-α-hydroxylase enzyme and are in various stages of clinical testing.[16]

Mass spectrometry as an analytical toolset has been so successful in its most obvious application (*i.e.* molecular identification) that information developed through this technique has frequently been years ahead of other aspects of biological research. The case of bile acids is one such example. The complexity of bile acid biosynthesis involving a wide array of very similar compounds (Figure 8.3) was largely worked out decades ago with straightforward mass spectrometry analyses.[17] Their analysis in plasma was routine decades ago.[18] Yet, annotating these various compounds for their functions has lagged behind and only in recent years have the biological properties of the various bile acids begun to be understood.[19] This disparity is a vivid indication of the importance of analytics to biological research. Mass spectrometry drove the science of molecular characterization forward rapidly but there were no parallel technologies to take these structures to function. Only now with complex biological tools such as gene addition/elimination have the transporters, receptors and transcription factors that transduce bile acids into cellular signals been recognized and their widespread functions in biological regulation revealed.[20]

However, translating this research into diagnostics of normal human metabolism needed an approach different from hepatic biopsy. The ability of plasma measures of bile acid concentrations to accurately reflect bile acid synthesis

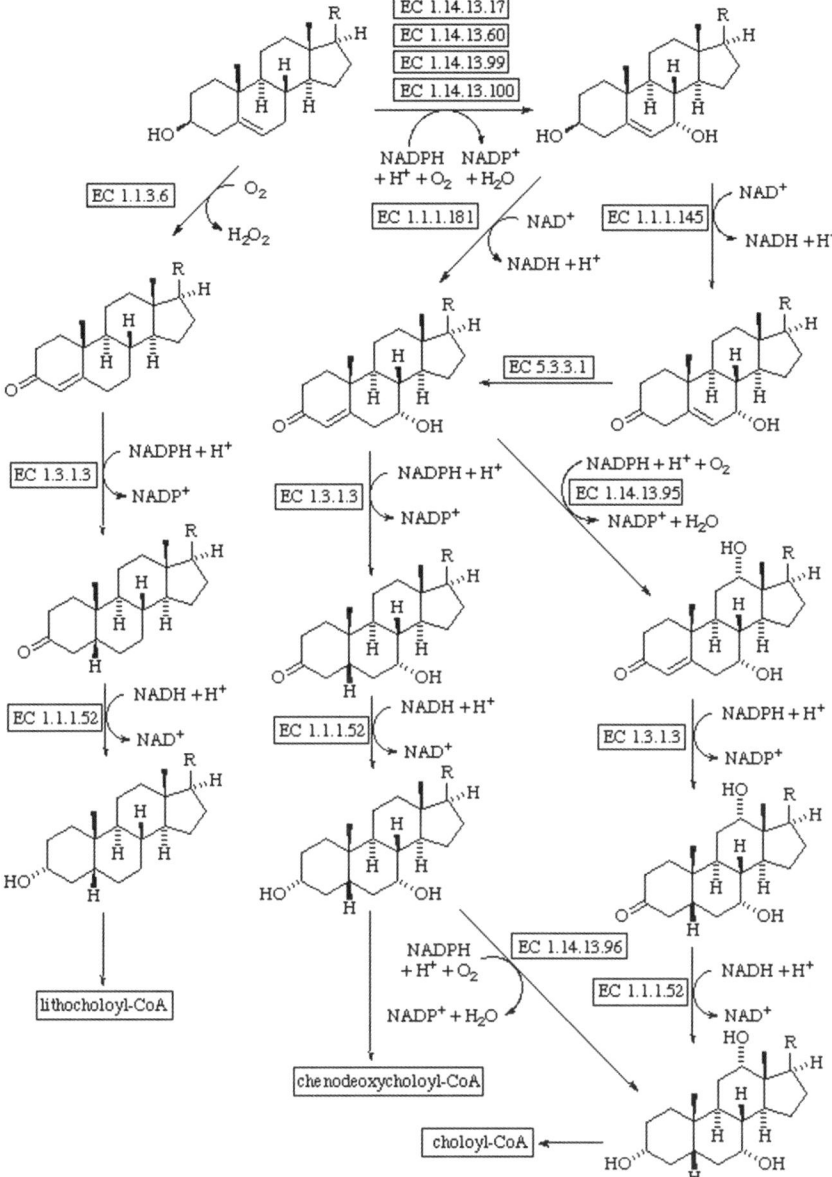

Figure 8.3 The remarkable diversity of bile salt derivatives formed in the liver, kidney and intestine. The structures of all these intermediates were solved decades before their functions were recognized.

in vivo was a breakthrough in the diagnostics of hepatic metabolism and the use of isotope dilution mass spectrometry was crucial to this research.[21] Though rare, genetic disorders of bile acid metabolism occur and are now routinely diagnosed by high-throughput analyses of the various intermediates in the pathway measured from blood using high-throughput mass spectrometry of blood spots.[22]

8.2.5 Excretion of Bile Salts

The transport of bile salts from the liver into the gall bladder, excretion from the gall bladder into the intestine, absorption from the intestine back into portal blood circulation and finally reuptake into the liver is a highly regulated process accomplished by a series of transporters that have been identified and characterized only over the past few years.[23] Once again, the disparity between enabling technologies has been an important block in understanding these transport pathways. Protein biochemistry has made remarkable achievements in identifying the catalytically active proteins and assembling them into metabolic pathways. The tools to acquire this knowledge relied upon the ability of biochemists to identify proteins for their ability to act as chemical catalysts, assaying the conversion of substrate to product. Analyzing the functions of proteins whose sole function is to bind amphiphilic lipids, however, is a much more difficult process. As a result, the large families of lipid transport proteins are only now being revealed with painstaking slowness. Their functions are being defined largely by the complex process of gene elimination, gene addition experiments which must examine suitable models for corresponding loss of function, gain of function outcomes. These studies are discouragingly more difficult than a mass spectrometric analysis.

8.2.6 Cholesterol Transport and Lipoproteins

Cholesterol is perhaps the most intriguing chemical abundant in biology. Cholesterol is ostensibly insoluble in water and at the same time insoluble in oil. Where is it soluble? Biology takes two strategies to resolve this apparent paradox. First, cholesterol is only soluble within the cellular compartment in which it functions, *i.e.* the biological membrane. Cholesterol can be dispersed in phospholipid bilayers almost to the point of equimolar mixtures. Hence, the majority of cholesterol within cells, tissues and fluids is dissolved within phospholipid bilayers and monolayers. Cholesterol is stored not as free cholesterol but as a fatty acid ester form in which it is relatively co-soluble in an oil phase with triglycerides. These solubility constraints mean that cholesterol is not free to diffuse across or between cells and tissues. This in turn means that the transport of cholesterol between cells requires explicit transport systems. Cholesterol moves within humans between cells as a portion of elaborate and compositionally diverse particles made up of lipids and proteins including lipoproteins, vesicles and globules (Figure 8.4). A vast scientific literature, which is beyond the scope of this review, describes these particles, their composition, synthesis, transport, disassembly and their regulation. However, in establishing the knowledge of these particles that transport and regulate cholesterol, the receptors that bind them and the enzymes that form and remodel

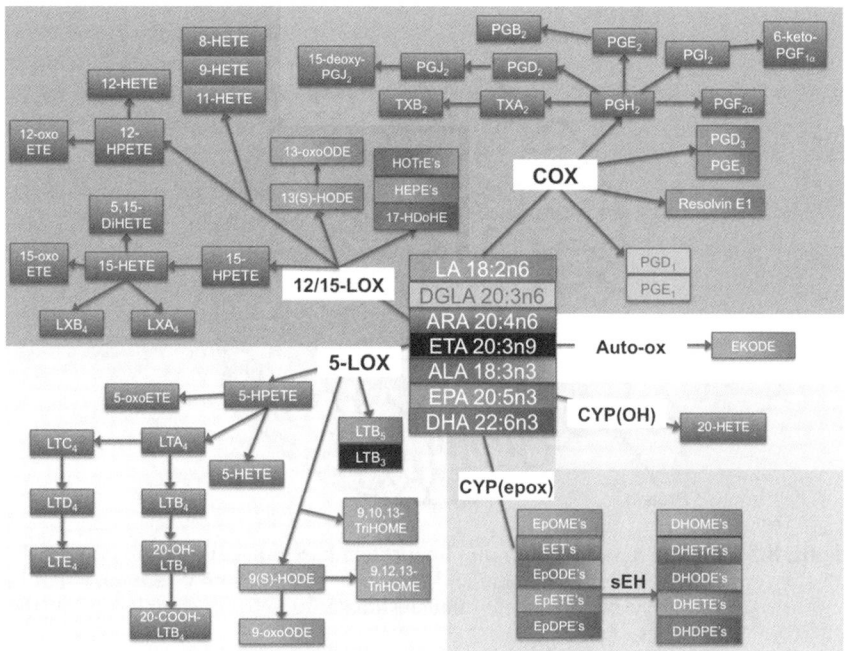

Figure 8.4 Oxylipid pathways. This figure illustrates just a small subset of the various compounds formed from the simple oxidation of polyunsaturated fatty acids by enzyme systems responding to stress. It has been documented that these compounds interact in their signalling actions: hence it is necessary to measure the full complement of molecules to assign a causal role to this largely pro-inflammatory pathway.

them, mass spectrometry has been a critical instrumentation platform in providing detailed chemical information.

Research today takes advantage of the spectacular increases simultaneously in the accuracy, sensitivity and speed of mass detection of mass spectrometry with sequence information of proteins to accomplish an almost unimagined capability—proteomics and lipidomics.[24–26] The ability of mass spectrometers to so accurately determine precise masses that mathematical computations can solve for sequence of proteins was a major scientific breakthrough. Other chapters in this book describe these techniques in detail. However, the daunting complexity of cells and most biofluids (blood) in terms of the total numbers and structures of proteins still exceeds the capability of proteomics to solve the structures of all proteins present. Nevertheless, proteomics capabilities are possible for subsets of cellular compartments. Thus, proteomics has been successfully applied to the task of measuring all the proteins associated with specific lipid particles.

The application of proteomic capability has had an immediate impact on the entire field of lipoprotein biology.[27] Formerly, lipoproteins were considered to be comparatively simple in their protein composition, dominated by a single

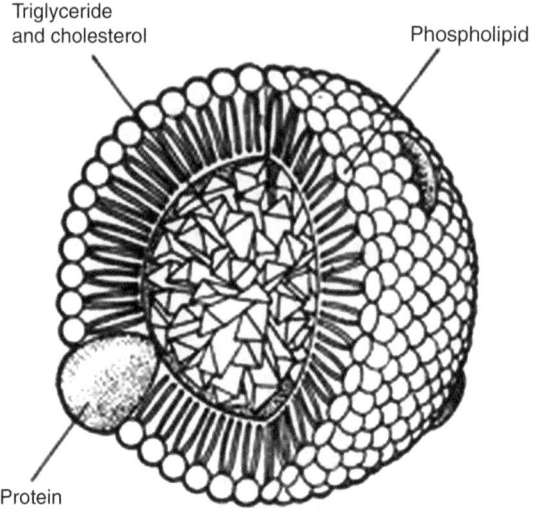

Triglyceride
and cholesterol   Phospholipid

Protein

Figure 8.5   Simple representation of a lipoprotein based on early perceptions of their composition. Proteomic and lipidomic measures now recognize that this simple view drastically underestimated the true complexity of these particles.

apoprotein—apoB or apoA. The figures created to represent lipoproteins are clearly dated renditions of a state of ignorance of their complexity (Figure 8.5). The very first proteomics studies have exploded this perception. The number of proteins and peptides associated with lipoproteins in plasma is not only quite complex but highly variable depending on the individual and their metabolic state sampled. The proteome of the smallest particle class, high density lipoproteins (HDL) (Table 8.1), illustrates the much more complex protein content of HDL. Because all of these proteins cannot be arrayed on all particles, the number of proteins found implies not only a wide range of different particles but a wide range of additional functions associated with the lipoproteins in general.[28] Indeed, lipid particles must now be considered to be more like dynamic docking stations orchestrating various life processes. Scientists are rebuilding the databases of blood lipoproteins, now including the apoprotein composition of the various lipoproteins in discrete populations and as a function of diverse phenotypes and physiological states. Mass spectrometry based proteomics is literally making this entire field possible.

Surprisingly, the application of the enhanced accuracy, sensitivity and efficiency of mass spectrometry was not applied to cataloguing the lipid composition of cells and biofluids as lipidomics until relatively recently.[29] Analytical strategies coupling different modalities—separation science and mass spectroscopy, notably liquid chromatography and mass spectrometry—into complementary platforms have been most successful. Nonetheless, even a single complex platform is unable to solve the varying analytical challenges of

Table 8.1 Proteome of the smallest particle class, HDL.

apoC-IV
PON1
Complement C3
apoA-IV
apoE
apoL1
SAA1
β-2-glycoprotein1
Complement C4B
apoC-II
apoM
LCAT
CETP
α-2-antiplasmin
α-1-acid-glycoprotein 2
PLTP
Angiotensinogen
apoB-100
α-2-HS-glycoprotein
apoF
Complement C4
apoD
α-1-antitrypsin
apoC-I
apoC-III
apoA-I
SAA
apoA-II
SAA2
Serum albumin
Vitronectin
HRP
Clusterin

the different molecular species of lipids. The enabling capabilities of high-performance liquid chromatography (HPLC) coupled to electrospray ionization-mass spectrometry (ESI-MS) to simultaneously identify and analyze the full range of phospholipids from a complex mixture was the first proof of principle of lipidomics.[29–31]

Mass spectrometry is now able to identify molecules by virtue of precise molecular mass measurement, thus providing the analytical power needed to identify large numbers of complex lipids. Some important problems remain, however. The isomeric diversity of fatty acids in biology still frustrates mass analyzers because the absolute mass of some of the fatty acid positional isomers is identical. Because these fatty acid isomers are importantly different biologically, this limitation still needs to be addressed for the field of lipidomics. Second, mass spectrometer ion sources are still inconsistent. As a result mass spectrometry is not as quantitatively accurate as metabolite measurements need to be. For many applications, measuring the subtle changes associated with early stages of disease development is necessary.

A recent approach to resolve lipid species with overlapping molecular weights of glycerolipids with identical mass ion molecular weights because of isomeric differences in the fatty acid structures on the glycerol backbone uses HPLC coupled directly to tandem mass spectrometry (MS/MS).[32,33] Normal phase HPLC coupled to electrospray ion-trap mass spectrometry can simultaneously examine changes in multiple phospholipid classes (phosphatidylethanolamine, phosphatidylinositol, phosphatidylcholine, lysophosphatidylcholine and sphingomyelin) and the individual molecular species within each class without prior derivatization, thin layer chromatography (TLC) or solid phase extraction (SPE) of the bulk lipid extract. Tandem liquid chromatography coupled to mass spectrometry has also been used to separate complex lipid species prior to ionization and detection by mass spectrometry.[34]

The strategy of combining tandem analytical methods to perform the sensitive, high-throughput, quantitative and comprehensive analysis of lipid metabolites for high throughput of very large numbers of molecules is driving all of the fields of metabolism. This option is propelling the field of lipid biochemistry forward rapidly in describing which lipids are present in different tissues, cells, subcellular organelles and lipoproteins. Nonetheless, the problem of quantitation is critical for understanding biological processes with the accuracy necessary to distinguish the early stages of the slow, incremental deterioration of the vasculature seen in atherogenesis and the development of heart disease.

Quantification using mass spectrometry requires that each analyte is compared to an internal standard and the actual amounts determined by peak area ratios.[35] These methods are clearly problematic for highly complex samples with dozens of different species of lipids as they require such a large number of internal standards. Sacrificing quantification of metabolites for higher throughput is possible when only chemical identification is needed; however, heart disease research is more challenging. Quantitative analyses are needed to detect the small differences that are the basis of varying states of metabolic health underlying the causes of the different phases of atherosclerosis leading to heart disease. Hence, the decision to forego quantitative precision limits the future utility of the integrated databases of metabolites that are produced by these analyses. The success of the extensive experimentation conducted on the population of Framingham, Massachusetts and the diet and lifestyle causes of heart disease was predicated upon banking quantitative data, and the limitations of this and other large databases lie in the lack of quantitative detail for many input and outcome variables.[36] It will be necessary then to revisit these studies to rebuild quantitative databases against which individual profiles can be compared to distinguish important differences (*i.e.* diagnostic applications, unintended side effects of therapeutics).

8.2.7 Atherogenic Lipoproteins

Decades of research on the normal biology of lipoproteins failed to resolve why this aspect of metabolism, when only slightly elevated, led to such an increase in the risk of disease. Part of the problem, of course, is that it is so slow. The pathologic processes within the coronary arteries leading ultimately to atherosclerotic cardiovascular disease develop progressively over years, if not decades. The atherosclerotic tissue itself is not healthy, but characterized by markers of inflammation and fibroproliferative responses of cells within the vessel wall.[37] This inflamed tissue is also chemically distinct. Compositional hallmarks of the disease process consist of subendothelial lipid deposits, primarily as cholesterol, cholesteryl esters and in later stages, calcium. The most conspicuous manifestations of early atherosclerosis-impaired endothelial function are lipid-laden macrophages. These cells, first identified histologically as foam cells form the distinctive property of fatty streaks and which, after necrosis, leave cholesterol crystals within the arterial wall.

Substantial chemical evidence accumulated using straightforward mass spectrometry analysis of diseased artery tissue implicated that the source of arterial cholesterol deposits as plasma low density lipoprotein (LDL). By simple compositional comparisons, it was clear that LDL was accumulating within these cells. A major question for researchers was why did LDL accumulate in monocytes within the artery wall? The answer was not obvious. In fact, it had been known for decades that neither monocytes nor macrophages could be shown to take up native LDL, even when presented with high concentrations of this lipoprotein.[38] Thus the mechanism underlying the strong association between elevated plasma LDL cholesterol and cardiovascular mortality remained one of science's intriguing puzzles. Hints were there. In the same earlier work documenting the fact that monocyte/macrophages did not take up normal LDL, Goldstein *et al.* had showed that, *in vitro*, chemical modification of LDL to a non-native form led macrophages to rapid massive, receptor-mediated uptake.[38] This receptor was ultimately identified as scavenger receptor type A.[39] But what modified the LDL *in vivo*? Several lines of research indicated that, *in vivo*, the most physiologically plausible modification to LDL was oxidative modification of lipid moieties.[40] This basic mechanism changed the course of heart disease research and much of nutrition research. The theory that chemical oxidation was central to the development of the processes underlying the atherogenicity of LDL *in vivo* explained a wide range of epidemiological observations, including those related to diet.[41]

Once it was proposed that chemical oxidation was a key to the development of heart disease *via* the conversion of native benign LDL to modified, atherogenic LDL chemists immediately began the process of studying LDL oxidation, analyzing LDL oxidation products, *in vitro* and in atherosclerotic plaque, and exploring the variables that could accelerate oxidation *in vivo*. Describing the *in vitro* chemical destruction of the lipid moieties of isolated LDL, triglyceride-rich lipoprotein (TRL), or model liposomes from oxidation caused by several different initiating species prompted a genuine renaissance of the field of lipid oxidation. Furthermore, if oxidation was a key propellant of heart disease, the logical corollary was that inhibiting LDL oxidation would lower the risk of heart disease. Such a promise led to a literal explosion in scientific research and a global race to research and develop antioxidant strategies to enhance the protection of LDL and lower their oxidation as the means to slow heart disease.[42,43] Mass spectrometry was an enabling technology for this pursuit, first in establishing the oxidation products of LDL oxidation and secondly in the determination of the structures of compounds capable of acting as antioxidants. The composition of LDL was shown to alter its oxidation, both susceptibility and the nature of products.[44] Then the race was on to modify the susceptibility of LDL to oxidation *in vitro* and to translate protection mechanisms to protection *in vivo*. The former has been met with unqualified success. The latter still awaits confirmation.

The chemical nature of the oxidative processes within LDL were studied in detail both to determine their rate and the specific components that led to the conversion of native LDL into particles that were recognized by monocyte scavenger receptors.[45] Oxidation is a highly complex process with many different chemical products. The processes of oxidation of the lipids of LDL yield many reactive compounds, oxidized cholesterol and many forms of fatty acid hydroperoxides. Many of these decompose into reactive aldehydes, both small fragments of larger lipids and the inert truncated phospholipids, triglycerides and cholesterol esters. Modification of the LDL particle can occur *via* direct adduct formation between lipid oxidation products such as hydroxynonenal and the apoB protein through Micheal addition reactions to histidines on the surface.[46] Alternatively, oxidized complex lipids—particularly arachidonyl-containing phosphatidylcholine—create a distortion of the lipid surface sufficient to create a non-native LDL particle.[47]

As the chemical products of oxidation were discovered, the destructive nature of oxidation of LDL *in vivo* became appreciated since the products formed in this process caused a variety of effects in the monocytes that take them up and accelerate lesion formation. Evidence that antioxidants slowed this process was further evidence that oxidation was a causal mechanism of atherogenesis and that antioxidants could present a preventive strategy.[48] In addition to the composition of the LDL themselves, various processes that could promote oxidation varied from exogenous oxidants (*e.g.* smoking, pollution and dietary hydroperoxides) to endogenous oxidants including free metals and stimulated phagocytes; even endothelial cells themselves were capable of promoting LDL oxidative modification.[49]

The conditions necessary for the initial oxidation of LDL *in vivo* are believed to arise from retention of LDL or other lipoproteins within the cage-like structures of fibres and fibrils that are secreted by cells of the arterial wall into the subendothelial space.[50] Altered endothelial permeability and enhanced retention of circulating LDL are among the very earliest indications of vascular damage. In later stages of atherosclerotic lesion formation, altered endothelial metabolism and function, inflammation and necrotic changes all stem from the generation of reactive-oxygen species (ROS), lipid peroxides and resultant protein modifications on a variety of targets. To no surprise to immunologists, this chaos of chemical damage is a recipe for inflammation.

8.2.8 Inflammation

Whereas the scientific knowledge of cholesterol metabolism is relatively mature with well-defined pathways complete with annotated functions and translated into biomarkers of disfunction, the same cannot be said of inflammation. Growing evidence suggests that sustained, low level inflammation is linked to risk of developing a variety of degenerative diseases and especially heart disease.[51] However, that evidence is not accompanied by a mature scientific knowledge base to explain it. Inflammation is discouragingly typified by poorly understood pathways, perplexing dynamics, a lack of functional annotation and unresponsive biomarkers.

One of the major problems is that inflammation is itself not yet well defined. In general, inflammation can be considered to be the collateral damage to tissues, cells and molecules that results from endogenous immune activities. The difficulties of study are largely related to the immune system itself. The immune system plays a lesser role in normal physiology, but instead lies dormant awaiting external stress. In simplified terms, success for the immune system is maintaining surveillance, in recognizing and in reacting aggressively to eradicate an invading pathogen, eliminate a deranged cell or detoxify an inappropriate toxin, and then returning to its surveillance state.

Studying immune functions requires scientists to abandon the normal model of examining systems at rest and instead perturb biological models with standardized exogenous stressors. This approach has been very informative where applied successfully, but it is difficult to standardize stress in humans. One of the themes emerging nonetheless in study after study is that, with aging, control of the immune system is lost. Even in the absence of any apparent exogenous stress, various components of the immune system's surveillance are inappropriately activated by chemical damage. Instead of benignly responding to and removing stress, the immune system is chronically activated even by endogenously created damage. To the extent that these inappropriate immunologic processes have been characterized, chronic inflammation is associated with almost all stages of heart disease.[37]

Three lines of evidence are particularly convincing that chronic inflammation is causally involved with the development and severity of heart disease:

- The presence of elevated markers of immune activation, including C-reactive protein (CRP), has been demonstrated in multiple clinical, epidemiologic and prospective trials to be associated with increased risk of disease.[52]
- The paradoxical property of aspirin, a drug that inactivates the prostaglandin biosynthetic pathway (an entire arm of immune signalling) in lowering heart disease rates demonstrated in a large number of studies.[53]
- The consumption of long chain omega-3-polyunsaturated fatty acids, chemically more susceptible to oxidation and, all things being equal, a potentially more damaging dietary family of fatty acids for the destruction of lipoproteins has nonetheless been repeatedly associated epidemiologically in

prospective trials and a variety of animal models with lowered rates of heart disease.[54]

In view of this evidence, scientific research is racing to understand the processes of inflammation and to develop accurate, mechanistically linked biomarkers of inflammation and interventions that can ameliorate inflammation without causing a disabling of the normal functions of immunity. It can genuinely be stated that, without advances in mass spectrometry, these goals cannot be achieved.

8.2.9 Inflammation Targets and Diagnostics

The hallmark of inflammatory diagnostics is C-reactive protein (CRP), a plasma protein produced by immune cells that after decades of measurements in various populations has been repeatedly demonstrated to be elevated in those at risk of heart disease.[55] However, CRP does not appear to be causally related to heart disease and thus does not provide a diagnostic that would respond successfully to treatment and accurately predict improved risk. Hence, more specific targets of immune action and chronic activation are necessary.[56]

Currently, inflammation is largely characterized (diagnosed) by the measurement of chronically elevated concentrations of intermediates or control points of immune activation that would otherwise be detectable immediately during a period of infection or similar event.[57] For many years, the field of inflammation pursued a single biomarker as a telltale indicator of inflammation. However, the immune system is a spectacularly complex system, with multiple independent and co-ordinated processes acting in ways varying depending on the nature of an exogenous stress and the stage of its resolution.[58] Hence scientists have begun to develop a broader concept of inflammation and its detection by multiple biomarkers. In this pursuit, mass spectrometry innovations have proven indispensable. In keeping with its role as a rapid response system, the immune system in addition to synthesizing new signal molecules, in many cases uses fragments of existing cellular biomolecules as the first wave of surveillance. Thus, the wide diversity of hydrolyzed phospholipids and massively complex families of oxidized fatty acids are a part of the language of communication in the early phase of immune activation.[59] Scientists are beginning to apply metabolomics technologies to describing these pathways during immune activation and during inflammation.

Oxidized polyunsaturated fatty acids are an important class of signal molecules for immune cell activation; their excess production has been linked causally to various events in heart disease and their inhibition has been linked to protection from heart disease and catastrophic events in heart disease.[60] The demonstration that genetic polymorphisms in the genes related to the production of these oxidized lipid pathways is evidence that the molecules are indeed causally related to disease and hence appropriate targets for intervention, diagnostics and follow-on theranostics.[61] These molecules are thus

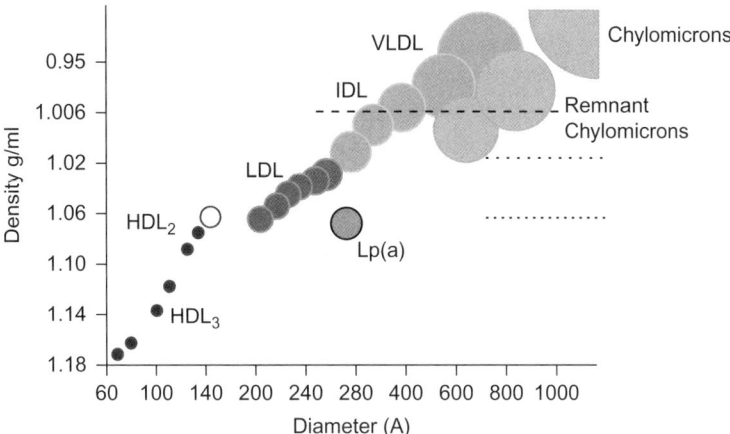

Figure 8.6   Diversity of sizes of lipoproteins. Just taking the perspective of size and density alone, lipoproteins are a heterogeneous population of particles. HDL = high density lipoprotein; IDL = intermediate density lipoprotein; LDL = low density lipoprotein; VLDL = very low density lipoprotein.

excellent candidates as mechanistic markers of inflammation and its role in atherosclerosis. However, their measurement is proving to be discouragingly difficult for several reasons. First is their complexity: the subset of oxidized lipid families shown in Figure 8.6 illustrates the sheer diversity of compounds within these signalling pathways (Figure 8.6). Second, their levels vary widely depending on a variety of acute physiological processes going on within individuals including eating.[62,63]

8.3   Next Steps

The remarkable demands that we make on our circulatory system all day, every day has an expected outcome. The human heart is our defining weak link. As a consequence of this rather singular cause of death in humans, a greater understanding of the biology of this system will continue to return on this investment with improved human health and longevity. There are three complementary research areas in which mass spectrometry as a toolset will continue to contribute:

- basic research—understanding the destructive processes within the vasculature;
- diagnostics—accurately assessing and predicting individual risk; and
- bioactive interventions—components that, when ingested, are capable of matching individual risk to specific biological solutions.

The most immediate challenges to the heart health research community are to understand the complex interactions between the structures and compositions of lipoproteins and the endothelial cell layer lining the vasculature.

The tools of proteomics and lipidomics made possible by mass spectrometry have established that lipoproteins are much more complex, diverse and dynamic than previously considered. These particles don't just transport lipids. They apparently carry out more diverse biological activities both within the vascular compartment and within cells and tissues. Yet there is a very poor understanding of their structure–function relationships. The composition of lipoproteins ranges widely during the day and across different individuals. How some of the lipoproteins are apparently far from benign carriers of insoluble lipids but instead become pro-inflammatory complexes of damaging constituents capable of propelling normal vascular epithelia towards disease needs to be understood.

Mass spectrometry empowered proteomics and metabolomics by virtue of simultaneous mass accuracy and analytical throughput. A similar step must be taken to accomplish high-throughput colloidal description of the genuine complexity within entire populations of lipoproteins. The "average" lipoproteome is not enough. Mass spectrometry will propel the next generation detailed, structure function description of the true nature of lipoproteins. Similarly, understanding how the cells of the vascular lining meet these particles is necessary. Endothelia encountering lipoproteins on their surface either react with a robust, healthy and ultimately protective response or they begin a long-term debilitating process of disease. The surface biology of these cells and the signalling that the surfaces produce must be described in chemical and structural detail. The continuing analytical growth of techniques such as MS imaging will be invaluable in providing a truly functional map of the endothelial surface.

Once this key knowledge is acquired, measures of the properties of lipoproteins and the health of the vasculature must be brought into practical measurements as diagnostics of lipoprotein and vascular health.

8.4   Personalized Health and Medicine

Linus Pauling first suggested that complex metabolite profiles could diagnose disease using non-invasive urine vapour and breath condensates analyzed by gas chromatography.[66] Now advances in analytical and computation technology are beginning to bring this concept to practice. What even Linus Pauling was hesitant to suggest was that such tools could obtain and integrate far more information than breath condensate patterns linked to disease and could address the overall goal of developing an accurate picture of the absolute health of each individual discretely.

The rise of heart disease as the leading cause of death in the world prompted a fundamental change in human healthcare: diagnostics and prevention with the introduction of routine cholesterol testing and cholesterol education programmes to guide individuals to personal strategies to lower risk. That basic principle of assessment and prevention is about to enter its next generation. The medical practices that embrace the genetic and biochemical diversity of patients have been termed personalized medicine. The benefits of individualized genetic and metabolic signatures will invariably become central to an overall systems approach to disease assessment and management. The view of individualized medicine is that patients will be treated with personally defined pharmaceutical needs. A global vision for individualized medicine was described by Sander in which patients would be genotyped early in life to define their metabolic baseline and then re-assessed throughout their lifetime using biochemical analyses to monitor environmental and nutritional influences.[67] Personalized medicine is already viewed as capable of providing dramatic advantages to:

- patients, who will benefit from improved health outcomes; and
- the provider sector from pharmaceutical companies to hospitals, which despite fragmented markets, will achieve greater efficacy, safety and efficiency.

Such a complementary approach highlights the strengths of both genomics and metabolomics, yet with a more personal view of medicine, phenotype measuring will be more critical than genotyping. Eventually, genotyping will help predict the metabolic predisposition of an individual toward disease or their potential intolerance to drugs or diet. Nonetheless, phenotyping is not only necessary to assessing current health status, it is the only means to document the efficacy of therapeutic solutions irrespective of genotypic prediction.

One more intriguing element of personal health assessment is on the horizon—personal response to challenge. While accurate assessment technologies will produce a new generation of diagnostics with broad utility in measuring phenotype, chemical measurements will also have a role in experiment-oriented clinical diagnostics. Many questions about individual metabolism are best addressed by challenging an individual or a surrogate of an individual (*e.g.* their own cultured cells) with substrates, drugs, nutrients or other compounds and by monitoring their metabolic response. Willard *et al.* employed this approach with success in identifying a highly unusual deficiency of δ-6-desaturase activity in a young patient.[66] Skin fibroblasts taken from the patient were cultured and challenged with a series of labelled fatty acids that comprised both precursors and products of the major lipid metabolism enzymes. The results indicated a defect in the δ-6-desaturase activity of the patient with no apparent secondary defects. On the basis of these results, the patient was able to supplement their dietary lipid intake with products of the δ-6-desaturase activity to mitigate the phenotypic consequences of the genetic deficiency. While this experiment was performed in series, with each fatty acid challenge following the last, it is easy to envision the broader application of the concept. Individuals could be challenged with a broad range of stressors simultaneously and their co-ordinated response followed as a part of health assessment.

This principle has been applied to dietary challenges.[67] Since heart disease susceptibility is now recognized to be driven by post-prandial events, the next generation of cholesterol education may well include a dietary challenge to recognize those individuals whose dietary responsiveness is the basis of increased risk. Thus, the future of mass spectrometry to the ultimate personalization of health, diet and lifestyle can already be imagined.[68] Combining personal measurements with designed interventions will provide each individual with the means to not only know more about their health but to intervene in its control—personally. Such a future is a fitting contribution by chemistry to the human condition.

References

1. R. Paoletti, A. Poli and A. Cignarella, *Expert Rev. Cardiovasc. Ther.*, 2006, **4**, 385.
2. US Department of Health and Human Services, *The Surgeon General's report on Nutrition and Health*, US Government Printing Office, Washington, DC, 1988, DHHS (PHS) Publication No. 88–50210.
3. H. Iso, D. R. Jacobs Jr, D. Wentworth, J. D. Neaton and J. D. Cohen, *N. Engl. J. Med.*, 1989, **320**, 904.
4. A. W. Alberts, *Am. J. Cardiol.*, 1998, **62**, 10J.
5. E. Pouteau, C. Piguet-Welsch, A. Berger and L. B. Fay, *Isotopes Environ. Health. Stud.*, 2003, **39**, 247.
6. G. Gremaud, C. Piguet, M. Baumgartner, E. Pouteau, B. Decarli, A. Berger and L. B. Fay, *Rapid Commun. Mass Spectrom.*, 2001, **15**, 1207.
7. M. K. Hellerstein and R. A. Neese, *Am. J. Physiol.*, 1992, **263**, E988.
8. W. N. P. Lee, E. A. Bergner and Z. K. Guo, *Biol. Mass. Spectrom.*, 1992, **21**, 114.
9. J. K. Kelleher and T. M. Masterson, *Am. J. Physiol. Endocrinol. Metab.*, 1992, **262**, E118.
10. J. K. Kelleher, A. T. Kharroubi, T. A. Aldaghlas, I. B. Shambat, K. A. Kennedy, A. L. Holleran and T. M. Masterson, *Am. J. Physiol.*, 1994, **266**, E384.
11. M. K. Hellerstein, M. Christiansen, S. Kaempfer, C. Kletke, K. Wu, J. S. Reid, N. S. Hellerstein and C. H. L. Shackleton, *J. Clin. Invest.*, 1991, **87**, 1841.
12. M. K. Hellerstein, *Metab. Eng.*, 2004, **6**, 85.
13. J. J. Clarenbach, B. Lindenthal, M. T. Dotti, A. Federico, J. K. Kelleher and K. von Bergmann, *Metabolism*, 2005, **54**, 335.
14. D. M. Klass, K. Bührmann, G. Sauter, M. Del Puppo, J. Scheibner, M. Fuchs and E. F. Stange, *Aliment. Pharmacol. Ther.*, 2006, **23**, 895.
15. K. Einarsson, B. Angelin, S. Ewerth, K. Nilsell and I. Bjorkhem, *J. Lipid Res.*, 1986, **27**, 82.

16. P. V. Luoma, *Eur. J. Clin. Pharmacol.*, 2008, **64**, 841.
17. W. J. Griffiths, *Mass Spectrom. Rev.*, 2003, **22**, 81.
18. K. D. Setchell and A. Matsui, *Clin. Chim. Acta*, 1983, **127**, 1.
19. D. J. Parks, S. G. Blanchard, R. K. Bledsoe, G. Chandra, T. G. Consler, S. A. Kliewer, J. B. Stimmel, T. M. Willson, A. M. Zavacki, D. D. Moore and J. M. Lehmann, *Science*, 1999, **284**, 1365.
20. F. Kuipers, J. H. Stroeve, S. Caron and B. Staels, *Curr. Opin. Lipidol.*, 2007, **18**, 289.
21. M. Axelson, A. Aly and J. Sjovall, *FEBS Lett.*, 1988, **239**, 324.
22. S. S. Sundaram, K. E. Bove, M. A. Lovell and R. J. Sokol, *Nat. Clin. Pract. Gastroenterol. Hepatol.*, 2008, **5**, 456.
23. H. N. Ginsberg, *Endocrinol. Metab. Clin. North Am.*, 1998, **27**, 503.
24. A. Kosters and K. J. Karpen, *Xenobiotica*, 2008, **38**, 1043.
25. A. J. Link, J. Eng, D. M. Schieltz, E. Carmack, G. J. Mize, D. R. Morris, B. M. Garvik and J. R. Yates, *Nat. Biotechnol.*, 1999, **17**, 676.
26. S. P. Gygi, B. Rist, S. A. Gerber, F. Turacek, M. H. Gelb and R. Aebersold, *Nat. Biotechnol.*, 1999, **17**, 994.
27. J. B. German, L. A. Gillies, J. T. Smilowitz, A. M. Zivkovic and S. M. Watkins, *Curr. Opin. Lipidol.*, 2007, **18**, 66.
28. M. Heller, E. Schlappritzi, D. Stalder, J. M. Nuoffer and A. Haeberli, *Mol. Cell. Proteomics*, 2007, **6**, 1059.
29. A. N. Hoofnagle and J. W. Heinecke, *J. Lipid Res.*, 2009, **50**, 1967.
30. X. Han and R. W. Gross, *J. Lipid Res.*, 2003, **44**, 1071.
31. X. Han and R. W. Gross, *Mass Spectrom. Rev.*, 2005, **24**, 367.
32. M. Pulfer and R. C. Murphy, *Mass Spectrom. Rev.*, 2003, **22**, 332.
33. D. Pacetti, E. Boselli, H. W. Hulan and N. G. Frega, *J. Chromatogr. A.*, 2005, **1097**, 66.
34. M. Masoodi and A. Nicolaou, *Rapid Commun. Mass Spectrom.*, 2006, **20**, 3023.
35. A. Takatera, A. Takeuchi, K. Saiki, T. Morisawa, N. Yokoyama and M. Matsuo, *J. Chromatogr. B.*, 2006, **838**, 31.
36. D. Sun, M. G. Cree and R. R. Wolfe, *Anal. Biochem.*, 2006, **349**, 87.
37. S. Kathiresan, A. K. Manning, S. Demissie, R. B. D'Agostino, A. Surti, C. Guiducci, L. Gianniny, N. P. Burtt, O. Melander, M. Orho-Melander, D. K. Arnett, G. M. Peloso, J. M. Ordovas and L. A. Cupples, *BMC Med. Genet.*, 2007, **8**, S1–S17.
38. P. Libby, G. Hansson and J. Pober, in *Molecular Basis of Cardiovascular Disease: A Companion to Braunwald's Heart Disease*, ed. K. R. Chien, Saunders, Philadelphia PA, 1999. p. 349.
39. J. L. Goldstein, Y. K. Ho, S. K. Basu and M. S. Brown, *Proc. Natl. Acad. Sci. U. S. A.*, 1979, **76**, 333.
40. T. Kodama, M. Freeman, L. Rohrer, J. Zabrecky, P. Matsudaira and M. Krieger, *Nature*, 1990, **343**, 333–337.
41. D. Steinberg and J. Witzum, in *Molecular Basis of Cardiovascular Disease: A Companion to Braunwald's Heart Disease,* ed. K. R. Chien, Saunders, Philadelphia PA, 1999. p. 458.

42. R. L. Walzem, S. Watkins, E. N. Frankel, R. J. Hansen and J. B. German, *Proc. Natl. Acad. Sci. U. S. A.*, 1995, **92**, 7460.

43. E. N. Frankel, J. Kanner, J. B. German, E. Parks and J. E. Kinsella, *Lancet*, 1993, **341**, 454.

44. A. Lapointe, C. Couillard and S. Lemieux, *J. Nutr. Biochem.*, 2006, **17**, 645.

45. D. Steinberg, *J. Lipid Res.*, 2009, **50**, S376.

46. E. J. Parks, J. B. German, P. A. Davis, E. N. Frankel, C. T. Kappagoda, J. C. Rutledge, D. A. Hyson and B. O. Schneeman, *Am. J. Clin. Nutr.*, 1998, **68**, 778.

47. D. Steinberg, S. Parthasarathy, T. E. Carew, J. C. Khoo and J. L. Witztum, *N. Engl. J. Med.*, 1989, **320**, 915.

48. B. A. Bruenner, A. D. Jones and J. B. German, *Chem. Res. Toxicol.*, 1995, **8**, 552.

49. M. Navab, J. A. Berliner, A. D. Watson, S. Y. Hama and M. C. Territo, *Arterioscler. Thromb. Vasc. Biol.*, 1996, **16**, 831.

50. H. Sies, W. Stahl and A. R. Sundquist, *Ann. N. Y. Acad. Sci.*, 1992, **669**, 7.

51. R. Stocker and J. F. Keaney Jr, *Physiol. Rev.*, 2004, **84**, 1381.

52. K. J. Williams and I. Tabas, *Arterioscler. Throm. Vasc. Biol.*, 1995, **15**, 551.

53. M. Hulsmans and P. J. Holvoet, *J. Cell. Mol. Med.*, 2009, Epub ahead of print.

54. A. M. Kampoli, D. Tousoulis, C. Antoniades, G. Siasos and C. Stefanadis, *Trends Mol. Med.*, 2009, **15**, 323.

55. Antithrombotic Trialists' Collaboration. *Br. Med. J.*, 2002, 324, 71.

56. C. Wang, W. S. Harris, M. Chung, A. H. Lichtenstein and E. M. Balk, *Am. J. Clin. Nutr.*, 2006, **84**, 5.

57. P. M. Ridker, *Clin. Chem.*, 2009, **55**, 209.

58. G. K. Hansson, *N. Engl. J. Med.*, 2005, **352**, 1685.

59. R. R. Packard and P. Libby, *Clin. Chem.*, 2008, **54**, 24.

60. B. D. Levy, C. B. Clish, B. Schmidt, K. Gronert and C. N. Serhan, *Nat. Immunol.*, 2001, **2**, 612.

61. W. Siess, K. J. Zangl, M. Essler, M. Bauer, R. Brandl, C. Corrinth, R. Bittman, G. Tigyi and M. Aepfelbacher, *Proc. Natl. Acad. Sci. U. S. A.*, 1999, **96**, 6931.

62. M. W. Buczynski, D. S. Dumlao and E. A. Dennis, *J. Lipid Res.*, 2009, **50**, 1015.

63. R. N. Lemaitre, K. Rice, K. Marciante, J. C. Bis, T. S. Lumley, K. L. Wiggins, N. L. Smith, S. R. Heckbert and B. M. Psaty, *Atherosclerosis*, 2009, **204**, e58.

64. W. C. Tsai, Y. H. Li, C. C. Lin, T. H. Chao and J. H. Chen, *Clin. Sci.*, 2004, **106**, 315.

65. M. Lankinen, U. Schwab, A. Erkkilä, T. Seppänen-Laakso, M. L. Hannila, H. Mussalo, S. Lehto, M. Uusitupa, H. Gylling and M. Oresic, *PLoS One*, 2009, **4**, e5258.

66. L. Pauling, A. B. Robinson, R. Teranishi and P. Cary, *Proc. Natl. Acad. Sci. U. S. A.*, 1971, **68**, 2374.
67. C. Sander, *Science*, 2000, **287**, 1977.
68. A. M. Zivkovic, M. M. Wiest, U. Nguyen, M. L. Nording, S. M. Watkins and J. B. German, *Metabolomics*, 2009, **5**(2), 209.

CHAPTER 9
Nutrition and Immunity

MARTIN KUSSMANN[a,b]

[a] Nestlé Research Centre, Vers-chez-les-Blanc, PO Box 44, CH-1000 Lausanne 26, Switzerland; [b] Faculty of Science, Aarhus University, Ny Munkegade, Building 1521, DK-8000 Aarhus C, Denmark

9.1 The Immune System

As reviewing the essentials of the immune system would expand beyond the scope of this chapter, the reader shall be briefly reminded that the immune system is a complex ensemble of biological entities (cells, tissues, organs) and processes (inflammation, immune tolerance) that protect the integrity of the organism from external and internal threats.[1] External threats include microorganisms (bacteria, viruses, parasites), their toxic products (exotoxins, endotoxins, enzymes), and air- and food-borne allergens. The immune system also responds to severe trauma such as burns and physical injuries in a manner similar to the shock response that occurs with an overwhelming bacterial infection. Internal threats include:

- microorganisms that are otherwise normal commensals in the gut, respiratory and urogenital system and on the skin;
- abnormal cells (cancer); and
- the tendency of the immune system to attack itself (autoimmunity).

The two major functional components of the immune system are innate immunity and adaptive immunity. Innate immunity is by definition present prior to exposure to antigen and therefore can not be customized (adapted).

RSC Food Analysis Monographs No. 9
Mass Spectrometry and Nutrition Research
Edited by Laurent B. Fay and Martin Kussmann
© The Royal Society of Chemistry 2010
Published by the Royal Society of Chemistry, www.rsc.org

It is phylogenetically conserved in all multicellular organisms.[2] The innate immune system is designed to recognize a few highly conserved motifs present in microorganisms. In contrast, the adaptive immune system is highly specific and increases both in specificity and magnitude with repeated exposure to antigen.[3] Unlike the innate immune system that recognizes pathogens through specific molecular markers, the cells of the adaptive immune system need to be educated to discriminate between "self" and "non-self".

9.2 Nutrition and Immunity

9.2.1 Macro- and Micronutrients

Nutritional status plays an important role in the functioning of the immune system.[4] Dietary proteins, carbohydrates and fats, as well as micronutrients (vitamins and minerals), interact with immune cells systemically in blood, regional lymph nodes and in the specialized gastrointestinal immune system.[5] The role of specific macro- and micronutrients in immune function has been extensively discussed in the literature in a dedicated issue of *Nutrition*. For example, Vanderhoof summarized the importance of carbohydrates, primarily seen as a source of energy, in immuno-nutrition.[6] Various amino acids such as glutamine,[7] arginine,[8] taurine[9] and sulfur-containing amino acids[10] have been reviewed in terms of their immunomodulatory properties. In addition, (poly)-unsaturated fatty acids have an impact on immune status and the role of ω-3 fatty acids has been specifically discussed by Alexander.[11]

Declines in both specific and non-specific immunity have been reported in association with under-nutrition and protein deficiency.[12] There is also considerable evidence that deficiencies of trace elements such as iron, zinc, selenium and copper, and vitamins A, B6, B12, folic acid, C, D and E are associated with impairments in immune function.[4,13–16] While natural food has the potential to supply most of the essential macro- and micronutrients, dietary supplements and/or enriched foods might be of great value in stress situations such as premature life, ageing or disease, or extreme conditions like exercise.

In terms of nutritional support and promotion of a healthy immunity, three levels of care are being pursued. The primary level consists of the provision of all key micro- and macronutrients to sustain immune cells and functions. The second level corresponds to the modulation of the immune system to appropriately respond to specific but broad areas of concern—an example would be proper management of inflammation. The tertiary level reflects nutritional interventions tailored to the individual immune disposition and situation and is hence part of preventive and personalized nutrition.[17]

9.2.2 Malnutrition, Under-nutrition and Immunity

The causal relationship between famine and pestilence has been known for millenna.[18] Malnutrition and infection are the two major obstacles for health, development and survival worldwide, with malnutrition being the commonest cause of immunodeficiency worldwide.[14] While infection and malnutrition aggravate each other, nutrition does not impact all infections to the same extent.[19] Nutritional deficiency is commonly associated with impaired immune responses, especially cell-mediated immunity, cytokine production, secretory antibody response and affinity.[14] The proper consumption and absorption of micronutrients is essential for optimal immune responses (*e.g.* zinc, iron, selenium, vitamin A, pyridoxine, vitamin E).[20] But macronutrient balance also plays a role: animal proteins are generally superior to vegetable proteins in maintaining immunity. Moreover, there are subtle differences in immune responses of animals fed casein- and whey-based diets.[18] During periods of stress and illness, production of glutamine—the most abundant intracellular amino acid—is upregulated as branched chain amino acids are metabolized by skeletal muscle.[21] Glutamine is an important energy source for intestinal enterocytes and for rapidly proliferating cells such as immunocytes which react to challenges imposed by injury and illness.[22]

The immune system is undergoing permanent renewal and produces millions of immune cells daily. Immune cell renewal is elevated during infectious disease, and recovery depends on the rate of cell division between the invading microorganism and that of immune cells. The immune system uses both macro- and micronutrients involved in DNA, RNA and protein synthesis.[23] Thus, under-nutrition has a strong influence on the immune system at all ages but mainly in growing and aged humans, *i.e.* when the body's nutritional reserves are limited. At those life stages, under-nutrition is a major factor leading to immunodeficiency and thereby to higher infection rates.[23] This chapter there-fore dedicates a special section to nutrition and immune function in newborns and infants as well as in elderly.

9.3.   Mass Spectrometry in Immunology– Immunoproteomics

Immunoproteomics with a perspective from biomarker discovery to diagnostic applications was recently reviewed by Tjalsma *et al.*[24] The concept here is to refine, multiplex and accelerate mass spectrometry- and proteomics-based antibody analytics and diagnostics.

9.3.1   Mass Spectrometry in Immune-related Nutritional Intervention

In 2008 de Roos and McArdle presented their view on how to deploy pro-teomics as a platform for biomarker development in nutrition research.[25] This paper is probably the most comprehensive summary of (immune-related) nutritional studies as monitored by mass spectrometry and much of the work cited therein is either discussed directly below (with those being immune-rele-vant in a broader sense) or further on in this chapter. These studies are mostly based on the classical proteomic approach, *i.e.* protein separation by two-dimensional (2D) gel electrophoresis followed by protein spot excision, in-gel protein digestion and mass spectrometric protein identification, the latter mainly deploying matrix-assisted laser desorption ionization (MALDI) mass fingerprinting but also liquid chromatography tandem mass spectrometry (LC-MS/MS).

For example, de Roos and McArdle demonstrated by MS-based proteomics that two structurally very similar dietary conjugated linoleic acid (CLA) isomers had divergent mechanistic effects on atherosclerosis development and insulin resistance in apoE 2/2 mice.[26] Equally relying on proteomics, their group could furthermore show that:[26,27]

- the consumption of dietary fish oil and *trans*10, *cis*12 CLA induced differential expression of long-chain acyl-CoA thioester hydrolase protein as an indicator of fatty acid β-oxidation in the liver; and
- the consumption of dietary fish oil, olive oil and *trans*10, *cis*12 CLA induced differential expression of adipophilin protein as an indicator of selective hepatic lipid accumulation and triglyceride secretion.

Arbones-Mainar and colleagues followed a proteomics-rooted approach to better understand the mechanisms by which olive oil fatty acids, or its minor antioxidant constituents, may affect hepatic metabolic pathways, oxidative stress and, eventually, atherogenesis.[28]

Mitchell *et al.* evaluated matrix-assisted laser desorption ionization time-of-flight (MALDI-ToF) mass spectrometry as a method for revealing protein biomarkers of an immune-modulating diet.[29] They identified α-2–HS glycoprotein B-chain as a biomarker of fruit and vegetable intake; during separate feeding periods, 38 participants ate a basal diet devoid of fruits and vegetables and a basal diet supplemented with cruciferous (broccoli) family vegetables. At the end of each seven-day feeding period, serum samples were obtained and abundant proteins were depleted. MALDI-ToF spectra were analyzed using peak picking algorithms and logistic regression models. Two significant mass peaks could classify participants based on diet (basal *vs.* cruciferous) with 76% accuracy. One peak was identified as the B-chain of α-2–HS glycoprotein, a serum protein previously found to vary with diet and be involved in immune function and insulin resistance.

A 2D gel- and MS-based proteome study published by the Daniel group also aimed to reveal protein biomarkers of dietary intake; they identified alterations in peripheral blood mononuclear cell (PBMC) proteins of healthy males ingesting flaxseed for a week. PBMCs from the same study subjects were also exposed *ex vivo* to physiological concentrations of enterolactone (a metabolite produced from dietary lignans by colonic microflora) to assess whether similar effects on the proteome could be observed as those caused by dietary flaxseed. A fairly robust change in 16 PBMC proteins was observed upon flaxseed consumption. Four out of these 16 protein changes were similar to those found in blood mononuclear cells exposed *ex vivo* to enterolactone:[30]

- enhanced levels of peroxiredoxin;
- decreased levels of long-chain fatty acid β-oxidation multi-enzyme complex proteins; and
- levels of glycoprotein IIIa/II.

However, most of the more traditional nutritional interventions reported to date that aim to improve immune condition have not (yet) deployed mass spectrometry to assess status, bioavailability and metabolism of the nutrient(s) or dietary antigens of interest and their effects at molecular level.[31,32] Rather, few- or single-point read-outs were performed with classical assays based for example on high performance liquid chromatography (HPLC) and internal standards.[33] Immune studies are often based on mononuclear cells cultured in standardized systems applying a chosen stimulus and a single endpoint. The analytical methods to assess immune response to nutritional or other stimuli have been reviewed[34] and encompass mainly immune cell-based assays (*in vitro* models or *ex vivo* samples), cytokine measurements, flow cytometry and delayed-type skin hypersensitivity testing.

While it is reasonable to link nutrient intake with immunological outcomes, it would be desirable to make measurements in between the very beginning and the very end of an intervention study, namely the known quantity of the nutrient as orally taken in and physiological endpoints. One of the rare examples of such an investigation is the study by Woelkart *et al.* who looked at bioavailability and pharmacokinetics of *Echinacea purpurea* and their interaction with the immune system.[35] *Echinacea* is a widely used herbal remedy for the prevention and treatment of the common cold. In order to compare the bioavailability of alkamides (the main lipophilic *Echinacea* constituents) from liquid and tablet preparations of *E. purpurea* in humans and to study the effects on *ex vivo* stimulated blood cells, a randomized, single dose, crossover study was performed. Liquid chromatography coupled to electrospray ionization ion-trap mass spectrometry (ESI-IT-MS) was used to determine the content of alkamides in serum. Both *E. purpurea* preparations led to the same effects on the immune system according to the concentration of pro-inflammatory cytokines.

Modern nutritional intervention studies should try to follow the fate of the nutrient at molecular level, and then add bioavailability and bioefficacy data to intake information in order to establish a better causality between the nutrient and its effect.[36] Mass spectrometry is ideally suited for nutrient bioavailability, bioefficacy and metabolism studies thanks to the intrinsic sensitivity and information-rich spectra it can deliver for almost all organic compounds.[36] Apart from these assets for targeted nutrient and metabolite analysis—especially in the highly sensitive and selective single reaction monitoring (SRM) and multiple reaction monitoring (MRM) mode as performed on triple quadrupole (QqQ) machines for proteomic purposes[37,38]—MS can empower nutritionists "to be prepared for the unexpected": it can elucidate nutrient metabolism in a holistic way and enables metabolite discovery.[39] The latter aspect is of particular importance to molecular nutritional research because the desired health effects of nutrients as enriched or "remixed" in functional food must not be compromised by less desirable side effects. In other words, health promotion through adapted nutrition "must get everything right".

9.3.2 Mass Spectrometry in Discovery of Immune Markers and Targets

Markers to measure immunomodulation in human nutrition intervention studies have been reviewed by Albers *et al.*[40] These markers do not descend from "omic" approaches but rather reflect targeted measurements of biomolecules or read-outs from cellular assays, typically performed in (pre-) clinical settings. The role of proteomics deployed for the discovery of biomarkers in gastro-intestinal diseases has been outlined by Song and Hanash,[41] who describe protein microarrays, mass spectrometry-based proteomic tools and guidelines for biomarker development. The authors state that inflammatory bowel disease (IBD) and irritable bowel syndrome (IBS) represent diseases for which biomarkers are still pending and that proteomics may help in identifying them. Within IBD, better markers are needed to distinguish between Crohn's disease and ulcerative colitis, and to improve diagnosis and prediction of therapy.

A review by Purcell and Gorman[42] on mass spectrometry-based studies of immune responses discusses the role of proteomics in:

- elucidation of the cytotoxic T lymphocytes;
- T cell—B cell co-operation and antibody secretion;
- defining targets of T cell immunity;
- discovery of T cell epitopes;
- analysis of antigen presenting cell (APC) surface proteins; and
- sequencing of major histocompatibility complex (MHC)-bound peptides.

Addressing a more specific immune context, Weingarten *et al.* discussed the application of mass spectrometric protein analysis to biomarker and target finding for immunotherapy.[43] Their article focuses on regulatory T cells that play a central role in maintaining the immunological balance and inhibiting T cell activation both *in vivo* and *in vitro*. The enhancement of suppressor cell function is suggested as a target for immunotherapeutic treatment of immune-mediated disorders such as multiple sclerosis and Crohn's disease. The proposed method of choice to elucidate the still unclear effector functions of regulatory T cells is differential proteomics of human and murine T cell populations. Applying such an approach, the same group at Protagen AG plus other colleagues have assessed the human CD4+ CD25+ regulatory T cell proteome and identified galectin-10 as a novel marker essential for their anergy and suppressive function.[44]

Cereals are the most important nutritional component in the human diet. Food-induced allergic reactions to these substances therefore have serious implications and exhaustive diagnosis is required. Such diagnosis is still difficult

because of the incomplete knowledge about major cereal allergens. In particular, few food-induced allergic reactions to maize have been reported and no
information on the allergenic proteins is available. Having observed several
anaphylactic reactions to maize, Pastorello *et al.*[45] aimed to identify major
maize allergens and their cross-reactivity with other cereals, as well as to peach,
because the majority of patients also reacted to *Prunoideae* fruits. Twenty-two
patients that showed systemic symptoms, positive skin prick tests and serum-
specific immunoglobulin E (IgE) antibodies after maize ingestion were selected.
The IgE reactivity pattern was identified by sodium dodecyl sulfate polyacrylamide gel electrophoresis (SDS-PAGE) and immunoblotting. The major
allergen identified was then purified by HPLC and characterized by mass
spectrometry.

Proteomics in humans with auto-immune diseases has been reviewed by
Chan and Utz[46] with a discussion of associated effects in inflammation. The
diagnostic and therapeutic potential of glycans in inflammation has been
assessed by Dube *et al.*,[47] with particular emphasis on glycosylation changes
resulting from chronic inflammation. The review emphasizes the challenge of
glycomics and glycoproteomics (*i.e.* the analogue of proteomics at glycan and
glycoprotein level); glycan biosynthesis is not template- but enzyme-dependent
(glycosyltransferases and glycosidases form glycans on lipid and protein scaf-
folds), and therefore renders the global, quantitative analysis of glycan
expression a daunting task. However, the functions of glycans found at sites of
chronic inflammation are relatively well-defined compared to, for example,
cancer-associated glycans and their changes.

Aguiar *et al.* presented a mass-spectrometry based, clinically relevant assay
for the quantification of C-reactive protein (CRP), a well-established and
clinically relevant marker of inflammation.[48] Exact quantities of two synthetic
[13]C-labeled CRP tryptic peptides were added as internal standards to the
sample prior to chemical treatment, tryptic digestion and LC-MS quantification *ex vivo*. The method was applied to the quantification of urinary CRP from
a study of drug-induced nephrotoxicity.

9.4   Proteomics of Intestinal Epithelial Cells

A series of recent publications focused on intestinal epithelial cells (IECs)
harvested *ex vivo* or cultivated *in vitro* for functional studies of inflammation-
related gut disorders. An *in vitro* proteome analysis of intestinal epithelial cells
has demonstrated the cytokine-induced synthesis of proteins involved in the
amplification of the inflammatory response such as heterogeneous nuclear
ribonucleoprotein JKTB, interferon-induced 35-kDa protein proteasome sub-
unit LMP2 or arginine metabolism-related enzymes (tryptophanyl- tRNA
synthase, indoleamine-2,3-dioxygenase and arginosuccinate synthetase).[49]

Shkoda *et al.* presented protein expression profiles in the intestinal epithelium from patients with inflammatory bowel disease.[50] The scientific rationale
behind this work was IEC function alteration shown to be critical in initiation

and progression of chronic intestinal inflammation in the genetically susceptible host. The 2E MALDI-MS proteomic study compared ileal and colonic primary IECs from patients with Crohn's disease, ulcerative colitis to those from non-inflamed controls. Among the 21 proteins found regulated relative to the normal IECs, nine reached statistical significance and the most pronounced changes were detected for programmed cell death protein 8 and annexin 2A. Moreover, changes in expression of proteins implicated in signal transduction, stress response and energy metabolisms were found in IBD patients. A further interesting observation was the differential expression of the signal transduction regulator Rho GDIα, an inhibitor of cell cycle progression and mediator for pro-apoptotic mechanisms. The induction of Rho GDIα has been associated with the destruction of epithelial cell integrity and increase in intestinal permeability.

The same group deployed a proteomics approach to investigate the role of interleukin-10 (IL-10) to block endoplasmic reticulum stress in IECs.[51] Primary IECs from IL-10 − / − mice and IBD patients revealed increased expression levels of the glucose-regulated endoplasmic reticulum stress protein (grp)-78 under conditions of chronic inflammation. Primary IECs from both inflamed IL-10 − / − mice and IBD patients demonstrated activated endoplasmic reticulum stress responses in the intestinal epithelium. One anti-inflammatory mechanism of IL-10 seems to root in the inhibition of inflammation-induced ER stress response by modulating ATF-6 nuclear recruitment to the grp-78 gene promoter. The authors concluded that loss of regulation with respect to endoplasmic reticulum responses in the epithelium may contribute to the pathogenesis of chronic intestinal inflammation.

The Déchelotte group compared the proteomes of human intestinal epithelial HCT-8 cells *in vitro* after glutamine supplementation under non-stimulated and inflammatory[52] and apoptotic conditions.[53] Glutamine (Gln) is an important amino acid for the enterocytes. It promotes intestinal growth and maintains gut structure and function, especially during inflammation, where the endogenous Gln stores are rapidly depleted. Increased gut proteolysis, in addition to a reduction of mucosal protein synthesis, may lead to mucosal atrophy in the absence of adequate nutritional supply. Two-dimensional gel and MS-based proteomics were utilized to characterize glutamine effects on the human intestinal epithelial HCT-8 cell line under non-treated and pro-inflammatory conditions.[52] Under non-stimulated conditions, 24 proteins were differentially expressed in response to Gln. Half of these proteins are implicated in protein biosynthesis or proteolysis and 20% in membrane trafficking. Under pro-inflammatory conditions, 27 proteins were up- or downregulated by Gln. Among these, 40% are involved in protein biosynthesis or proteolysis, 16% in membrane trafficking, 8% in cell cycle and apoptosis mechanisms, and 8% in nucleic acid metabolism.

The influence of glutamine on intestinal proteome expression in apoptotic conditions was also studied in HCT-8 cells.[53] The pharmaconutritional effects of glutamine were determined under 2 mM (physiological concentration) and 10 mM (pharmaconutritional concentration) conditions. Among 1800 protein spots revealed in both conditions, 28 proteins were differentially expressed in response to an increased glutamine concentration in the culture medium, with 24 identified by mass spectrometry. Of these, 34% are involved in cell cycle and apoptosis, 17% in signal transduction, and 13% in cytoskeleton organization. The proteome-based findings are relevant to establish the effects of glutamine on intestinal barrier function and inflammatory responses, and to open new mechanistic approaches to optimize nutritional support under specific conditions.

Intestinal epithelial cell protrusions referred to as microvilli or brush border membranes (BBMs) are specialized in digestion, uptake and transport of nutrients from intestinal lumen into the circulation. Native protein complexes in murine intestinal BBMs have been recently described.[54] The blue native PAGE (BN-PAGE) technique combined with LC-MS/MS was recruited to separate and identify native digestive protein complexes in BBMs in order to better understand the physiology and pathology of digestion and absorption. Twenty-three distinct protein complexes were found and their protein composition was determined. Overall, 55 individual proteins were identified including peptidases, enzymes of carbohydrate metabolism, membrane transporters, cytoskeletal proteins, chaperones and regulatory enzymes.

9.5   Inflammation and Nutrition

9.5.1   Definition of Inflammation

Inflammation is a basic process whereby tissues of the body respond to injury. Inflammation has been described as purposeful, timely, powerful and, as a consequence, also as dangerous, if resolution is not initiated.[55] The normal outcome of the acute inflammatory programme is successful resolution and repair of tissue damage, rather than persistence of the inflammatory response.[56] Emerging evidence suggests that a co-ordinated programme of resolution initiates during the first few hours after an inflammatory response begins. Natural resolution of inflammation is a highly complex, multifactorial and tightly controlled process driven by removal of the initial stimulus, decrease in pro-inflammatory mediators (mainly cytokines, chemokines), elimination of damaged and inflammatory cells, and promotion of repair.[57,58] Although inflammation is essential for tissue homeostasis, prolonged inflammation is a hallmark of many chronic diseases such as inflammatory bowel disease and auto-immunity. Moreover, chronic inflammation has been shown to be implicated in critical conditions such as atherosclerosis, arthritis, cancer and asthma—all leading to tissue destruction, fibrosis and impairment or loss of organ function.

9.5.2   Inflammation and Nutrition

Pro-inflammatory cytokines and oxidant molecules produced during the inflammatory response following infection and injury may be beneficial or detrimental to the patient, depending on the amounts and contexts in which they are produced. Aberrant or excessive production is implicated in inflammatory disease. Systems exist for the control of cytokine production and oxidant actions. The former include the hormones of the hypothalamo–pituitary–adrenal axis, acute phase proteins and endogenous inhibitors of interleukin (IL)-1 and tumour necrosis factor (TNF). The latter encompass endogenously synthesized antioxidants (*e.g.* glutathione) and dietary antioxidants (*e.g.* tocopherols, ascorbates and catechins). Nutrients change cytokine production and potency by influencing tissue concentrations of molecules implicated in cytokine biology (for a review see ref. 59). Monounsaturated fatty acids and ω-3 polyunsaturated fatty acids (PUFAs) suppress TNF and IL-1 production and actions, while n-6 PUFAs exert the opposite effect. Low antioxidant intake results in enhanced cytokine production and effects. The anorexia that follows infection and injury may be purposeful to permit substrate release from endogenous sources to support and control the inflammatory process. Therefore, prior as well as concurrent nutrient intake co-determines the outcome of the inflammatory response.[59]

Figure 9.1 displays schematically how inflammation can be managed by nutritional means. Probiotics, TGF-β caseinate, antioxidants and lipids influence intestinal inflammation, gut integrity, tissue repair and oxidative stress. Antioxidants and lipids also modulate acute phase proteins and the glutathione (GSH)-based redox system in the liver. Moreover, free amino acids feed into muscle catabolism and can influence the oxidative stress in the muscle.

The Daniel group applied a 2D gel- and MALDI mass spectrometry-based strategy to reveal proteomic biomarkers of dietary response in human PBMCs;

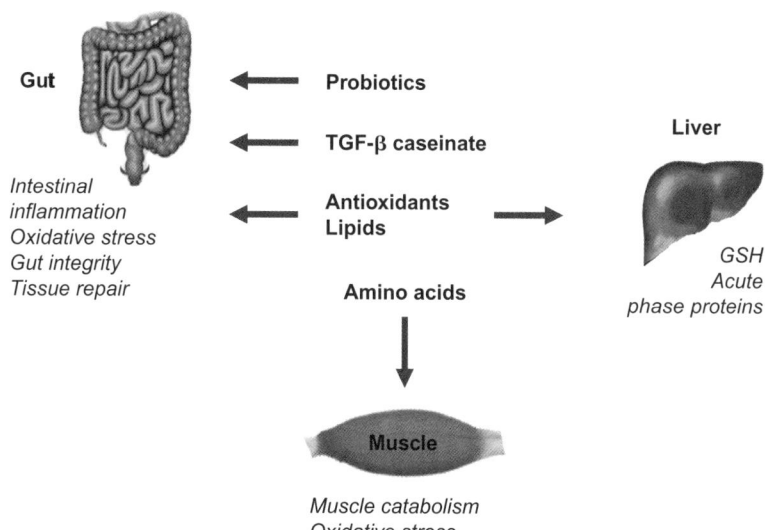

Figure 9.1 Managing inflammation by nutritional means.

postmenopausal women received a supplementation with an isoflavone extract for eight weeks.[60] Twenty-nine proteins—including several involved in the anti-inflammatory response—showed altered expression in the mononuclear blood cells following the soy-isoflavone intervention. As no overall anti-inflammatory activity of the soy intervention was observed at the level of clinically relevant inflammation markers in plasma, the PBMC proteome was suggested to be a more sensitive target to detect inhibition of inflammatory processes and to possibly respond earlier than those plasma markers classically assessed.

9.5.3   Intestinal Inflammation

Celiac disease and Crohn's disease are prototypic disorders of gastrointestinal mucosal immune function.[61] Crohn's disease is characterized by chronic inflammation of the gastrointestinal tract and associated with multiple genetic mutations, at least one of which has been clearly implicated in innate immunity. Moreover, the disease appears to involve abnormal immune responses to gut microbiota.[61] Celiac disease is a disorder of the small intestine characterized by chronic inflammation of the mucosa caused by loss of tolerance to dietary antigens. Among the associated cofactors identified are antigenic peptides in wheat, rye and barley diets. Most patients have complete remission after elimination of these cereals. The immune system cannot afford to err on the side of caution because failure to mount effective and vigorous immune responses will be exploited by pathogens. This can be exemplified by celiac disease, in which the high prevalence of HLA-DQ2 in the general population suggests an evolutionary advantage of this allele against infection, even when facing the negative effects of the coincidental affinity of gluten peptides for HLA-DQ2 to cause celiac disease. Because of these conflicting interests of the immune system, it may be unrealistic to prevent chronic inflammatory (gut) diseases and hence new treatments should be based on a molecular understanding of the disease.[62]

Inflammatory bowel disease arises in part from a genetic predisposition through the inheritance of contributory genetic polymorphisms. These gene variants may be associated with an abnormal response to normal luminal bacteria. In view of these findings, Ferguson *et al.* presented a nutrigenetic review on IBD and dietary exposure/intervention.[63] *In vivo* models of inflammatory bowel disease elucidate important mechanisms of chronic inflammation. Roy *et al.* applied nutrigenomics to an animal model of inflammatory bowel disease.[64] However, their investigation of the effects of diets enriched with eicosapentaenoic acid (EPA) and arachidonic acid (ARA) remained at transcriptomic level and did not deploy mass spectrometry.

The Bendixen group has presented one of the few mass spectrometry-based *in vivo* proteomic studies in the context of intestinal inflammation.[65] Acquisition of passive immunity by endocytosis of intact immunoglobulins (Ig) from colostrum is critical for preventing intestinal and systemic diseases in neonatal mammals. Therefore the group compared proteome patterns of healthy and inflamed gut tissues harvested from pre-term piglets to investigate the effect of inflammation on acquisition of passive immunity. A clear difference in the 2D gel electrophoretic protein patterns between healthy and inflamed intestinal tissues was revealed, suggesting that inflamed tissues failed to absorb and transfer Ig from colostrum to epithelial cells. Mass spectrometry identified isoforms of the IgA and IgG heavy chain and Ig κ and λ light chains as being absorbed by healthy intestinal tissues and indicated that colostrum protein uptake in the porcine gut is a selective process deranged in inflamed pre-term intestine.

Widening the context of intestinal inflammation, the mechanisms of salicylic acid modulating potentially pro-cancerous activity in the colon were investigated in a rat model of oxidative stress using MS-based proteomics.[66] Supplementation of salicylic acid resulted in expression changes of 55 cytosolic proteins extracted from the distal colon. The functions of these proteins related to redox balance, protein folding, protein transport, energy metabolism and cytoskeletal regulation.

9.5.4 Holistic Views of Inflammation

Innate immunity is the main mechanism for immediate responses to infection and cellular injury. Elements of innate inflammation are conserved in all multicellular organisms and predate the evolution of the adaptive immune system.[55] The recognition of "non-self" in combination with so-called "danger signals" (derived from bacteria or damaged cells) and the subsequent inflammatory response to this recognition comprises effector mechanisms of both innate and adaptive immunity.[67–69] In westernized countries, most infectious diseases of the gut are largely condemned, while gastrointestinal food allergies and idiopathic inflammatory conditions have dramatically increased: we seem to now have inflammation without infection. The absence of gut infection may have disturbed the balance between the normal bacteria that colonize the healthy gut and the mucosal immune system.[62]

Activation of the adaptive immune system is essential to mount (antigen)-specific, mainly Th1-driven, responses and to generate regulatory T cells. The latter are key players in the control of inflammation either by direct cell contact and/or secretion of immuno-regulatory cytokines such as IL-10 and/or TGF-β.[70,71]

In a network-based analysis of systemic inflammation in humans, Calvano *et al.* showed the genome-wide transcriptional response to systemic administration of bacterial lipopolysaccharides (LPS).[72] Transcriptomic analysis of PBMCs demonstrated the temporal activation of gene clusters implicated in innate immune responses, but also interconnected genes involved in cell cycle control, apoptosis, cytoskeleton protein synthesis and mitochondrial energy production. This example highlights the self-limiting character of the innate immune response in healthy conditions. By contrast, chronic inflammation manifests itself by a deregulation of functional modules interrelated in physiological conditions. The erosion of such functional networks documents the high degree of plasticity of immune cells to rapidly adapt to changing conditions such as injury, inflammatory insults or infections with the overall goal of re-establishing homeostasis. The complexity and flexibility of immuno-regulatory networks highlights the need for their holistic analysis, to which the omics sciences with chip-based transcriptomics and mass spectrometry-rooted proteomics are now beginning to contribute.

9.5.5 Mass Spectrometry in Inflammation

Despite the fragmentary understanding of inflammation networks and only emerging contributions from mass spectrometry-driven, holistic proteomic studies, inflammation is already a major target for dietary intervention with bioactive food ingredients. Several nutritional strategies, including n-3 PUFA, antioxidants vitamins, plant flavonoids, prebiotics and probiotics are being explored with the aim of dampening chronic inflammatory processes. However, nutritional studies still largely deploy cell cultures and animal models, and the potential of extrapolating to human nutrition remains limited. Therefore, more studies in human subjects and holistic, non- or minimally invasive readouts are urgently required. Mass spectrometry clearly has to expand its role here as the tool of choice to comprehensively interrogate easily accessible body fluids such as blood,[73,74] urine,[75] saliva,[76] tears[77] and nasal fluid.[78]

Although nutritional studies have focused on therapy of inflammatory conditions and appropriate nutrition may lower the risk of such conditions, strong molecular evidence of this effect is currently lacking.[79] This said, naturally occurring "nutraceuticals", especially antioxidant bioactives such as plant phenols, vitamins, carotenoids and terpenoids, have revealed benefits by tempering sustained inflammation accompanying chronic disease.[80] Targeted genes directly involved in inflammation encompass cyclo-oxygenase-2 (COX-2), TNF-α, IL-1, phospholipase A2, 5–lipoxigenase (LOX) and inducible nitric oxide synthase (iNOS). Almost a thousand plant extracts were screened for potential modulators of COX-2 expression.

9.6 Allergy and Nutrition

9.6.1 Definition of Allergy

The term allergy is understood as the overshooting, IgE-mediated response of an organism towards an allergen.[81] Atopy means, in more general terms, the disposition for the development of allergic symptoms.[82] Allergy can manifest in various forms such as neurodermitis (affecting the skin) and asthma (affecting the respiratory tract).[81] Accordingly, allergens can be classified according to their channels of interaction with the host: airborne allergens invade the respiratory system, food allergens are taken up by the gastrointestinal tract and contact allergens act through the skin. The channel of invasion does not necessarily correspond to the locus of allergy manifestation: some food allergens can, for example, provoke allergic reactions in the respiratory tract.[81]

Allergy is mainly governed by Th2 cells, which express the interleukins (ILs) 4, 5, 6, and 13. Autoimmunity is controlled by Th1 cells expressing IL-2, IL-12, IL-18, interferon (IFN) α and γ, as well as TNF-α and TNF-β. T-regulatory cells (Tregs), secreting IL-10 and TGF-β, control the balance between Th1 and Th2 cells, and regulate in this way the specific allergen response and maintain normal immunity.[83]

9.6.2 Allergy is a Public Health Issue

Over the last 25 years, the occurrence of allergy has dramatically increased and the World Health Organization (WHO) has declared it as one of the epidemics of the 21st century. The so-termed "hygiene hypothesis" proposes a paucity of microbial exposure during childhood as one of the causes of the allergy epidemic in Western countries.[84] Western infants show a delayed acquisition of several gut microbes and a reduced turnover of strains in the microbiota, indicating exposure to a low variety of environmental bacteria.[85]

9.6.3 Allergy Markers

Most of the to-date identified 20 or more allergy-associated genetic markers are rather indicators of inflammation than of allergy. Kornman *et al.* described a nutrigenomics strategy to better understand the associations between genetic variations, the susceptibility to inflammation and the nutritional intervention potential.[86] Important allergy markers accepted to date are IL-10,[87] TGF-β, TLRs,[88] PD-1 and CTLA-4.[89] Yet, specific IgE levels are successfully used as indicators for an allergic condition. Roughly 150 genes are suspected to be linked with the multiple phenomena of the three allergic diseases—atopic dermatitis, hay fever and asthma.[90] Today, only few gene trait links for allergy susceptibility and pre-disposition are established, one of which is the identification of a susceptibility locus for asthma-related traits on chromosome seven revealed by a genome-wide scan in a Finnish founder population.[91] Figure 9.2 shows a large-format 2D gel proteome display of PBMCs.

9.6.4 Food Allergy

Food allergy is an adverse reaction to food or food additives with an underlying immunological mechanism. Its incidence in young children and among adults is approximately 1.3% and 0.3%, respectively. Parental history of atopy is a significant causal factor and exposure to common allergenic foods in infancy increases risk. For these reasons, exclusive breastfeeding and maternal avoidance of peanut, egg, fish and dairy products during lactation have been recommended and shown to reduce the occurrence of food allergy.[93] The

MW
[kDa] 2 ←————————————————— **pI** —————————————————→ **11**

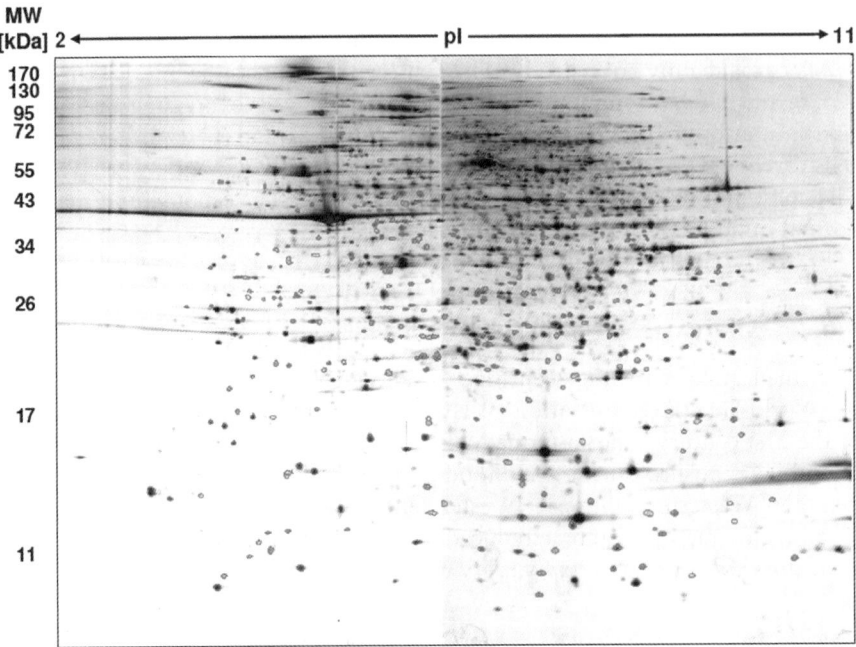

Figure 9.2   Large-format 2D gel proteome display of PBMCs. PBMCs represent an intensely assessed immune cell population because they are available in large numbers and by minimally invasive means, *i.e.* blood sampling. The gel spans a pI range from 2 to 11 and a Mr range from >10 to *ca.* 200 kDa. In this particular study on allergy biomarkers,[92] typically 2000 protein spots were detected per gel, ∼1200 spots were matched, and ∼700 spots were matched and quantified across all technical and biological replicates.

consequences of breastfeeding and early nutrition are discussed in more detail in a subsequent, dedicated section.

Many food allergens have been identified and these stimuli are often structurally well characterised, typically by mass spectrometry of the implied proteins and peptides. This source of risk necessitates detecting and monitoring (potential) allergens before, during and after food processing.[94] A list of the ten most sensitising proteins has been proposed. Although this may vary from country to country, these proteins basically derive from egg, fish, shellfish, milk, soy, wheat, peanuts, tree nuts, citrus fruits and sesame seeds. Most of these food allergens are glycoproteins and most in the range 14–40 kDa.[93] These physicochemical characteristics render them ideal analytes for mass spectrometry and proteomics, with their power to identify, sequence and quantify proteins and post-translational modifications such as glycosylation, and to differentiate between protein isoforms.[95,96] Cow milk protein allergy is still an increasing problem for infants. MALDI-ToF-MS is well suited to address this concern, as it has been extensively used to characterize allergens in cow milk.[97] Further examples for mass spectrometric efforts in protein allergen characterisation are:

- identification of the hazelnut 11–S allergen;[98]
- discovery of sesame seed allergens;[99] and
- immunological analysis of shrimp allergens.[100]

In contrast to the advanced level of understanding about allergen structures, the molecular mechanisms deciding on a normal or an allergic reaction (*i.e.* the consequences of allergen exposure for the host) are incompletely understood. Prediction of allergy risk and onset is mainly based on family history data. In terms of individual disposition, genetics and environmental influence are difficult to dissect. The environmental imprinting as a counter-player of the genetic determination is most important during pregnancy, the weaning period and in early childhood (see also Section 9.8 on "Early Nutrition").

Circulating leukocytes (or PBMCs) are good objects for proteomic studies of an individual's immune status.[101] They are available in large amounts from healthy and diseased subjects, can be harvested by minimally invasive means and cultured under near-physiological conditions. Moreover and importantly, PBMCs have a normal active metabolism.[102] Differential proteomics of PBMCs require a sufficient number of biological and technical replicates in order to understand the pronounced and meaningful inter-donor variability in protein profiles and to discern it from the undesired experimental variations.[103]

9.7 Gut Mucosal Immunity, Intestinal Microbiota and Probiotics

9.7.1 Gut Mucosal Immunity

Second to the respiratory tract, the gastrointestinal tract is the body's largest tissue boundary with a surface area of *ca.* 300 m^2.[21] It interacts with nutrients, exogenous compounds and gut microbiota, and its condition is influenced by these environmental factors and host genetics. Intestinal functions such as digestion, nutrient absorption, barrier integrity, motility and mucosal immunity are all under complex regulatory control.[104] Moreover, the intestine is the primary immune organ of the body represented by the gut-associated lymphoid tissue (GALT) exerting innate and acquired immunity. Three constituents are in permanent contact and dialog with each other—the microbiota (see Section 9.7.2), the mucosal barrier and the local immune system.[105] Figure 9.3 shows how mass spectrometry and omics tools can come into play to investigate the intestinal immune system; at the gut barrier, enteric bacteria enter into crosstalk with the mucosal immune system.[55]

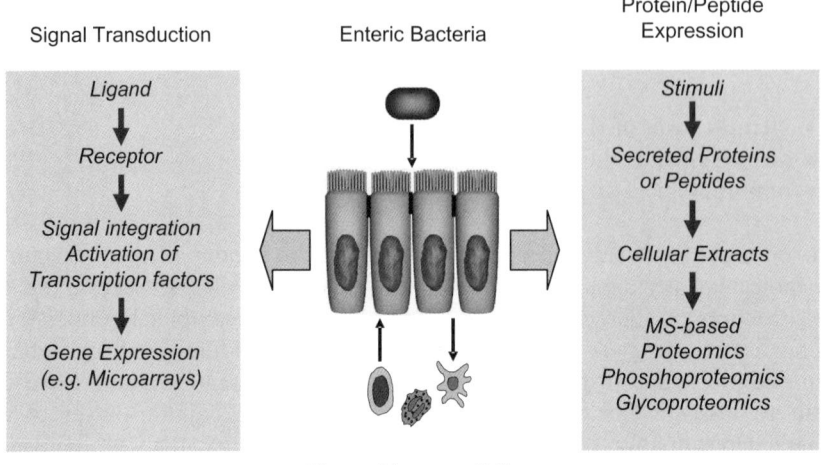

Mucosal Immune Cells

Figure 9.3   At the gut barrier, enteric bacteria enter into cross-talk with the mucosal
immune system.[58] This interface can be sampled by Omics techniques to
elucidate signal transduction and protein/peptide secretion. Signal transduction can be globally assessed by gene microarrays whereas MS-based
proteome analysis can reveal immune biomarkers.

9.7.2   Gut Microbiota

Humans and other mammals are colonized by a vast, complex and dynamic
consortium of microorganisms. In fact, adult humans are numerically more
prokaryotic than eukaryotic: estimates are that 90% of our cells are microbial,
whereas only 10% are human.[106] The impact of these indigenous microbial
communities on our physiology is likely to be most pronounced in the intestine
because this organ harbours the vast majority of our bacteria. Microbial
densities in the proximal and middle small intestine are relatively low but
increase dramatically in the distal small intestine ($\sim 10^8$ bacteria per g of
luminal contents) and colon (10^{11}–$10^{12}\,\mathrm{g}^{-1}$).[106]

Gut microbes conduct a multitude of biochemical reactions and can be
collectively thought of as a metabolically active "organ." This metabolic entity
plays a critical role in nutrition, degrading a number of dietary substances that
are otherwise non-digestible.[107] One "raison d'être" for this metabolically
active microbial society is to harvest energy from nutrients, especially
carbohydrates.[108]

The microbiota in the adult human body consists of an enormous biomass of
$> 100\,000$ billion bacteria spread over > 400 different species which generate
intense metabolic activity, mainly in the colon, and play an important physiological role in the host.[105] The microbiota has a major impact on gastrointestinal and mucosal immune functions. Colonization of the gut by
commensal bacteria has been shown to alter intestinal physiology of the host by
modulation of genes implicated in nutrient absorption, mucosal defences and

xenobiotic metabolism.[109,110] It is now established that one of the essential functions of the colonic microbiota is its ability to resist colonization by any new strain of bacteria from the exterior.[105]

9.7.3 Probiotics

Probiotics are live microbial food and feed supplements, which beneficially affect the host by improving its intestinal microbial balance.[111] Recent evidence indicates that probiotics (*e.g. Streptococcus thermophilus* and *Lactobacillus bulgaricus*) may influence both systemic and gut-associated immune responses.[112] Some probiotics enhance while others suppress immune responses.[113] Probiotics seem to act through stimulating regulatory T cells, which can activate both these responses.[114] Most of the immunobiological effects of probiotics are likely to take place in gut-associated lymphoid tissue, including Peyer's patches, in the small intestine. Due to comparable numbers of probiotic and resident bacteria at that location, probiotics may compete with luminal microbiota more successfully than in the colon, which is already heavily populated with indigenous bacteria. Furthermore, the cross-talk between probiotics and the small intestine may be different from that in the colon and it may be age-dependent.[114]

Certain probiotic strains are reported to control inflammation,[115] reduce the risk of allergy[116] and restore gut comfort in chronic painful conditions.[117] Probiotics have furthermore been reported to promote tolerance,[118] maintain intestinal immune homeostasis[119] and prevent atopy.[120] Lactobacilli are one of the most frequently used strains of probiotic bacteria in the management of gastroenteritis, inflammatory bowel diseases[121] and atopic diseases.[122] These probiotic bacteria are also suggested to regulate immunity and promote mucosal tolerance, which is in part mediated by Treg cells.

Clinical applications of probiotics for adults encompass inflammatory bowel disease, irritable bowel syndrome, *Helicobacter pylori* gastritis and improvement of intestinal transit. For infants and children, probiotics have been administered to fight acute diarrhoea, acute childhood constipation, *Helicobacter pylori* gastritis and intestinal bacterial overgrowth. For a review of these studies see Walker *et al.*[114]

Bifidobacteria, a class of probiotics, are important components of the human intestinal microbiota, in which they occur at concentrations of 10^9–10^{10} cells per gram of feces,[123] and of fermented milk products, to which they are added mainly because of their health promoting activities.[124] The entire genome of *Bifidobacterium longum* has been sequenced.[125] Champomier-Verges *et al.*[126] reviewed mass spectrometry-based proteomic studies dealing with lactic acid bacteria. Two research interests were pursued. The first aimed to establish a systematic protein map for taxonomy and function assignment of proteins. The second axis focused on proteins, the synthesis of which is induced by various environmental factors. Such studies may give new insights for the usefulness of bacteria in human health and in the struggle against bacterial pathogens.

Tolerance to digestive stresses is one of the main factors limiting the use of microorganisms as live probiotic agents. These effects as well as technological stresses (heat, pressure, shear) are major factors affecting viability and thus the efficiency of probiotic microorganisms in food products. Proteomic analyses have shown that pre-treatment of the probiotic strain *Propionibacterium freudenreichii* with a moderate concentration of bile salts greatly increased its survival rate in subsequent challenges.[127,128] Marvin-Guy *et al.* have published a rapid identification of stress-related fingerprints from whole cells of *Bifidobacterium lactis* using MALDI-MS.[129] Guillaume *et al.* have found markers of heat shock resistance by comparing the proteomes of two *Bifidobacterium longum* strains.[130] The proteomic data were compared to and corroborated by a related gene expression study.[131]

9.7.4 Prebiotics and Synbiotics

Prebiotics have been defined as non-digestible food ingredients that beneficially affect the host by selectively stimulating the growth and/or activity of probiotic bacteria in the colon and thus improve host health.[132] As a logical extension of the probiotic concept backed up by prebiotic ingredients, combined symbiotic approaches are being pursued[133] that aim to define the right combination of beneficial gut bacteria and food ingredients which foster the growth and activity of the latter.[134] For example, one study evaluated the effects of six weeks' consumption of a symbiotic product containing *Lactobacilli* and fructo-oligosaccharides (FOS) on intestinal microbiota, self-reported intestinal function and the immune function of generally healthy adults.[135] Although no differences in self-reported improvement were found with treatment of mild gastrointestinal symptoms present at baseline, there was a significant improvement overall in symptoms and in motility in the symbiotic group compared with the placebo group. Intestinal microbiota did not change as a result of symbiotic consumption.

Liquid chromatographic separation coupled to mass spectrometric detection is the platform of choice to structurally characterize prebiotics and study their metabolic fate. This has been pursued, for example, in the context of (milk) oligosaccharides by Lebrilla *et al.*[136,137] and is further discussed in the milk-related section of this chapter. LoCascio *et al.* utilized a HPLC-chip ToF mass spectrometric approach to glycoprofiling of bifidobacterial consumption of human milk oligosaccharides (HMOs).[137] HMOs were separated from pooled human breast milk samples and several bifidobacterial strains grown on them. The oligosaccharides were isolated and purified from the supernatant and analyzed on a high-resolution ESI Q hybrid Fourier transform ion cyclotron resonance (FTICR) mass spectrometer and on a HPLC-chip-ToF-MS system. LoCascio *et al.* demonstrated strain-specific, preferential consumption of small-chain glycans secreted in early human lactation.

Apart from this latter application, mass spectrometry remains under-deployed for such purposes. However, as a key analytical element of functional genomics, mass spectrometry is expected to serve in unravelling inter-dependencies between prebiotics and probiotics, commensal host–bacterial relationships in the gut[109] as well as naturally occurring and designed symbiotic relationships.[138–140]

9.7.5 Gut Ecology

NMR- and MS-based metabolomics is uniquely suited and increasingly deployed to capture the metabolic interplays between the host metabolism, and its symbiotic and parasitic microorganisms in the colonic flora.[141,142] The characterization and mathematical modelling[143] of this metabolic cross-talk between microbiota and host should result in a better understanding of the long-term health consequences associated with an optimal or impaired microbiotic activity. Metabolomics bears also great potential for investigating the effects of dietary ingredients that target the colonic flora such as probiotics (mainly lactobacteria) or prebiotics (mainly soluble fibres).

It is extremely difficult to simulate the complex bacterial–mucosal immune interaction using *in vitro* models. Nicholls *et al.* deciphered metabolic events associated with acclimatization of germ-free rats to standard laboratory conditions.[144] Martin and colleagues modelled transgenomic metabolic effects consecutive to the inoculation of non-adapted human faecal flora in a mouse model.[141] In order to elucidate gut microbial effects under relevant conditions, animals with *a priori* sterile gastrointestinal tract and monocolonised with probiotics are now used as a suitable model, especially the gnotobiotic mouse,[109,110,145] but also germ-free piglets.[146] The latter model was deployed to investigate the effects of bacterial colonization on the porcine intestinal proteome by mass spectrometry.[146] Small intestinal protein expression patterns in gnotobiotic pigs maintained germ-free or mono-associated with either *Lactobacillus fermentum* or non-pathogenic *Escherichia coli* were studied. A common reference combined with stable isobaric tags (iTRAQ) for relative protein quantification revealed that bacterial colonization differentially affected proteolysis, epithelial proliferation and lipid metabolism, which corroborated studies of other germ-free animal models.

Our molecular understanding of how members of the intestinal microbiota degrade complex polysaccharides derives from studies of *Bacteroides thetaiotaomicron*, a prominent and genetically changeable component of the normal human and mouse gut. Colonization of germ-free mice with *B. thetaiotaomicron* (Btheta) has shown how this anaerobe modifies many aspects of intestinal cellular differentiation/gene expression to benefit both host and microbe.[108,145,147] The Btheta proteome encompasses specific functions for polysaccharide acquisition and hydrolysis, and an environment sensing system.[148] The same group undertook a combined gene expression and GC-MS-based metabolomics study; GC-MS was performed on the standard mouse chow diet and on the total caecal contents recovered from sterile and

Btheta-colonized animals. Sonnenburg *et al.* found that the predominant *in vivo* responses to Btheta-association reflected glycobiome activation:[149]

- Btheta glycosyl hydrolases correspond to the most prominent sugars in the environment;
- Btheta prefers the monosaccharides that can be metabolized most efficiently; and
- Btheta is able to degrade both plant- and host-derived polysaccharides.

Compared to the gut commensal and probiotic Btheta, *Bifidobacterium longum,* a minor member but a commonly used probiotic, has a more restricted glycan-degradation machinery but a larger repertoire of transporters[108] suggesting that *B. longum* may directly benefit from Btheta's "upstream" polysaccharide degradation.[114] To address this latter hypothesis, the Gordon group colonized germ-free mice with *B. thetaiotaomicron* and *B. longum.* Simultaneous whole genome transcriptional profiling of both bacterial species in their gut habitat and of the intestinal epithelium, combined with mass spectrometric analysis of habitat-associated carbohydrates, revealed that *B. longum* expanded the diversity of polysaccharides targeted for degradation by *B. thetaiotaomicron* (*e.g.* mannose- and xylose-containing glycans) and induces host gene expression involved in innate immunity. Although the overall transcriptome expressed by *B. thetaiotaomicron* when it encounters *B. longum* in the caecum depends upon the genetic background of the mouse, Btheta's expanded capacity to utilize polysaccharides occurs independently of host genotype and is also observed with a fermented dairy product-associated strain, *Lactobacillus casei.* Hence, this gnotobiotic mouse model provides a controlled case study of how a resident symbiont and a probiotic species mutually adapt their substrate utilization, and illustrates both the generality and specificity of the relationship between a host, a component of its microbiota and intentionally consumed microbes.[149]

The pioneering work by Gordon *et al.* documents two things:

- gut ecology is extremely complex and it takes an ecosystem approach to understand the health impact of the intestinal microbiota including probiotics;[145] and
- genomics as well as mass-spectrometry-rooted proteomics and metabolomics are the tools of choice to provide holistic mechanistic insights into this host–microbe cross-talk.[138]

9.8. Early Nutrition and Immunity

9.8.1 Immune Development around Birth

Profound immunological changes occur during pregnancy, involving a polarization of T helper (Th) cells towards a dominance of Th2 and regulatory T cell effector responses in both mother and foetus. This situation is important to maintain pregnancy through avoidance of the rejection of the immunologically incompatible foetus. During the third trimester of human pregnancy, foetal T cells are able to mount antigen-specific responses to environmental and food-derived antigens and antigen-specific T cells are detectable in cord blood in virtually all newborns indicating *in utero* sensitization. If the neonatal immune system is not able to downregulate the pre-existing Th2 dominance effectively, an allergic phenotype may develop.

Important changes occur also around birth so that the neonate's immune system becomes competent and functional, and the gut is colonized with bacteria. Mucosal immune response is primed at birth and responses generated at this time support specific immunity in later life.[150] Infants are born with a practically sterile gut, which is rapidly colonized. The predominant source for initial colonization is the maternal flora, followed by the environmental flora.[105] Exposure to bacteria during birth and from the mother's skin, and the provision of immunological factors in breast milk are among the key events that promote maturation of the infant's gut and the gut-associated as well as systemic immune systems. The maturing small intestine of the newborn is initially exposed to a large number of colonizing bacteria acquired while passing through the birth canal. In the absence of mature intestinal function (mucus production, peristalsis, *etc.*), large numbers of bacteria colonize the small intestine. This contrasts with the mature intestine, in which large numbers of colonizing bacteria are only present in the distal ileum, caecum and colon. The early exposure of the small intestine to colonizing flora is an important step in the appropriate maturation of mucosal immune system.[114] A compositional comparison of the intestinal microflora between healthy and allergic infants, for example, showed that the latter had fewer Lactobacilli and Bifidobacteria, but more Clostridia and coliform bacteria.[151] Probiotics can downregulate, namely minimize, an IgE-mediated allergic response[152] and are involved in re-establishing oral tolerance to food allergens even after sensitisation.[153]

The introduction of infant formula and solid foods exposes the baby to novel food antigens and affects the gut flora. Nutrition is the source of antigens to which the immune system must become tolerant. Nutrition provides factors, including nutrients, that themselves might modulate immune maturation and delivers compounds that influence the intestinal microbiota, which in turn affects antigen exposure, immune maturation and responses. Through these mechanisms, nutrition early in life influences and even "programmes" later immune competence, *i.e.* the ability to both mount an appropriate immune response upon infection, and develop a tolerogenic response to "self" and to benign environmental antigens.[154]

9.8.2 Milk as the Ideal Early (Immuno-) Nutrition

Figure 9.4 shows the major human milk proteins. Human milk mainly consists of caseins, α-lactalbumin, lactoferrin, albumin and various immunoglobulins.

These predominant proteins account for >99% of the milk protein mass. However, the remaining <1% encompass a complex blend of bioactive proteins and peptides, which is still far from being fully exploited at both analytical and functional level.

The mammary gland has a large metabolic potential including the large-scale synthesis of milk proteins, carbohydrate and lipids. Peng and colleagues carried out a proteomic analysis of mammary tissue to discover proteins affecting lipid metabolism.[155] Unfractionated microsomes from lactating bovine mammary tissue were separated with 1D-PAGE and identified by LC-ESI-MS/MS, yielding 703 proteins including 160 predicted transmembrane proteins. More than 50 proteins were associated with cellular uptake, metabolism and secretion of lipids. This database provides a proteomic view of the metabolic potential of the mammary gland. In a related study, the Smith group characterized the human mammary epithelial cell (HMEC) proteome;[156] they reported on a cysteinyl peptide enrichment (CPE) approach, which improved both protein sequence and overall proteome coverage. The combined analyses of HMEC tryptic digests with and without CPE resulted in ~4300 different proteins with an estimated 10% gene coverage of the human genome. CPE contributed roughly an additional 1000 relatively low abundant proteins, resulting in a further increase in proteome coverage. Almost 1400 proteins were observed with increased sequence coverage. Comparative protein distribution analyses revealed that the CPE method is not biased with regard to protein molecular weight (Mr), isoelectric point (pI), cellular location or biological function.

Secretory immunoglobulins, lysozyme, interferon and growth factors are known to confer immunological advantages to breast milk. Inhibition of bacterial pathogens and permissive growth of a protective colonic microbiota are partly promoted by breast milk.[157] Besides providing nutrition to the newborn, milk also protects the neonate and the mammary gland against infection. Breast-fed newborns have been shown to experience a lower incidence of gastrointestinal infections and inflammatory, respiratory and allergic diseases. This finding has been attributed to a diversity of protective factors in breast milk. One specific biological activity in mother's milk was reported to be the one of soluble CD14 (sCD14).[158,159] The study indicated a central role for sCD14 during bacterial colonization of the gut and suggested sCD14 to be involved in modulating local innate and adaptive immune responses, thus controlling homeostasis in the neonatal intestine. Another related study revealed an interaction between soluble Toll-like receptor 2 (sTLR2) and sCD14 in plasma and milk, proposing the existence of a novel innate immune mechanism regulating microbially induced TLR triggering.[160] A particular fraction of human milk, generated by a special chromatography based on restricted access material (RAM), was characterized by 2D LC-MS/MS in order to elucidate the protein composition and to discover novel molecules that potentially interact with sCD14.[161]

Differences were observed in the composition of intestinal bifidobacterial species depending on the type of milk fed. While *B. breve* is one of the

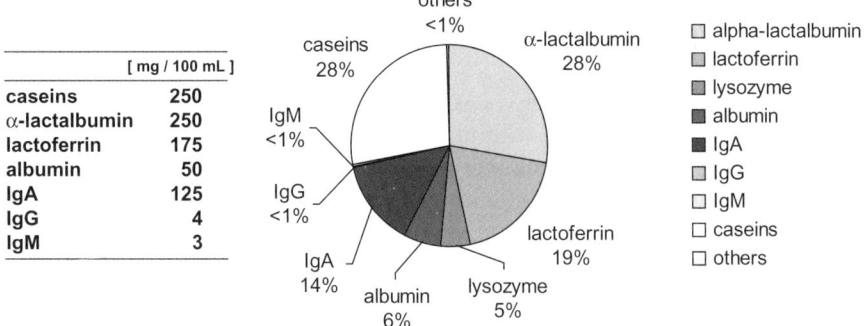

	[mg / 100 mL]
caseins	250
α-lactalbumin	250
lactoferrin	175
albumin	50
IgA	125
IgG	4
IgM	3

Figure 9.4   Major human milk proteins.

predominant species of the gut microbiota of breast-fed babies, *B. catenulatum* and *B. adolescentis* are characteristic of that of formula-fed infants.[162] Puertollano *et al.* assessed the differential effect of the bifidobacterial species identified in the intestinal microbiota of breast-fed and formula-fed infants on cytokine production by PBMCs.[163] The effects of different bifidobacterial species were analyzed individually and in combinations representing their proportions in infants under both feeding types. The effects of breast-fed and formula-fed bifidobacterial species combinations on cytokine production were not significantly different. These results suggest that the presence or absence of particular bifidobacterial species and the overall composition of the bifido-bacterial population in the infant gut could be key factors defining the immunomodulatory effect of the gut microbiota in early life.

9.8.3   Mal-/Under-nutrition and Immune Imprinting

Moore summarized the context of nutrition, immunity, and the fetal and infant origins of disease in developing countries.[164] For instance, events in early life strongly influence the adult survival prospects of rural Africans:[165] nutritional status depends highly on season in rural settings as can be found in Gambia. There, individuals born during periods of seasonal nutritional deprivation are more susceptible to mortality from infectious diseases in adult life. A permanent negative imprinting of the immune system during fetal growth caused by malnutrition appears to be a likely explanation.[165] A related study on long-term effects of perinatal nutrition was published by Ghattas *et al.*, who investigated T lymphocyte kinetics of young Gambian men depending on the nutritional status of their mothers at the time of their birth.[166] This study rooted analytically in stable-isotope labelling of T cell subsets combined with GC-MS and revealed the astonishing finding that, in healthy young Gambian men, T lymphocyte homeostasis is extremely robust regardless of perinatal nutritional compromise.

9.9. Mass Spectrometry in the Analysis of Milk

9.9.1 Milk and Health

Milk is a rich source of bioactives beneficial for human health. It is the only nutrition that has co-evolved with mankind and is therefore particularly relevant and suited to support healthy growth and development of the neonate and infant, including the maturation and maintenance of a balanced immune system.[167] Milk bioactives derive from the protein and peptide,[168] the lipid[169] and the oligosaccharide complement[136] of milk from diverse mammalian species.

The protein complement of human milk roughly splits into caseins and whey with a 50 : 50 weight/weight ratio. Bovine milk consists of 80% caseins and 20% whey proteins;[168] a review of bioactive peptides and proteins present in milk and dairy products has been published by Severin and Xia.[168] For example, caseins serve as ion carriers and precursors of bioactive peptides, whereas whey proteins have major functions in immune modulation and defence.[170]

Specific milk fractions have been shown to alleviate immune deregulations like inflammation and osteoarthritis but without addressing molecular mechanisms or directly identifying ingredients responsible for the observed effects. More recently, milk peptides and hydrolysates moved into the focus of studies of bioactive compounds. Peptides with opioid, antihypertensive, antithrombotic, immunomodulating and metal-binding activities have been described in the review by Severin and Xia.[168]

9.9.2 Milk Analytics

Due to its analytical versatility and power for structure elucidation and quantification of larger biomolecules, mass spectrometry has developed into the major contributor to comprehensive biomolecule characterisation in milk, nowadays known under the more recently coined terms milk proteomics/peptidomics, lipidomics and glycomics. Casado *et al.* recently released a comprehensive review of the protein/peptide, lipid and carbohydrate complement of human and animal milk as assessed by various mass spectrometric approaches.[171] Fong *et al.* presented an update on bovine whey protein fractionation and characterisation by proteomic techniques including chromatography, gels and mass spectrometry.[172]

Traditionally, milk and its fractions have been assessed in terms of composition, physicochemical properties and biological functions, and mass spectrometry is largely contributing to various areas of milk and dairy research (reviewed in ref. 173 and ref. 174). These studies include identification of milk protein variants and glycoforms, falsification of milk with non-dairy ingredients and identification of peptides in dairy products. Moreover, mass spectrometry has become an indispensable technique for the quality assessment of milk- and dairy based products (reviewed in ref. 175).

Most relevant to the scope of this chapter, Smolenski *et al.* have characterised host defence proteins in milk deploying a dual proteomics approach;[176] they applied both classical 2D gel electrophoresis and MALDI-ToF-MS, and a shotgun LC-MS/MS technique to bovine skim milk, whey and milk fat globule membrane (MFGM) fractions. Milk from peak lactation as well as during colostrum formation and mastitis was analyzed. In total, 2903 peptides were detected by LC-MS and 2770 protein spots by 2D gel electrophoresis. From these, 95 distinct gene products were identified, comprising 53 identified through direct LC-MS/MS and 57 through 2D gel electrophoresis and mass spectrometry. The latter stemmed from a total of 363 spots analyzed, with 181 being identified. At least 15 identified proteins are involved in host defence.

9.9.3 Breast Milk and Substitutes

Human breast milk is still considered the gold standard for neonate and infant nutrition. D'Auria *et al.* have undertaken a proteomic evaluation of different mammalian species for the potential to optimize infant formulas classically based on bovine milk and to recruit possible alternative sources for human breast milk substitutes.[177] Goat, horse, donkey and water buffalo milk were compared to human and bovine samples by 2D gel electrophoresis and mass spectrometry. In a milk protein hydrolysis study, the release of β-casomorphin-5 (BCM5) and β-casomorphin-7 (BCM7) was investigated during simulated gastrointestinal digestion (SGID) with pepsin of bovine β-casein variants, commercial milk-based infant formulas and experimental infant formulas.[178] β-Casein variants were extracted from raw milks derived from Holstein–Friesian and Jersey cow breeds. Identification and quantification of BCMs involved HPLC coupled to tandem MS. In view of the ongoing debate on β-casein health benefits, these data from SGID of infant formulas provide information for the evaluation of the potential bioactivity of bovine milk protein used in the manufacturing of infant formulas.[178] In another infant milk product-related study, lactosylated proteins of infant formula powders were investigated and this resulted in the identification of α-lactalbumin with five lactosylated peptides.[179] These may serve as protein markers to detect chemical modification induced by milk processing and/or storage.

Purification and characterization of novel peptide antibiotics from human milk has been described by Liepke and co-workers.[180] Digestion of human milk by infants was simulated by using pepsin under acidic conditions to generate peptides with antimicrobial activity. LC fractionation followed by MALDI-MS analysis allowed the identification of novel casein- and lactoferrin-derived fragments, which inhibited the growth of bacteria and yeasts.

9.9.4 Colostrum

Human colostrum (*i.e.* early breast milk) is an important source of protective, nutritional and developmental factors for the newborn. Colostrum (and other fractions) from different species were investigated mass spectrometrically by several groups mainly resulting in protein catalogues of these samples[177,181] including low-abundance proteins.[182] A recent profiling of human colostrum revealed, after immunodepletion of high abundant milk proteins, a list of 151 low abundant proteins, 83 of which have not been previously reported in human colostrum or milk.[183]

9.9.5 Milk Fat Globule Membrane

The milk fat globule membrane (MFGM) is derived from the apical region of the mammary gland epithelial cells and budded off around the milk lipids, the latter being secreted by the mammary gland cells. MFGM is considered to be similar to any other eukaryotic cell membrane and accounts for 2–4% of the total human milk protein content.[184] MFGM may therefore contain—in addition to molecules previously described to be associated with this membrane (mucins, lactadherin, adipophilin, CD 36 and butyrophilin)—other factors to date thought to be exclusively found in cellular membranes. Figure 5 shows the physical organisation and major membrane-anchored proteins of the milk fat globule membrane.[185]

Nutritional and technological aspects of MFGM material have been recently reviewed.[185] Argov *et al.* recently investigated the particle size-dependent lipid content of human milk fat globules by Raman spectroscopy and reviewed milk fat globule composition, size and distribution.[186] Despite the large body of knowledge about its unusual biochemical structure, little is known about the physiological function of MFGM for the nursing infant. As such, it bears great potential for the identification of new proteins in milk and the exploitation of these proteins for dairy product development.[187] While MFGM proteins have a low nutritional value in classical terms, they play important roles in cellular processes and defence mechanisms for the newborn. MFGM is a particularly rich source of bioactive peptides and proteins.[188] Smolenski *et al.* compared the host defence proteome in MFGM, whey and skimmed milk by direct LC-MS/MS and 2D gels plus MALDI-ToF-MS.[176] Milk samples from peak lactation, during colostrum formation and during mastitis were analyzed resulting in a total of ~2900 peptides detected by LC-MS and ~2800 protein spots resolved by 2D gel electrophoresis. Of these, 95 distinct gene products were identified, comprising 53 unravelled by the shotgun and 57 through the gel approach. At least 15 proteins were found to be involved in host protection against infection.

Several Italian groups teamed up to chart the human colostral MFGM proteome and established a 2D gel electrophoresis MFGM protein database.[184] Reinhardt and co-workers analyzed the composition of bovine MFGM by 1D-PAGE and nano LC-MS/MS proteomics and identified 120 proteins, 71% of which were membrane associated.[189] Pursuing an iTRAQ-based shotgun proteomic approach, these authors also investigated developmental changes in the bovine MFGM proteome during the transition from colostrum to milk.[190] They identified 138 proteins, with 26 being upregulated and 19 downregulated in day 7 MFGM compared with colostral MFGM. Mucin-1 and mucin-15 were upregulated in MFGM from day 7 milk. Adipophilin, butyrophilin and

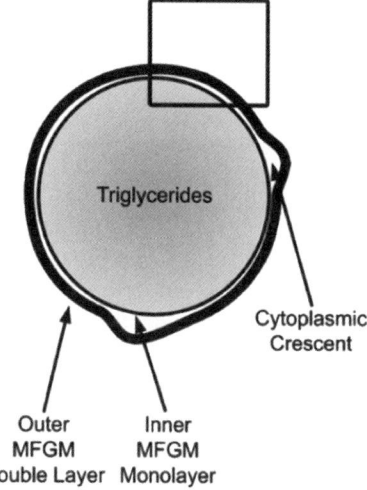

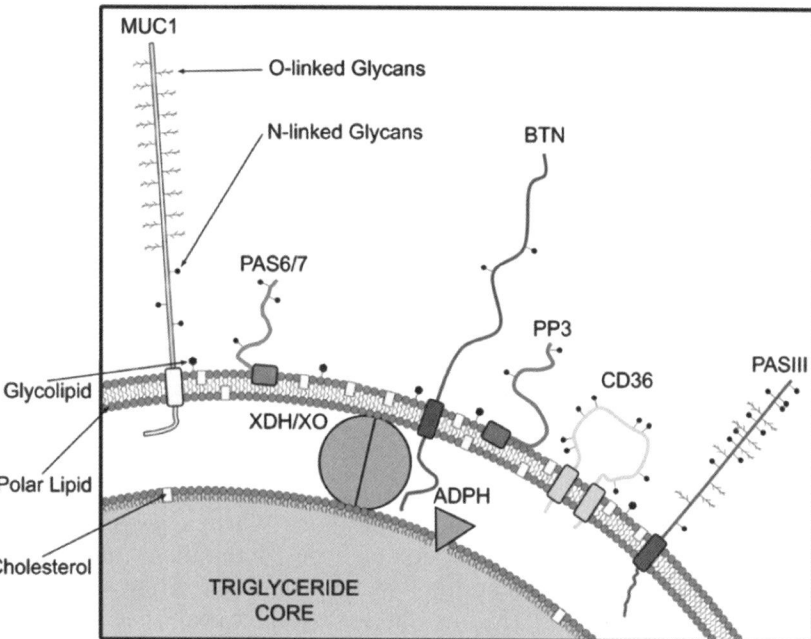

Figure 9.5 Physical organisation (top) and zoom (below) major membrane-anchored proteins of the milk fat globule membrane (MFGM): lactadherin (PAS6/7) protects from viral gut infection; mucin-1 (MUC1) and mucin-15 (PASIII) protect against rotavirus infection; adipophilin (ADPH) is involved in fatty acid/triacylglyceride (TAG) uptake and transport; fatty acid binding protein (FABP); xanthine dehydrogenase (XDH/XO) is bactericidal and anti-inflammatory; butyrophilin (BTN) suppresses multiple sclerosis; and platelet glycoprotein 4 (CD36). Modified from ref. 188.

xanthine dehydrogenase were individually upregulated in day 7 MFGM compared to the colostral fraction. Proteins associated with lipid transport, synthesis and secretion were also upregulated in day 7 MFGM. In contrast, apolipoproteins A1, C-III, E and A-IV were downregulated in day 7 MFGM relative to colostral MFGM.

Affolter *et al.* took a more targeted proteomic approach to compare MFGM-enriched milk fractions from different sources.[191] Applying a strategy based on multiple reaction monitoring (MRM) and labelled, proteotypic peptides as internal standards, they quantified seven bioactive MFGM proteins in absolute terms, namely lactoferrin, α-lactoglobulin, mucin, fatty acid binding protein (FABP), lactadherin, xanthine dehydrogenase/oxidase, adipophilin and butyrophilin.

Wilson *et al.* elucidated differences in sugar epitopes on human and bovine MFGM.[192] Their data indicate that human milk may provide different innate immune protection against pathogens compared to bovine milk as evidenced by the presence of Lewis b epitope (a target for *Helicobacter pylori*) on human but not bovine MFGM mucins.

9.9.6   Milk Protein Modifications

A considerable effort of MS-based milk protein research has focused on the elucidation of post-translational modifications such as glycosylation and phosphorylation.[193,194] For instance, Kjeldsen *et al.* aimed to completely characterize post-translational modification (PTM) sites in the bovine milk protein PP3 by tandem mass spectrometry with electron capture dissociation (ECD) as the last stage.[195] In their approach, termed "reconstructed molecular mass analysis" (REMMA), the molecular mass distribution of the intact protein is measured first, revealing the extent and heterogeneity of modifications. The protein is then digested, peptides are separated by reversed phase (RP) HPLC and analyzed by Fourier transform mass spectrometry (FTMS). Vibrational excitation (collisional or infrared) or electron capture dissociation of peptide ions provides protein identification. When a measured peptide molecular mass suggests the presence of a post-translational modification, vibrational excitation determines the type and structure of the modification, while ECD determines the PTM site. Chromatographic peak analysis continues until full sequence coverage is reached, after which the molecular mass is reconstructed and compared with the measured value. Agreement indicates that the PTM characterization is complete. This procedure has been applied to the bovine milk PP3 protein containing 25% modifications by weight and yielded all known modifications (five phosphorylations, two *O*- and one *N*-glycosylation) as well as a previously unreported *O*-linked NeuNAc-Hex-[NeuNAc]HexNAc group at Ser^{60}. FTMS-based REMMA can serve as the basis for high-throughput, high-sensitivity PTM characterization.

Protein alterations[196,197] and covalent complexes between milk proteins (*e.g.* caseins and β-lactoglobulin)[198] have also been investigated. Casein micelles, for example, formed by interaction of milk caseins and calcium phosphate, maintain a supersaturated calcium phosphate concentration in milk, providing the newborn with sufficient calcium phosphate for the mineralization of the rapidly growing calcified tissues. The phosphorylation state of caseins plays an important role in the interaction with calcium phosphate and thereby in the organization of the micelles.[199] Other trace elements associated with milk proteins, such as iron in lactoferrin,[200] are important constituents to provide the newborn with essential nutrients. Inductively coupled plasma (ICP) MS has the analytical potential for "element-tagged" proteomics of milk, resulting in quantitative information on multi-element distribution patterns in different milk sources.[201]

9.9.7 Cryptomes

Bioactivities of peptides encrypted in major milk proteins are latent until released and activated, *e.g.* during gastrointestinal digestion or food processing. Bioactive peptides can be produced *in vivo* following intake of milk proteins. Moreover, the proteolytic system of bacterial species used in fermentation (*e.g.* yogurt, cheese) can contribute to the liberation of bioactive peptides or precursors thereof.[202] A wide range of proteins contain concealed functional units that can be liberated to generate novel bioactivities. Autelitano *et al.* term this "hidden" subset of peptides, residing within the proteome, the "cryptome",[203] and it is suggested to represent a vast array of cryptic peptides or "crypteins", with manifold bioactivities that can be liberated from the parent protein *via* proteolytic cleavage. Mass spectrometry is the obvious and powerful tool to study those "cryptomes".

9.9.8 Milk Allergens

Despite all these health beneficial effects of milk proteins, milk is also a source of protein allergens. Natale and co-workers characterized milk allergens by 2D gel electrophoresis immunoblotting and mass spectrometry.[204] The serum from 20 milk-allergic subjects was searched for major cow's milk allergens followed by MALDI-ToF-MS identification of the proteins. Zeece *et al.* investigated the effect of high-pressure treatment on *in vitro* digestibility of β-lactoglobulin (β-LG) under simulated gastric conditions using pepsin.[205] The proteomic study, based on one-dimensional (1D) gels and MALDI-ToF-MS, concluded that high-pressure treatment increased the digestibility of β-LG and represents a promising processing technology for reducing the allergenicity of known allergens in a wide variety of food materials.

9.9.9 Human Milk Oligosaccharides

Human milk is often the sole dietary source for the first few months in life. It contains all the nutrients necessary for the infant to thrive, but also ingredients that may provide health benefits beyond those of traditional nutrients. Human milk oligosaccharides (HMOs) represent an abundant and diverse component of human milk, even though they have no direct nutritive value to the infant.[136] One litre of mature human milk contains approximately 5–10 g unbound oligosaccharides, and >130 different HMOs have been identified. Both their high amount and structural diversity are unique to human milk. Only trace amounts of these oligosaccharides are present in mature bovine milk and, as a consequence, in bovine milk-based infant formula. The potential health benefits of HMOs uncovered over the years may affect breast-fed infants both locally and systemically.[206] A recent hypothesis proposes that they could be substrates for the development of intestinal microbiota and the mucosal immune system.[137,207] Kunz and Rudloff recently reviewed the health-promoting aspects of milk oligosaccharides with reference to:[207]

- (prebiotic) oligosaccharides as growth factors for Bifidobacteria;
- anti-adhesion effects of milk oligosaccharides;
- systemic effects;
- leukocyte-endothelial interactions;
- plant-derived prebiotic oligosaccharides (PBOs) *vs.* HMOs;
- linkage specificity between monosaccharides in HMOs and PBOs; and
- benefits of milk oligosaccharides compared to fructo- and galacto-oligosaccharides.

Lebrilla's group is one of the pioneers of the quantitative and structural analysis of mammalian milk oligosaccharides[208] which they separate from the lipids and proteins of individual human milk samples and analyse by a combination of microchip LC-MS and MALDI-FTICR-MS.[136] Accurate mass measurements obtained through an orthogonal time-of-flight (o-ToF) MS provides oligosaccharide composition for *ca.* 200 individual molecular species. Comparison of microchip LC-MS profiles from different women revealed inter-individual, lactation phase-dependent and even daily[209] variations in milk oligosaccharide composition. While microchip LC-MS profiling provides routine identification of milk oligosaccharides, tandem MS in combination with exoglycosidase digestion distinguishes structural isomers.[136]

9.9.10   Milk Lipids

Milk fat is a remarkable source of energy, fat-soluble nutrients and bioactive lipids for mammals. The composition and content of lipids in milk fat vary widely among mammalian species. Milk fat is not only a source of bioactive lipid components; it also serves as an important delivery medium for nutrients, including the fat-soluble vitamins. Bioactive lipids in milk include triacylglycerides, diacylglycerides, saturated and polyunsaturated fatty acids, and phospholipids. Beneficial activities of milk lipids include antimicrobial, anti-inflammatory and immuno-suppressive properties. The major mammalian milk consumed by humans as a food commodity is that from cows, whose milk fat composition is distinct due to their diet and the presence of a rumen. As a result of these factors, bovine milk fat is lower in polyunsaturated fatty acids and higher in saturated fatty acids than human milk, and the consequences of these differences are being researched.[169]

Odham's group has published an LC-MS/MS study on sphingomyelins as found in an enriched sample of polar lipids from bovine milk.[210] Intact sphingomyelins were separated by normal-phase HPLC and detected by positive mode ESI-MS for structural information. In atmospheric pressure chemical ionisation (APCI), in-source fragmentation of sphingomyelin ions led to the formation of ceramide ions. With the latter as precursors, ions representative of both the long-chain base and the fatty acid parts were detected in APCI-MS/MS *via* collision-induced dissociation (CID). At least 36 protonated molecules of intact sphingomyelin were detected in the bovine milk sample.

Precht *et al.* published comparative studies of isomeric 18 : 1 acids in cow, goat and ewe milk fats by low-temperature high-resolution capillary gas-liquid chromatography, but without deploying mass spectrometry as a detector.[211] The same groups also investigated individual isomeric 18 : 1 acids in cow, goat and ewe milk fats by low-temperature high-resolution capillary gas-liquid chromatography[212] as well as individual *trans-* and *cis*-16 : 1 isomers in the same sources applying the same GC/LC technique.[213]

9.10 Conclusions

Nutrition has a strong influence on immune status, development and decline. Consequently, nutritional modulation of immunity is a major axis in nutrition and health research with the objectives to favourably "programme" neonate immunity, maintain immune homeostasis throughout life and reinforce immunity in elderly. Modern immune-modulating nutrition accompanies consumers through their life stages and styles.

An area of immunology and nutrition where mass spectrometry is already a well-established working horse is allergen detection, identification and characterisation. Immune relevant food sources like milk have been extensively investigated by MS in terms of their bioactives complement. A few nutritional interventions have been monitored by MS regarding their immune effects, mainly assessing the PBMC proteome—the latter serving in general as an accessible and relevant immune cell population readily amenable to mass spectrometric proteomics. Intestinal cells have served as a model to study gut immunity by MS means. Moreover, MS is rapidly emerging as the platform complementary to NMR in metabolomic investigations of host–microbe interactions and gut microbiota characterisation.

While mass spectrometry is certainly a most powerful tool to assess immune status and nutritional immune modulation, it is to date largely under-deployed. As the mature and diverse technology platform delivers quantitative, information-rich data and is highly accurate and sensitive, mass spectrometry in immunology and nutrition means for today and tomorrow:

- discovery of biomarkers for immune status and nutritional intervention;
- mass spectral monitoring of nutritional intervention and bioavailability/ bioefficacy studies.

Extending the rather traditional and few molecule-directed bioavailability studies to comprehensive, mass spectrometry-based investigation of metabolism and combining such approaches with MS-rooted proteomics paves the way to proceed from single nutrient bioavailability to multiple-nutrient bioefficacy studies. As mass spectrometry is a central platform to both proteomics and metabolomics, this technology will rapidly expand its role in holistic nutritional biomarker discovery. The nutrition community today largely sticks to traditional proteomic workflows based on 2D gels, but the array of deployed tools will increasingly include stable-isotope and label-free techniques, both enabling a higher throughput.

The complexity and subtlety of improving human health through nutrition requires holistic and sensitive approaches. Due to its versatility, sensitivity, accuracy, information richness and holistic nature, a rapidly expanding business for the application of mass spectrometry to nutrition and health is predicted.

References

1. M. E. Gershwin, J. B. German and K. L. Keen, *Nutrition and Immunology: Principles and Practice,* Humana Press, Totowa, NJ, 2000.
2. J. A. Hoffmann, F. C. Kafatos, C. A. Janeway and R. A. Ezekowitz, *Science*, 1999, **284**, 1313.
3. K. Takeda, T. Kaisho and S. Akira, *Annu. Rev. Immunol.*, 2003, **21**, 335.
4. N. S. Scrimshaw and J. P. SanGiovanni, *Am. J. Clin. Nutr.*, 1997, **66**, 464S.
5. S. Cunningham-Rundles, *Nutr. Rev.*, 1998, **56**, S27–S37.
6. J. A. Vanderhoof, *Nutrition*, 1998, **14**, 595.
7. D. W. Wilmore and J. K. Shabert, *Nutrition*, 1998, **14**, 618.
8. D. Evoy, M. D. Lieberman, T. J. Fahey III and J. M. Daly, *Nutrition*, 1998, **14**, 611.
9. H. P. Redmond, P. P. Stapleton, P. Neary and D. Bouchier-Hayes, *Nutrition*, 1998, **14**, 599.
10. R. F. Grimble and G. K. Grimble, *Nutrition*, 1998, **14**, 605.
11. J. W. Alexander, *Nutrition*, 1998, **14**, 627.
12. B. M. Lesourd, *Am. J. Clin. Nutr.*, 1997, **66**, 478S.
13. R. K. Chandra, *Am. J. Clin. Nutr.*, 1997, **66**, 460S.
14. R. K. Chandra, *Am. J. Clin. Nutr.*, 1991, **53**, 1087.

15. W. R. Beisel, R. Edelman, K. Nauss and R. M. Suskind, *JAMA*, 1981, **245**, 53.

16. S. N. Meydani and A. A. Beharka, *Nutr. Rev.*, 1998, **56**, S49–S58.

17. M. Kussmann and L. B. Fay, *Per. Med.*, 2008, **5**, 447.

18. R. K. Chandra, *Proc. Natl. Acad. Sci. U. S. A.*, 1996, **93**, 14304.

19. R. K. Chandra, *Lancet*, 1983, **1**, 688.

20. G. T. Keusch, *J. Nutr.*, 2003, **133**, 336S.

21. C. D. Johnson and K. A. Kudsk, *Clin. Nutr.*, 1999, **18**, 337.

22. W. W. Souba, *Annu. Rev. Nutr.*, 1991, **11**, 285.

23. B. Lesourd, *J. Nutr. Health Aging*, 2004, **8**, 28.

24. H. Tjalsma, R. M. J. Schaeps and D. W. Swinkels, *PROTEOMICS Clin. Appl.*, 2008, **2**, 167.

25. B. de Roos and H. J. McArdle, *Br. J. Nutr.*, 2008, **99**(Suppl 3), S66–S71.

26. B. de Roos, G. Rucklidge, M. Reid, K. Ross, G. Duncan, M. A. Navarro, J. M. Arbones-Mainar, M. A. Guzman-Garcia, J. Osada, J. Browne, C. E. Loscher and H. M. Roche, *FASEB J.*, 2005, **19**, 1746.

27. B. de Roos, I. Duivenvoorden, G. Rucklidge, M. Reid, K. Ross, R. J. Lamers, P. J. Voshol, L. M. Havekes and B. Teusink, *FASEB J.*, 2005, **19**, 813.

28. J. M. Arbones Mainar, K. Ross, G. J. Rucklidge, M. Reid, G. Duncan, J. R. Arthur, G. W. Horgan, M. A. Navarro, R. Carnicer, C. Arnal, J. Osada and B. de Roos, *J. Proteome Res.*, 2007, **6**, 4041.

29. B. L. Mitchell, Y. Yasui, J. W. Lampe, P. R. Gafken and P. D. Lampe, *Proteomics*, 2005, **5**, 2238.

30. D. Fuchs, R. Piller, J. Linseisen, H. Daniel and U. Wenzel, *Proteomics*, 2007, **7**, 3278.

31. R. K. Chandra, *Lancet*, 1992, **340**, 1124.

32. J. D. Bogden, A. Bendich, F. W. Kemp, K. S. Bruening, J. H. Shurnick, T. Denny, H. Baker and D. B. Louria, *Am. J. Clin. Nutr.*, 1994, **60**, 437.

33. S. N. Meydani, M. P. Barklund, S. Liu, M. Meydani, R. A. Miller, J. G. Cannon, F. D. Morrow, R. Rocklin and J. B. Blumberg, *Am. J. Clin. Nutr.*, 1990, **52**, 557.

34. S. Cunningham-Rundles, *Nutr. Rev.*, 1998, **56**, S27–S37.

35. K. Woelkart, E. Marth, A. Suter, R. Schoop, R. B. Raggam, C. Koidl, B. Kleinhappl and R. Bauer, *Int. J. Clin. Pharmacol. Ther.*, 2006, **44**, 401.

36. M. Kussmann, M. Affolter, K. Nagy, B. Holst and L. B. Fay, *Mass Spectrom. Rev.*, 2007, **26**, 727.

37. L. Anderson and C. L. Hunter, *Mol. Cell Proteomics*, 2006, **5**, 573.

38. S. A. Gerber, J. Rush, O. Stemman, M. W. Kirschner and S. P. Gygi, *Proc. Natl. Acad. Sci. U. S. A.*, 2003, **100**, 6940.

39. K. Nagy, M. C. Courtet-Compondu, B. Hoist and M. Kussmann, *Anal. Chem.*, 2007, **79**, 7087.

40. R. Albers, J. M. Antoine, R. Bourdet-Sicard, P. C. Calder, M. Gleeson, B. Lesourd, S. Samartin, I. R. Sanderson, J. Van Loo, F. W. Vas-Dias and B. Watzl, *Br. J. Nutr.*, 2005, **94**, 452.

41. K. Song and S. Hanash, *Gastroenterology*, 2007, **131**, 1375.
42. A. W. Purcell and J. J. Gorman, *Mol. Cell Proteomics*, 2004, **3**, 193.
43. P. Weingarten, P. Lutter, A. Wattenberg, M. Blueggel, S. Bailey, J. Klose, H. E. Meyer and C. Huels, *Methods Mol. Med.*, 2005, **109**, 155.
44. J. Kubach, P. Lutter, T. Bopp, S. Stoll, C. Becker, E. Huter, C. Richter, P. Weingarten, T. Warger, J. Knop, S. Mullner, J. Wijdenes, H. Schild, E. Schmitt and H. Jonuleit, *Blood*, 2007, **110**, 1550.
45. E. A. Pastorello, L. Farioli, V. Pravettoni, M. Ispano, E. Scibola, C. Trambaioli, M. G. Giuffrida, R. Ansaloni, J. Godovac-Zimmermann, A. Conti, D. Fortunato and C. Ortolani, *J. Allergy Clin. Immunol.*, 2000, **106**, 744.
46. S. M. Chan and P. J. Utz, *Ann. N. Y. Acad. Sci.*, 2005, **1062**, 61.
47. D. H. Dube and C. R. Bertozzi, *Nat. Rev. Drug Discov.*, 2005, **4**, 477.
48. M. Aguiar, R. Masse and B. F. Gibbs, *Anal. Biochem.*, 2006, **354**, 175.
49. S. Barcelo-Batllori, M. André, C. Servis, N. Lévy, O. Takikawa, P. Michetti, M. Reymond and E. Felley-Bosco, *Proteomics*, 2002, **2**, 551.
50. A. Shkoda, T. Werner, H. Daniel, M. Gunckel, G. Rogler and D. Haller, *J. Proteome Res.*, 2007, **6**, 1114.
51. A. Shkoda, P. A. Ruiz, H. Daniel, S. C. Kim, G. Rogler, R. B. Sartor and D. Haller, *Gastroenterology*, 2007, **132**, 190.
52. S. Thebault, N. Deniel, R. Marion, R. Charlionet, F. Tron, D. Cosquer, J. Leprince, H. Vaudry, P. Ducrotte and P. Déchelotte, *Proteomics*, 2006, **6**, 3926.
53. N. Deniel, R. Marion-Letellier, R. Charlionet, F. Tron, J. Leprince, H. Vaudry, P. Ducrotte, P. Déchelotte and S. Thebault, *Mol. Cell Proteomics*, 2007, **6**, 1671.
54. M. Babusiak, P. Man, J. Petrak and D. Vyoral, *Proteomics*, 2007, **7**, 121.
55. M. Kussmann and S. Blum-Sperisen, *Endocrin. Metab. Immune. Disord. Drug Targets*, 2007, **7**, 271.
56. P. M. Henson, *Nat. Immunol.*, 2005, **6**, 1179.
57. J. Han and R. J. Ulevitch, *Nat. Immunol.*, 2005, **6**, 1198.
58. C. N. Serhan and J. Savill, *Nat. Immunol.*, 2005, **6**, 1191.
59. R. F. Grimble, *Nutrition*, 1998, **14**, 634.
60. D. Fuchs, K. Vafeiadou, W. L. Hall, H. Daniel, C. M. Williams, J. H. Schroot and U. Wenzel, *Am. J. Clin. Nutr.*, 2007, **86**, 1369.
61. S. P. James, *J. Allergy. Clin. Immunol.*, 2005, **115**, 25.
62. T. T. Macdonald and G. Monteleone, *Science*, 2005, **307**, 1920.
63. L. R. Ferguson, A. N. Shelling, B. L. Browning, C. Huebner and I. Petermann, *Mutat. Res.*, 2007, **622**, 70.
64. N. Roy, M. Barnett, B. Knoch, Y. Dommels and W. McNabb, *Mutat. Res.*, 2007, **622**, 103.
65. M. Danielsen, T. Thymann, B. B. Jensen, O. N. Jensen, P. T. Sangild and E. Bendixen, *Proteomics*, 2006, **6**, 6588.
66. J. E. Drew, S. Padidar, G. Horgan, G. G. Duthie, W. R. Russell, M. Reid, G. Duncan and G. J. Rucklidge, *Biochem. Pharmacol.*, 2006, **72**, 204.
67. P. Matzinger, *Ann. N. Y. Acad. Sci.*, 2002, **961**, 341.

68. P. Matzinger, *Science*, 2002, **296**, 301.
69. R. J. Ulevitch, *Nat. Rev. Immunol.*, 2004, **4**, 512.
70. W. Strober, *J. Pediatr. Gastroenterol. Nutr.*, 2005, **40**(Suppl 1), S26.
71. W. Strober, I. Fuss and P. Mannon, *J. Clin. Invest.*, 2007, **117**, 514.
72. S. E. Calvano, W. Xiao, D. R. Richards, R. M. Felciano, H. V. Baker, R. J. Cho, R. O. Chen, B. H. Brownstein, J. P. Cobb, S. K. Tschoeke, C. Miller-Graziano, L. L. Moldawer, M. N. Mindrinos, R. W. Davis, R. G. Tompkins and S. F. Lowry, *Nature*, 2005, **437**, 1032.
73. N. L. Anderson and N. G. Anderson, *Mol. Cell Proteomics*, 2002, **1**, 845.
74. K. Rose, L. Bougueleret, T. Baussant, G. Bohm, P. Botti, J. Colinge, I. Cusin, H. Gaertner, A. Gleizes, M. Heller, S. Jimenez, A. Johnson, M. Kussmann, L. Menin, C. Menzel, F. Ranno, P. Rodriguez-Tome, J. Rogers, C. Saudrais, M. Villain, D. Wetmore, A. Bairoch and D. Hochstrasser, *Proteomics*, 2004, **4**, 2125.
75. J. Adachi, C. Kumar, Y. Zhang, J. V. Olsen and M. Mann, *Genome Biol.*, 2006, **7**, R80.
76. T. Guo, P. A. Rudnick, W. Wang, C. S. Lee, D. L. DeVoev and B. M. Balgley, *J. Proteome Res.*, 2006, **5**, 1469.
77. G. A. de-Souza, L. M. Godoy and M. Mann, *Genome Biol.*, 2006, **7**, R72.
78. B. Ghafouri, K. Irander, J. Lindbom, C. Tagesson and M. Lindahl, *J. Proteome Res.*, 2006, **5**, 330.
79. P. C. Calder, R. Albers, J. M. Antoine, S. Blum, R. Bourdet-Sicard, G. A. Ferns, G. Folkerts, P. S. Friedmann, G. S. Frost, F. Guarner, M. Lovik, S. Macfarlane, P. D. Meyer, L. M'Rabet, M. Serafini, W. van Eden, J. Van Loo, D. W. vas, S. Vidry, B. M. Winklhofer-Roob and J. Zhao, *Proc. Nutr. Soc.*, 2008, **67**, E9.
80. J. B. Hirsch and D. Evans, *Food Technol.*, 2005, **59**(24), 33.
81. C. Bindslev-Jensen, P. S. Skov, F. Madsen and L. K. Poulsen, *Ann. Allergy*, 1994, **72**(4), 317.
82. U. Wahn, E. von Mutius, S. Lau and and R. Nickel, *Nestle Nutr. Workshop Ser. Pediatr. Program*, 2007, **59**, 1.
83. C. A. Akdis, K. Blaser and M. Akdis, *Allergy*, 2004, **59**, 897.
84. H. Renz, N. Blumer, S. Virna, S. Sel and H. Garn, *Chem. Immunol. Allergy*, 2006, **91**, 30.
85. I. Adlerberth, E. Lindberg, N. Aberg, B. Hesselmar, R. Saalman, I. L. Strannegard and A. E. Wold, *Pediatr. Res.*, 2006, **59**, 96.
86. K. S. Kornman, P. M. Martha and G. W. Duff, Genetic variations and inflammation: a practical nutrigenomics opportunity. *Nutrition*, 2004, **20**, 44.
87. C. A. Akdis, A. Joss, M. Akdis and K. Blaser, *Int. Arch. Allergy Immunol.*, 2001, **124**, 180.
88. I. Sabroe, L. C. Parker, A. G. Wilson, M. K. Whyte and S. K. Dower, *Clin. Exp. Allergy*, 2002, **32**, 984.
89. R. Djukanovic, *Clin. Exp. Allergy*, 2000, **30**(Suppl 1), 46.
90. T. Illig and M. Wjst, *Paediatr. Respir. Rev.*, 2002, **3**, 47.

91. T. Laitinen, A. Polvi, P. Rydman, J. Vendelin, V. Pulkinnen, P. Salmi-kangas, S. Mäkelä, M. Rehn, A. Pirskanen, A. Rautanen, M. Zucchelli, H. Gullstén, M. Leino, H. Alenius, T. Petäys, T. Haahtela, A. Laitinen, C. Laprise, T. J. Hudson, L. A. Laitinen and J. Kere, *Science*, 2004, **304**, 300.

92. M. Blueggel, F. Spertini, P. Lutter, J. Wassenberg, R. Audran, Corthésy, B., S. Müllner, S. Blum, A. Wattenberg, A. Mercenier, M. Affolter and M. Kussmann, submitted.

93. R. K. Chandra, *Am. J. Clin. Nutr.*, 1997, **66**, 526S.

94. P. A. Eigenmann, *Allergy*, 2001, **56**, 1112.

95. M. Kussmann, M. Affolter and L. B. Fay, *Comb. Chem. High Throughput Screen.*, 2005, **8**, 679.

96. M. Kussmann and M. Affolter, *Curr. Opin. Clin. Nutr. Metab. Care*, 2006, **9**, 575.

97. S. W. Terheggen-Lagro, I. M. Khouw, A. Schaafsma and E. A. Wauters, *BMC. Pediatr.*, 2002, **2**, 10.

98. K. Beyer, G. Grishina, L. Bardina, A. Grishin and H. A. Sampson, *J. Allergy Clin. Immunol.*, 2002, **110**, 517.

99. K. Beyer, L. Bardina, G. Grishina and H. A. Sampson, *J. Allergy Clin. Immunol.*, 2002, **110**, 154.

100. C. J. Yu, Y. F. Lin, B. L. Chiang and L. P. Chow, *J. Immunol.*, 2003, **170**, 445.

101. J. A. Dunstan, L. Breckler, J. Hale, H. Lehmann, P. Franklin, G. Lyons, S. Y. Ching, T. A. Mori, A. Barden and S. L. Prescott, *Clin. Exp. Allergy*, 2006, **36**, 993.

102. J. J. Eady, G. M. Wortley, Y. M. Wormstone, J. C. Hughes, S. B. Astley, R. J. Foxall, J. F. Doleman and R. M. Elliott, *Physiol. Genomics*, 2005, **22**, 402.

103. B. de Roos, S. J. Duthie, A. C. Polley, F. Mulholland, F. G. Bouwman, C. Heim, G. J. Rucklidge, I. T. Johnson, E. C. Mariman, H. Daniel and R. M. Elliott, *J. Proteome Res.*, 2008, **7**, 2280.

104. M. A. Roberts, D. M. Mutch and J. B. German, *Curr. Opin. Biotechnol.*, 2001, **12**, 516.

105. P. Bourlioux, B. Koletzko, F. Guarner and V. Braesco, *Am. J. Clin. Nutr.*, 2003, **78**, 675.

106. D. C. Savage, *Annu. Rev. Microbiol.*, 1977, **31**, 107.

107. D. C. Savage, *Annu. Rev. Nutr.*, 1986, **6**, 155.

108. L. V. Hooper, T. Midtvedt and J. I. Gordon, *Annu. Rev. Nutr.*, 2002, **22**, 283.

109. L. V. Hooper and J. I. Gordon, *Science*, 2001, **292**, 1115.

110. L. V. Hooper, M. H. Wong, A. Thelin, L. Hansson, P. G. Falk and J. I. Gordon, *Science*, 2001, **291**, 881.

111. R. Fuller, *J. Appl. Bacteriol.*, 2004, **66**, 365.

112. S. N. Meydani and W. K. Ha, *Am. J. Clin. Nutr.*, 2000, **71**, 861.

113. M. Boirivant and W. Strober, *Curr. Opin. Gastroenterol.*, 2007, **23**, 679.

114. W. A. Walker, O. Goulet, M. Morelli and J. M. Antoine, *Eur. J. Nutr.*, 2006, **45**, 1.

115. E. Isolauri, P. V. Kirjavainen and S. Salminen, *Gut*, 2002, **50**(Suppl 3), III54–III59.

116. J. A. Vanderhoof and R. J. Young, *Ann. Allergy Asthma Immunol.*, 2003, **90**, 99.

117. C. P. Tamboli, C. Caucheteux, A. Cortot, J. F. Colombel and P. Desreumaux, *Best Pract. Res. Clin. Gastroenterol.*, 2003, **17**, 805.

118. E. Isolauri, Y. Sutas, P. Kankaanpaa, H. Arvilommi and S. Salminen, *Am. J. Clin. Nutr.*, 2001, **73**, 444S.

119. S. Blum and E. J. Schiffrin, *Curr. Issues Intest. Microbiol.*, 2003, **4**, 53.

120. M. J. Miraglia-del-Giudice, M. G. De Luca and C. Capristo, *Dig. Liver Dis.*, 2002, **34**(Suppl 2), S68–S71.

121. G. E. Bergonzelli, S. Blum, H. Brussow and I. Corthesy-Theulaz, *Digestion*, 2005, **72**, 57.

122. E. Isolauri, *Curr. Allergy Asthma Rep.*, 2004, **4**, 270.

123. G. W. Tannock, *Int. Dairy J.*, 1995, **5**, 1059.

124. M. Saarela, G. Mogensen, R. Fonden, J. Matto and T. Mattila-Sandholm, *J. Biotechnol.*, 2000, **84**, 197.

125. M. A. Schell, M. Karmirantzou, B. Snel, D. Vilanova, B. Berger, G. Pessi, M. C. Zwahlen, F. Desiere, P. Bork, M. Delley, R. D. Pridmore and F. Arigoni, *Proc. Natl. Acad. Sci. U.S.A*, 2002, **99**, 14422.

126. M. C. Champomier-Verges, E. Maguin, M. Y. Mistou, P. Anglade and J. F. Chich, *J. Chromatogr. B*, 2002, **771**, 329.

127. P. Leverrier, J. P. Vissers, A. Rouault, P. Boyaval and G. Jan, *Arch. Microbiol.*, 2004, **181**, 215.

128. P. Leverrier, D. Dimova, V. Pichereau, Y. Auffray, P. Boyaval and G. Jan, *Appl. Environ. Microbiol.*, 2003, **69**, 3809.

129. L. F. Marvin-Guy, S. Parche, S. Wagniere, J. Moulin, R. Zink, M. Kussmann and L. B. Fay, *J. Am. Soc. Mass Spectrom.*, 2004, **15**, 1222.

130. E. Guillaume, B. Berger, M. Affolter and M. Kussmann, *J. Proteomics*, 2009, **72**, 771.

131. B. Berger, D. Moine, R. Mansourian and F. Arigoni, *J. Bacteriol.*, 2010, **192**, 256.

132. G. R. Gibson and M. B. Roberfroid, *J. Nutr.*, 1995, **125**, 1401.

133. S. Bengmark, *Nutrition*, 1998, **14**, 585.

134. K. C. Mountzouris, A. L. McCartney and G. R. Gibson, *Br. J. Nutr.*, 2002, **87**, 405.

135. E. Nova, B. Viadel, M. Blasco and A. Marcos, *Proc. Nutr. Soc.*, 2008, **67**, E4.

136. M. R. Ninonuevo, P. Youmie, Y. Hongfeng, Z. Jinhua, R. E. Ward, B. H. Clowers, J. B. German, S. L. Freeman, K. Killeen, R. Grimm and C. B. Lebrilla, *J. Agric. Food Chem.*, 2006, **54**, 7471.

137. R. G. LoCascio, M. R. Ninonuevo, S. L. Freeman, D. A. Sela, R. Grimm, C. B. Lebrilla, D. A. Mills and J. B. German, *J. Agric. Food Chem.*, 2007, **55**, 8914.

138. J. Xu and J. I. Gordon, *Proc. Natl. Acad. Sci. U. S. A.*, 2003, **100**, 10452.

139. J. Xu, M. K. Bjursell, J. Himrod, S. Deng, L. K. Carmichael, H. C. Chiang, L. V. Hooper and J. I. Gordon, *Science*, 2003, **299**, 2074.

140. J. Xu, M. A. Mahowald, R. E. Ley, C. A. Lozupone, M. Hamady, E. C. Martens, B. Henrissat, P. M. Coutinho, P. Minx, P. Latreille, H. Cordum, A. Van Brunt, K. Kim, R. S. Fulton, L. A. Fulton, S. W. Clifton, R. K. Wilson, R. D. Knight and J. I. Gordon, *PLoS. Biol.*, 2007, **5**, e156.

141. F. P. Martin, M. E. Dumas, Y. Wang, C. Legido-Quigley, I. K. Yap, H. Tang, S. Zirah, G. M. Murphy, O. Cloarec, J. C. Lindon, N. Sprenger, L. B. Fay, S. Kochhar, P. van-Bladeren, E. Holmes and J. K. Nicholson, *Mol. Syst. Biol.*, 2007, **3**, 112.

142. F. P. Martin, E. F. Verdu, Y. Wang, M. E. Dumas, I. K. Yap, O. Cloarec, G. E. Bergonzelli, I. Corthesy-Theulaz, S. Kochhar, E. Holmes, J. C. Lindon, S. M. Collins and J. K. Nicholson, *J. Proteome Res.*, 2006, **5**, 2185.

143. O. Cloarec, M. E. Dumas, A. Craig, R. H. Barton, J. Trygg, J. Hudson, C. Blancher, D. Gauguier, J. C. Lindon, E. Holmes and J. Nicholson, *Anal. Chem.*, 2005, **77**, 1282.

144. A. W. Nicholls, R. J. Mortishire-Smith and J. K. Nicholson, *Chem. Res. Toxicol.*, 2003, **16**, 1395.

145. P. G. Falk, L. V. Hooper, T. Midtvedt and J. I. Gordon, *Microbiol. Mol. Biol. Rev.*, 1998, **62**, 1157.

146. M. Danielsen, H. Hornshoej, R. H. Siggers, B. B. Jensen, A. G. van Kessel and E. Bendixen, *J. Proteome Res.*, 2007, **6**, 2596.

147. L. Bry, P. G. Falk, T. Midtvedt and J. I. Gordon, *Science*, 1996, **273**, 1380.

148. J. L. Sonnenburg, J. Xu, D. D. Leip, C. H. Chen, B. P. Westover, J. Weatherford, J. D. Buhlerv and J. I. Gordon, *Science*, 2005, **307**, 1955.

149. J. L. Sonnenburg, C. T. Chen and J. I. Gordon, *PLoS. Biol.*, 2006, **4**, e413.

150. S. Cunningham-Rundles, *J. Nutr. Health Aging*, 2004, **8**, 20.

151. B. Bjorksten, P. Naaber, E. Sepp and M. Mikelsaar, *Clin. Exp. Allergy*, 1999, **29**, 342.

152. K. Shida, K. Makino, A. Morishita, K. Takamizawa, S. Hachimura, A. Ametani, T. Sato, Y. Kumagai, S. Habu and S. Kaminogawa, *Int. Arch. Allergy Immunol.*, 1998, **115**, 278.

153. N. Sudo, S. Sawamura, K. Tanaka, Y. Aiba, C. Kubo and Y. Koga, *J. Immunol.*, 1997, **159**, 1739.

154. P. C. Calder, S. Krauss-Etschmann, E. C. de Jong, C. Dupont, J. S. Frick, H. Frokiaer, J. Heinrich, H. Garn, S. Koletzko, G. Lack, G. Mattelio, H. Renz, P. T. Sangild, J. Schrezenmeir, T. M. Stulnig, T. Thymann, A. E. Wold and B. Koletzko, *Br. J. Nutr.*, 2006, **96**, 774.

155. L. Peng, P. Rawson, D. McLauchlan, K. Lehnert, R. Snell and T. W. Jordan, *J. Proteome Res.*, 2008, **7**, 1427.

156. T. Liu, W. J. Qian, W. N. Chen, J. M. Jacobs, R. J. Moore, D. J. Anderson, M. A. Gritsenko, M. E. Monroe, B. D. Thrall, D. G. Camp and R. D. Smith, *Proteomics*, 2005, **5**, 1263.

157. J. Levy, *Nutrition*, 1998, **14**, 641.

158. M. O. Labeta, K. Vidal, J. E. Nores, M. Arias, N. Vita, B. P. Morgan, J. C. Guillemot, D. Loyaux, P. Ferrara, D. Schmid, M. Affolter, L. K. Borysiewicz, A. Donnet-Hughes and E. J. Schiffrin, *J. Exp. Med.*, 2000, **191**, 1807.

159. K. Vidal, M. O. Labeta, E. J. Schiffrin and A. Donnet-Hughes, *Acta Odontol. Scand.*, 2001, **59**, 330.

160. E. LeBouder, J. E. Rey-Nores, N. K. Rushmere, M. Grigorov, S. D. Lawn, M. Affolter, G. E. Griffin, P. Ferrara, E. J. Schiffrin, B. P. Morgan and M. O. Labeta, *J. Immunol.*, 2003, **171**, 6680.

161. A. Panchaud, M. Kussmann and M. Affolter, *Proteomics*, 2005, **5**, 3836.

162. M. Haarman and J. Knol, *Appl. Environ. Microbiol.*, 2005, **71**, 2318.

163. E. Puertollano, T. Pozo, I. Nadal, A. Marcos, Y. Sanz and E. Nova, *Proc. Nutr. Soc.*, 2008, **67**, E85.

164. S. E. Moore, *Proc. Nutr. Soc.*, 1998, **57**, 241.

165. S. E. Moore, T. J. Cole, E. M. Poskitt, B. J. Sonko, R. G. Whitehead, I. A. McGregor and A. M. Prentice, *Nature*, 1997, **388**, 434.

166. H. Ghattas, D. L. Wallace, J. A. Solon, S. M. Henson, Y. Zhang, P. T. Ngom, R. Aspinall, G. Morgan, G. E. Griffin, A. M. Prentice and D. C. Macallan, *Am. J. Clin. Nutr.*, 2007, **85**, 480.

167. J. B. German, C. J. Morganv and R. E. Ward, *Aust. J. Dairy Technol.*, 2003, **49**.

168. S. Severin and W. Xia, *Crit. Rev. Food Sci. Nutr.*, 2005, **45**, 645.

169. J. B. German and C. J. Dillard, *Crit. Rev. Food Sci. Nutr.*, 2006, **46**, 57.

170. A. R. Madureira, C. I. Pereira, A.-M. P. Gomes, M. E. Pintado and F. X. Malcata, *Food Res. Int.*, 2007, **40**, 1197.

171. B. Casado, M. Affolter and M. Kussmann, *J. Proteomics*, 2009, **73**, 196.

172. B. Y. Fong, C. S. Norris and K. P. Palmano, *Int. Dairy J.*, 2008, **18**, 23.

173. H. F. Alomirah, I. Alli and Y. Konishi, *J. Chromatogr. A*, 2000, **893**, 1.

174. P. Ferranti, *Eur. J. Mass Spectrom.*, 2004, **10**, 349.

175. P. A. Guy and F. Fenaille, *Mass Spectrom. Rev.*, 2006, **25**, 290.

176. G. Smolenski, S. Haines, F. Y. Kwan, J. Bond, V. Farr, S. R. Davis, K. Stelwagen and T. T. Wheeler, *J. Proteome Res.*, 2007, **6**, 207.

177. E. D'Auria, C. Agostoni, M. Giovannini, E. Riva, R. Zetterstrom, R. Fortin, G. F. Greppi, L. Bonizzi and P. Roncada, *Acta Paediatr.*, 2005, **94**, 1708.

178. I. de Noni, *Food Chem.*, 2008, **110**, 897.

179. L. F. Marvin, V. Parisod, L. B. Fay and P. A. Guy, *Electrophoresis*, 2002, **23**, 2505.

180. C. Liepke, H. D. Zucht, W. G. Forssmann and L. Standker, *J. Chromatogr. B, Biomed. Sci. Appl.*, 2001, **752**, 369.

181. G. Miranda, M. F. Mahe, C. Leroux and P. Martin, *Proteomics*, 2004, **4**, 2496.

182. M. Yamada, K. Murakami, J. C. Wallingford and Y. Yuki, *Electrophoresis*, 2002, **23**, 1153.

183. D. J. Palmer, V. C. Kelly, A. M. Smit, S. Kuy, C. G. Knight and G. J. Cooper, *Proteomics*, 2006, **6**, 2208.
184. D. Fortunato, M. G. Giuffrida, M. Cavaletto, L. P. Garoffo, G. Dellavalle, L. Napolitano, C. Giunta, C. Fabris, E. Bertino, A. Coscia and A. Conti, *Proteomics*, 2003, **3**, 897.
185. K. Dewettinck, R. Rombaut, N. Thienpont, T. T. Le, K. Messens and J. van Camp, *Int. Dairy J.*, 2008, **18**, 436.
186. N. Argov, S. Wachsmann-Hogiu, S. L. Freeman, T. Huser, C. B. Lebrilla and J. B. German, *J. Agric. Food Chem.*, 2008, **56**, 7446.
187. M. Cavaletto, M. G. Giuffrida, D. Fortunato, L. Gardano, G. Dellavalle, L. Napolitano, C. Giunta, E. Bertino, C. Fabris and A. Conti, *Proteomics*, 2002, **2**, 850.
188. J. Charlwood, S. Hanrahan, R. Tyldesley, J. Langridge, M. Dwek and P. Camilleri, *Anal. Biochem.*, 2002, **301**, 314.
189. T. A. Reinhardt and J. D. Lippolis, *J. Dairy Res.*, 2006, **73**, 406.
190. T. A. Reinhardt and J. D. Lippolis, *J. Dairy Sci.*, 2008, **91**, 2307.
191. M. Affolter, L. Grass, F. Vanrobaeys, B. Casado and M. Kussmann, *J. Proteomics*, 2009.
192. N. L. Wilson, L. J. Robinson, A. Donnet, L. Bovetto, N. H. Packer and N. G. Karlsson, *J. Proteome Res.*, 2008.
193. J. W. Holland, H. C. Deeth and P. F. Alewood, *Proteomics*, 2004, **4**, 743.
194. J. W. Holland, H. C. Deeth and P. F. Alewood, *Proteomics*, 2005, **5**, 990.
195. F. Kjeldsen, K. F. Haselmann, B. A. Budnik, E. S. Sorensen and R. A. Zubarev, *Anal. Chem.*, 2003, **75**, 2355.
196. J. Hau and L. Bovetto, *J. Chromatogr. A*, 2001, **926**, 105.
197. M. Galvani, M. Hamdanv and P. G. Righetti, *Rapid Commun. Mass Spectrom.*, 2001, **15**, 258.
198. G. Henry, D. Molle, F. Morgan, J. Fauquant and S. Bouhallab, *J. Agric. Food Chem.*, 2002, **50**, 185.
199. E. S. Sorensen, L. Moller, M. Vinther, T. E. Petersenv and L. K. Rasmussen, *Eur. J. Biochem.*, 2003, **270**, 3651.
200. P. Bratter, I. N. Blasco, d. B. Negretti and V. A. Raab, *Analyst*, 1998, **123**, 821.
201. A. Sanz-Medel, M. Montes-Bayon and M.-L. F. Sanchez, *Anal. Bioanal. Chem.*, 2003, **377**, 236.
202. H. Meisel, *Biofactors*, 2004, **21**, 55.
203. D. J. Autelitano, A. Rajic, A. I. Smith, M. C. Berndt, L. L. Ilag and M. Vadas, *Drug Discov. Today*, 2006, **11**, 306.
204. M. Natale, C. Bisson, G. Monti, A. Peltran, L. P. Garoffo, S. Valentini, C. Fabris, E. Bertino, A. Coscia and A. Conti, *Mol. Nutr. Food Res.*, 2004, **48**, 363.
205. M. Zeece, T. Huppertz and A. Kelly, *Innov. Food Sci. Emerging Technol.*, 2008, **9**, 62.
206. L. Bode, *J. Nutr.*, 2006, **136**, 2127.
207. C. Kunz and S. Rudloff, *Int. Dairy J.*, 2006, **16**, 1341.

208. M. R. Ninonuevo, R. E. Ward, R. G. LoCascio, J. B. German, S. L. Freeman, M. Barboza, D. A. Mills and C. B. Lebrilla, *Anal. Biochem.*, 2007, **361**, 15.

209. M. R. Ninonuevo, P. D. Perkins, J. Francis, L. M. Lamotte, R. G. LoCascio, S. L. Freeman, D. A. Mills, J. B. German, R. Grimm and C. B. Lebrilla, *J. Agric. Food Chem.*, 2008, **56**, 618.

210. A. A. Karlsson, P. Michelsen and G. Odham, *J. Mass Spectrom.*, 1998, **33**, 1192.

211. D. Precht, J. Molkentin, F. Destaillats and R. L. Wolff, *Lipids*, 2001, **36**, 827.

212. D. Precht and J. Molkentin, *Nahrung-Food*, 1999, **43**, 233.

213. F. Destaillats, R. L. Wolff, D. Precht and J. Molkentin, *Lipids*, 2000, **35**, 1027.

CHAPTER 10

Mass Spectrometry, Nutrition and Protein Turnover

MICHAEL AFFOLTER

Functional Genomics Group, Department of Bioanalytical Sciences, Nestlé Research Centre, Vers-chez-les-Blanc, PO Box 44, CH-1000, Lausanne 26, Switzerland

10.1 Introduction

Protein turnover is the result of synthesis of new and breakdown of old proteins in the body thereby providing a mechanism for the maintenance of optimally functioning proteins[1] (Figure 10.1). Over the last 60 years, scientists have attempted to quantify rates of turnover as it represents a fundamental biological process in all living organisms.[2] Sprinson and Ritterberg[3] introduced quantitative measurements of protein turnover in humans in 1949 using a constant infusion of a tracer and measurement of the excretion of the tracer in the urine. This approach, called the end-product method, depends on measurement of the tracer and the two end products, urea and ammonia, in the urine.[2] Initially, radioactive isotopically labelled amino acids were used in combination with scintillation counting, but due to potential radiation damage, these tracers became restricted for human studies and were replaced by stable-isotope tracer amino acids labelled with ^2H, ^{13}C or ^{15}N isotopes, for which mass spectrometric detection methods have been applied. Currently, the tracers most commonly used in whole-body protein turnover studies are [^{15}N]glycine,[4] L-[^{13}C]leucine[5] and L-[$^2H_5$]phenylalanine.[6]

RSC Food Analysis Monographs No. 9
Mass Spectrometry and Nutrition Research
Edited by Laurent B. Fay and Martin Kussmann
© The Royal Society of Chemistry 2010
Published by the Royal Society of Chemistry, www.rsc.org

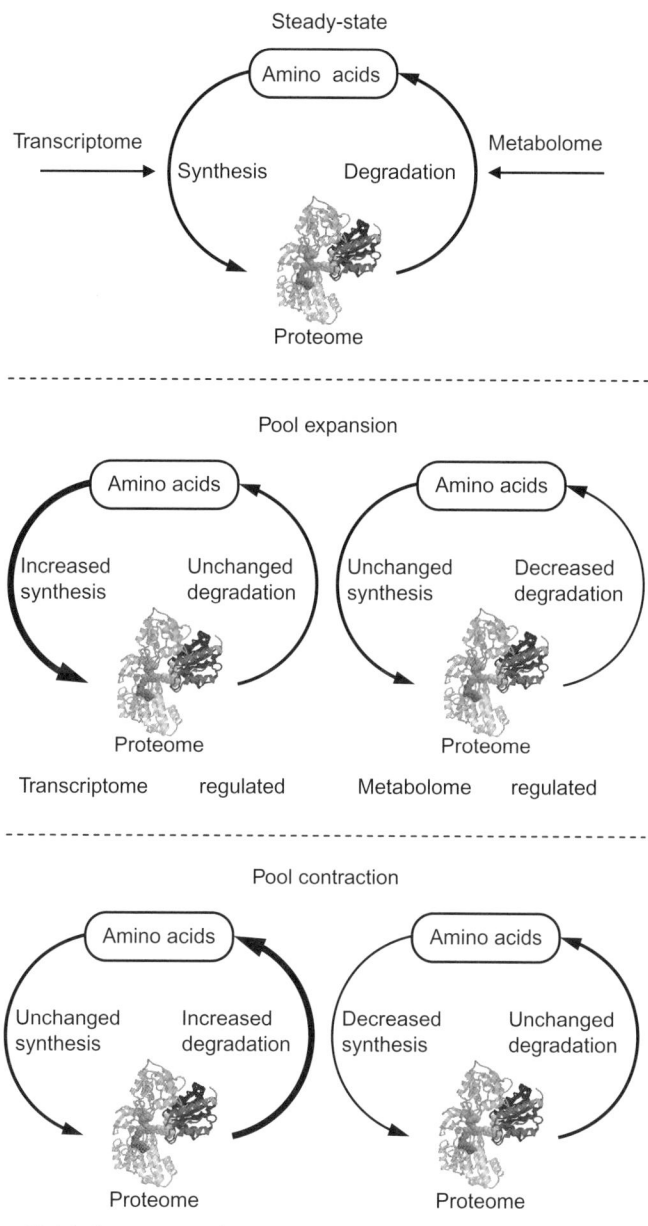

Figure 10.1 Protein turnover. The steady state reflects a balance between protein synthesis and degradation, and the concentration of the protein pool is not changing. Protein synthesis is linked to transcriptional processes whereas protein degradation is related to metabolic processes. Pool expansion can be induced by increased synthesis or decreased degradation while pool contraction is a result of increased degradation or decreased synthesis (adapted from ref. 21).

The end-product method is based on a simple two-pool model derived from the axiom that two products of the same precursor have the same activity under the condition that the precursor pool is the only source of the product.[7,8] Although quite popular in the 1980s, as it is non-invasive and easily applied, the end-product method was superseded by the "precursor" approach, in which a stable-isotope labelled amino acid is given by intravenous infusion for several hours until a steady state of enrichment is reached at plasma levels. Today, the precursor method is considered the gold standard for measuring whole body protein turnover in humans. Two different strategies—constant infusion and flooding dose approach—have been used to measure protein synthesis in humans.[9,10]

The constant infusion method is based on the infusion of labelled tracer amino acids until steady-state is reached, typically over 3–8 h. This prolonged labelling period renders the approach well-suited for the measurement of proteins with slow turnover rates such as muscle proteins. The synthesis rate of proteins with high turnover may be underestimated due to recycling of the tracer or possible secretion of newly synthesized proteins from the target organ, *e.g.* the liver or gut. Protein synthesis is calculated from the ratio of protein-bound and free amino acid isotope enrichment, with the additional advantage of concurrent determination of whole body and tissue-specific turnover rates.[11]

The flooding dose approach involves the injection of a large amount of unlabelled amino acid (tracee) along with the isotope labelled amino acid (tracer). This strategy rapidly increases the labelling of the intracellular amino acid pool, thus enabling the measurements of fast turnover rates, as typically found in liver proteins. It also equilibrates the free amino acid pools, such as the extracellular, intracellular and aminoacyl-transfer RNA pool, by providing an excess of the labelled amino acid.[12] The method is based on the assumption of equilibration between the precursor pools, which was demonstrated by Davis and coworkers,[13] namely that the free amino acid pools are equilibrated with aminoacyl-tRNA pools. Their results showed that 30 min after the administration of a flooding dose of phenylalanine, there was equilibration of the specific radioactivity of phenylalanine among the blood, tissue and aminoacyl-tRNA precursor pools. Equilibration of the specific radioactivity of the three precursor pools for protein synthesis occurred in both skeletal muscle and liver. Neither feeding nor hormonal fluctuations affected the equilibrium between the different precursor pools.[13]

Although the flooding dose method has substantial practical advantages, it is still debated whether the large dose of the tracee amino acid influences directly protein synthesis. Some studies suggest differences between flooding dose and constant infusion methods, *i.e.* stimulation of protein synthesis by a large dose of essential amino acids,[14] whereas others report similar synthesis rates obtained by the two approaches.[15] It seems that these concerns are less or not evident when short labelling periods of less than 30 min are studied. Whole body protein turnover or, more specifically, the fractional synthesis rate (FSR) of proteins, can be estimated either by the continuous infusion or flooding dose method.

10.2 Mass Spectrometry and Protein Turnover Analysis

The study of dynamic changes in the protein and amino acid pool requires methods to measure the incorporation or loss of a tracer. Historically, the tracer was an unstable (radioactive) isotope determined by scintillation counting. More recently, it was replaced or complemented by stable (non-radioactive) isotopes which are measured by mass spectrometry (MS). Two strategies exist to assess protein dynamics—to measure either the incorporation of label into protein (synthesis) or the loss of label from previously labelled protein (degradation). In order to determine protein synthesis by incorporation of labelled amino acids into proteins, it is necessary to know the extent to which the precursor pool is labelled, *e.g.* the relative isotope abundance (RIA). Although the true precursor pool for protein synthesis is the aminoacyl-tRNA pool, which is difficult to quantify in tissues or cells due to low abundance of these precursors, it is common to use free amino acids as surrogate for the aminoacyl-tRNA pool.[13]

Based on continuous infusion of labelled isotopes or through the flooding dose method, any study of protein turnover requires analytical tools to determine isotope enrichment in the sampled tissues or body fluids. An excellent review of analytical techniques for studying protein metabolism using radioactive or stable isotope tracers was published by Wolfe and Chinkes.[16] Owing to safe handling, stability and absence of radiation hazard, stable isotopes are preferred to radioisotopes in metabolic studies.[17,18] This comes, however, at the price of needing sophisticated MS systems and experienced operators, which is not the case for radioisotopes that are easily measured with liquid scintillation counters.

Not withstanding the cost factor, mass spectrometry is considered today the best technique to measure isotopic ratios and enrichments due to its precision, sensitivity and accuracy.[19] Two classes of MS systems are most suited to determine very small differences in isotope ratios:

- isotope ratio MS (IRMS); and
- atmospheric pressure MS.

Both are commonly operated with gas chromatography but more recently also liquid chromatography.

Considering carbon as one of the most commonly used element, the precision for the measurement of $^{13}C/^{12}C$ ratios is in the range of 0.0002% for GC-IRMS and 0.05% for GC-MS.[19] Typically, the IRMS value is expressed relative to a standard in ‰ ($\delta^{13}C$).

Another difference between the two MS devices is the type of species measured; IRMS determines the isotopic ratio after conversion of the organic molecules into ionized CO_2 whereas lower precision MS measures isotopomer ratios of ionized molecules (Figure 10.2). Baseline resolution of analytes is critical for IRMS as it distinguishes only CO_2, but less decisive for MS as

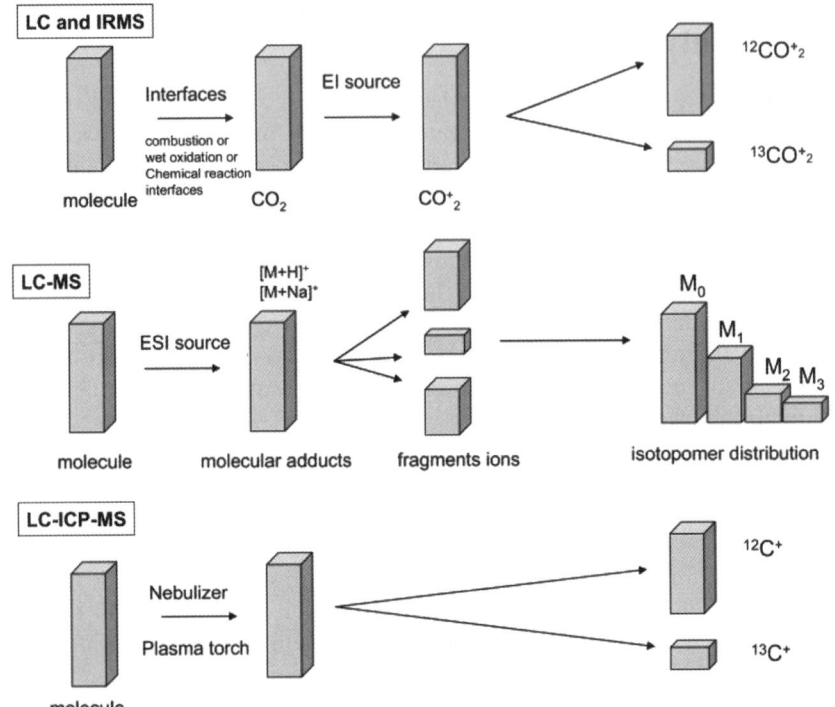

Figure 10.2 $^{13}C/^{12}C$ isotopic ratio determination using IRMS, MS, or ICP-MS devices. According to the interfaces used, the organic molecules are transformed either as ionized CO_2 gas, ionized molecules, or ionized molecular atoms. For LC-IRMS, $^{13}C/^{12}C$ isotopic ratio is measured using CO_2^+ as final species. For molecular MS, $^{13}C/^{12}C$ isotopic ratio is assessed through the isotopomer distribution in each of the fragment ions. For LC-ICP-MS, $^{13}C/^{12}C$ isotopic ratio is measured using $^{12}C^+$ and $^{13}C^+$-ionized atoms (reprinted with permission from ref. 19; Copyright 2009 Wiley Periodicals, Inc.).

co-elution of compounds can be discriminated in the analyzer. The precision of IRMS is ideally suited for determination of natural abundance variations and studies using stable-isotope enriched tracers, whereas the lower resolution MS is complementary to measure samples with higher isotope enrichment.[20]

10.2.1   Proteomic Technologies for Protein Turnover Analysis

Total protein turnover has been measured in a variety of organisms and tissues, either by the continuous infusion or the flooding dose method. Information on whole body turnover is valuable, especially in clinical application, but yields only a global view on the total protein pool either at whole tissue or organism level. Until the emergence of proteomics, limited effort has been invested in studying protein synthesis and degradation at the level of individual proteins.[21]

The recent development of new methods to address total protein metabolism and multiplexed protein analysis has allowed comprehensive profiling in the same labelling experiment.

Initially, proteomic technologies were developed for simultaneous (qualitative) identification of proteins in mixtures[22] and evolved more recently to (relative and absolute) quantitative characterization of whole proteomes.[23–26] Nevertheless, the "classical" approaches did not address the dynamics of the proteome in terms of steady-state, synthesis and degradation of individual protein species. It is the balance of synthesis and degradation that determines the concentration of any protein at steady state or a given moment in time ("proteomics snapshot"). The simple measurement of an increased or decreased expression of a given protein at a given time in a biological system does not allow the recognition of underlying mechanisms of synthesis or degradation. Despite this evident importance, the role of protein turnover has not been addressed for a long time in proteome studies. Some of the discrepancy observed between transcriptome and proteome data[27] might be explained by underlying variations of synthesis/degradation rates not captured by the traditional proteomic techniques.[28]

Addressing protein turnover dynamics represent some technical challenges for proteome-wide analysis. Whereas efficient and complete stable-isotope labelling of cells and tissue cultures has been shown to be feasible,[28,29] the high isotope enrichment needed for proteome-wide studies is still difficult to achieve in animals and humans. Modern MS systems used in proteomics can easily reach a resolution in the low parts per million (ppm) range, but the natural isotope abundance distribution of the unlabeled peptide would overlap with the profile of the labelled peptide. Ideally, mass differences of 4–10 Da introduced by a labelled precursor would greatly simplify distinguishing labelled and unlabelled forms.

A number of considerations have to be addressed in practice in order to select the "ideal" labelling precursor:[30]

- metabolic isolation (labelled atom centres do not distribute to any other amino acid);
- the cell should be auxotrophic for the amino acid (no dilution of precursor pool through *de novo* synthesis—this means an essential amino acid for humans);
- metabolically active precursor pool (rapid change in precursor isotope pool);
- abundant amino acid (high incorporation probability); and
- sufficient number of heavy atom centres (4–10 Da mass offset between labelled and unlabelled amino acid).

A very early implementation of proteome-wide turnover analysis was proposed by Cargile *et al.*,[29] named synthesis/degradation ratio mass spectrometry (SDMS), for the measurement of relative dynamic protein turnover. The approach included metabolic labelling of *Escherichia coli* with $[^{13}C_6]$-glucose

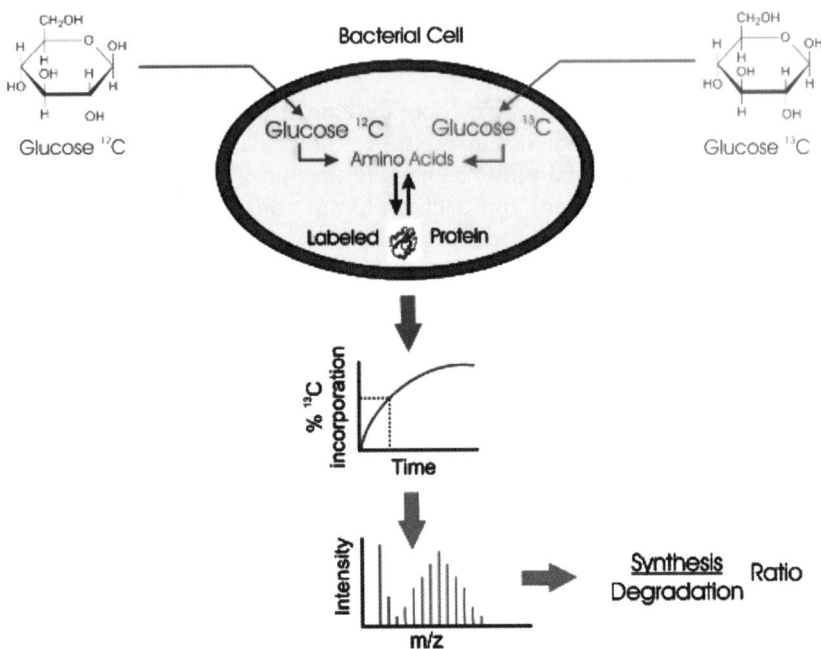

Figure 10.3 Experimental design of the synthesis/degradation ratio mass spectrometry experiment. The separation step and the Poisson modelling used for the S/D_{Rel} ratio calculation are the keys for high-throughput analysis (reprinted with permission from ref. 29; Copyright 2009 American Chemical Society).

followed by proteomic analysis—gel separation, in-gel digestion, matrix-assisted laser desorption ionization (MALDI) time-of-flight/time-of-flight (ToF/ToF) MS detection—and isotope data fitting to a Poisson distribution model in order to determine the overall turnover rate (Figure 10.3). This technique, although only applicable to cell cultures, enabled the simultaneous measurement of more than 20 proteins with synthesis/degradation ratios from 0.1 to 4.4.

A different approach was described by Pratt and co-workers,[28] in which yeast cells were grown in [^{2}H$_{10}$]-leucine over seven doubling times to obtain fully labelled cells. The chase experiment with unlabelled leucine and frequent sampling over 48 h resulted in degradation curves, from which protein turnover rates were calculated by non-linear curve fitting. Approximately 50 proteins from a two-dimensional (2D) gel were reported with corresponding degradation rates, indicating that "classical" proteomic techniques such as 2D gel separation and MALDI-MS analysis in combination with stable-isotope labelling are compatible with turnover determinations.

The true challenge in proteome-wide turnover studies is the incorporation of stable isotopes into animals and humans. First attempts have been successful in completely labelling rats with ^{15}N by feeding the animals a diet containing algal cells labelled with ^{15}N.[31] Long-term metabolic labelling with a diet enriched in

[15]N did not result in adverse health consequences. A labelling period of 44 days in a male rat resulted in a mean ^{15}N atomic enrichment of >90% in liver and plasma. Although this study did not specifically address protein turnover but rather quantitative changes in protein expression, the approach shows the feasibility of incorporating stable-isotope labelling in vertebrates.

Another study measured proteome dynamics in chicken fed with a semi-synthetic diet containing [^{2}H$_8$]-valine at a calculated relative isotope abundance (RIA) of 0.5.[32] The RIA was stable over an extended labelling window and enabled calculation of the rates of synthesis and degradation of individual proteins. By calculating the partition between newly synthesized (new) and pre-existing (old) components, and factoring in the total pool size as well as assumptions about tissue expansion, a detailed time course for the replacement of eight individual muscle proteins, reflecting muscle growth, was generated. This study demonstrated, for the first time, the possibility of addressing the analysis of turnover of individual proteins in intact animals.

Measurement of turnover in the human proteome was reported more recently using dynamic incorporation of stable isotopes in cell cultures with amino acids (dynamic SILAC).[33] Almost 600 proteins from human adeno-carcinoma cells were characterized for time-dependent changes thanks to the incorporation of [^{13}C$_6$]-arginine in a classic label-chase experiment. Although a large number of proteins was analyzed and turnover rates were deduced, the approach depends on cell cultures and thus excludes proteome-wide assessment in whole animals. Nevertheless, it shows the potential of modern proteomic technologies to measure the synthesis and degradation rates of individual proteins in the proteome.

The aforementioned approaches all require high enrichment of labelled proteins, which can be achieved in animals by feeding for long periods with highly enriched isotopes. In humans, this strategy is not practical and thus was adapted to allow the measurement of fractional synthesis rates of multiple plasma proteins after a meal that contained intrinsically labelled milk proteins.[34] To provide labelled proteins for metabolic studies, Boirie and co-workers developed a methodology some 15 years ago to produce large amounts of milk proteins intrinsically labelled with [^{13}C]-leucine.[35] The enrichment achieved in the milk proteins (casein and whey protein fraction) was between 10 and 20% ([^{13}C]-leucine atom percent excess), which was sufficient for human protein metabolism studies.

Jaleel *et al.* measured the incorporation rate of amino acids liberated during digestion of the labelled food protein and incorporation into plasma proteins was measured in humans.[34] The approach combined proteomic technologies (plasma protein purification, gel separation, LC-MS/MS-based protein identification) with traditional GC-MS analysis for isotope enrichment ([^{13}C$_6$]-phenylalanine) analysis. Twenty-nine individual plasma proteins were identified and their corresponding postprandial fractional synthesis rates were calculated based on the rate of [^{13}C$_6$]-Phe incorporation (Figure 10.4), showing a 30-fold difference in synthesis rates. The authors stress the point that con-centration of proteins between two study conditions may not change if both

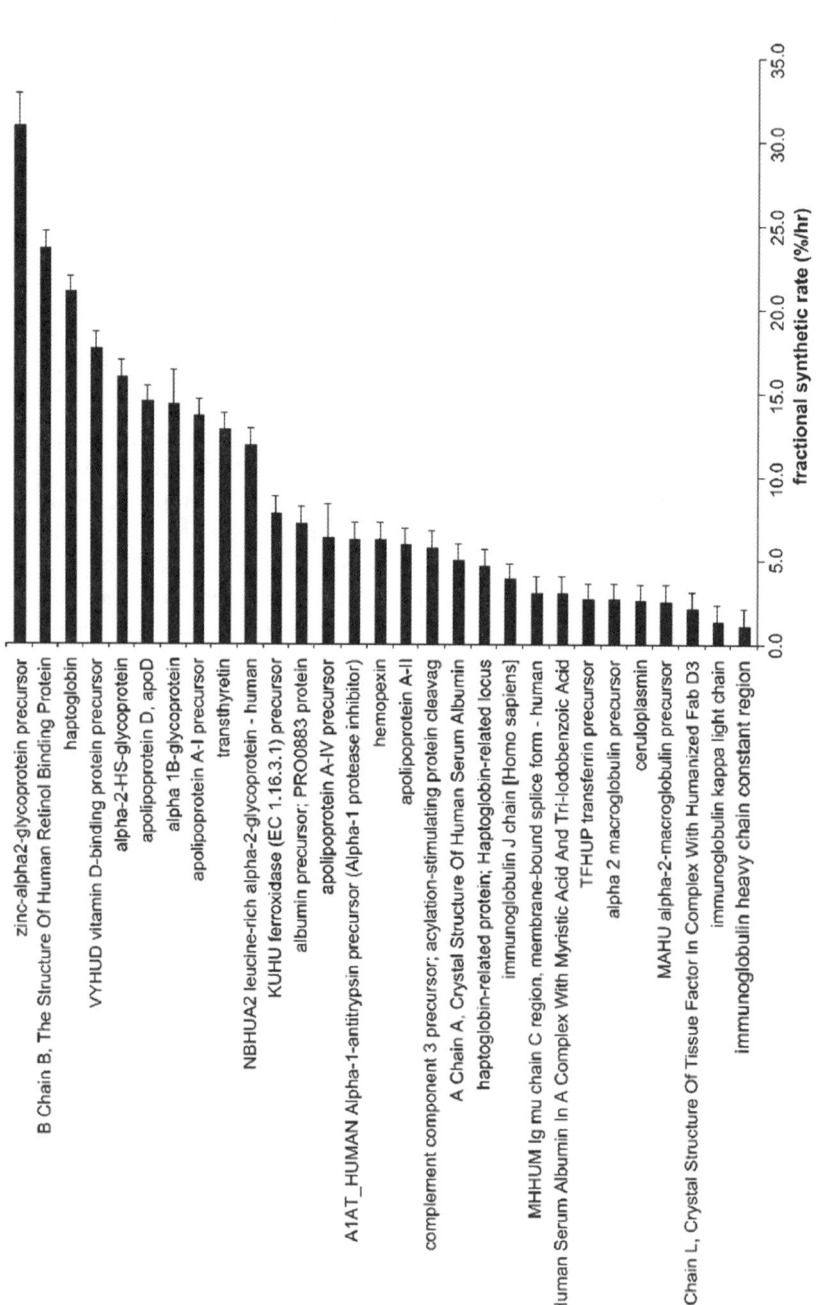

Figure 10.4 Fractional synthesis rates (FSR) of 29 plasma proteins. Proteins are arranged in hierarchical order of their FSR, showing the highest value for zinc-α2-glycoprotein and with immunoglobulin heavy-chain constant region showing the lowest. The FSR of each protein was calculated using precursor pool enrichment from either hepatic vein for proteins of hepatic origin or femoral artery for proteins of non-hepatic origin (reprinted with permission from ref. 34; Copyright 2009 American Physiological Society).

synthesis and degradation rates change in the same direction at the same magnitude, although an increased metabolism of a single protein or a cluster of proteins may have major functional consequences.

In a more recent study, the same authors report a methodology to measure synthesis rates of multiple muscle mitochondrial proteins.[36] The synthesis rates of 68 mitochondrial and 23 non-mitochondrial proteins isolated from a skeletal muscle mitochondria fraction showed a ten-fold difference between the lowest and highest rates. This approach represents a good measurement of translation rate of gene transcripts and thus offers a potential opportunity to understand the regulation of specific genes and the correlation of transcriptome and proteome results.

10.3   Nutrition and Protein Turnover

Nutrition plays a key role in body protein metabolism. Since stable-isotope methods became available, a great deal of research has been performed to study the nutritional influences on protein synthesis and degradation. Even before then, nutritional influences on nitrogen balance were studied focussing on the minimum requirements for protein and individual essential amino acids.[37] It was shown that nitrogen balance is negative after a fasting period, also called post-absorptive phase, *i.e.* when a person has not eaten for a few hours.[38] Circulating amino acids, glucose and lipids are derived from endogenous stores rather than from food being absorbed from the gut. Post-absorptive studies are generally performed after overnight fasting for 10–12 hours. The negative protein balance is explained mainly by loss of amino acids through oxidation, even if most of the amino acids derived from protein breakdown are re-incorporated into newly synthesized proteins. Therefore, the total protein balance is negative because no amino acids are available from food. The protein balance becomes positive after a meal if an adequate amount of protein is consumed. Hence, fluctuations between negative protein balance during the post-absorptive state are compensated by the positive balance during the fed state, if an adequate diet is consumed (Figure 10.5).

The relations between nutrition and protein metabolism are so multifaceted that not all aspects of this interesting field can be discussed in this chapter. The focus is therefore on nutritional modulation of skeletal muscle synthesis in the context of exercise and of skeletal muscle degradation in the context of ageing and sarcopenia.

10.3.1   Exercise and Protein Metabolism

Muscle plays a central role in whole-body protein metabolism[39] as it serves as the main reservoir for amino acids to maintain protein synthesis. In healthy humans, skeletal muscle comprises approximately 50% of total body mass[40] and accounts for almost 80% of the basal metabolic rate.[41] In addition, skeletal muscle accounts for about 80% of postprandial glucose uptake.[42]

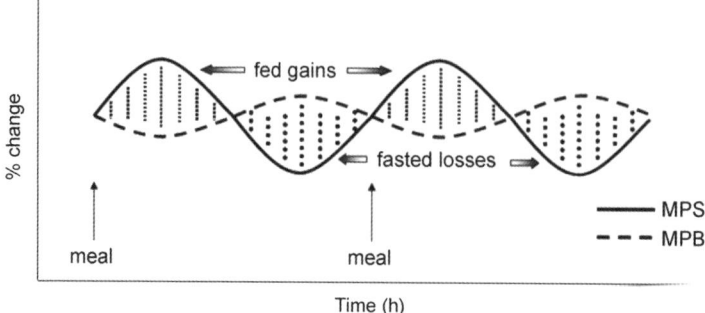

Figure 10.5   Changes in muscle protein synthesis (MPS) and muscle protein breakdown (MPB) in response to feeding, *i.e.* amino acids (reprinted with
permission from ref. 53; Copyright 2009 American Physiological
Society).

The constant mechanical and metabolic demand results in a continuous and
active remodelling of muscle, implicating tuned degradation and re-synthesis of
muscular proteins such as myofibrillar contractile as well as sarcoplasmatic
metabolic proteins. Both protein synthesis and degradation are affected by
different factors such as (mal)nutrition, physical (in)activity and mechanical
(un)loading, and the hormonal changes that parallel these processes.[40]
Approximately 50% of whole protein synthesis in humans originates from
skeletal muscle synthesis and is doubled in the fed compared to the fasted
state.[43] As shown by using the flooding dose approach with a series of essential
and non-essential amino acids, and MS-based analysis of [^{13}C]-leucine incorporation, amino acids are the main food components responsible for an
increased rate of protein synthesis. Essential amino acids (phenylalanine,
threonine) were identified to significantly induce muscle protein synthesis
whereas non-essential amino acids (arginine, serine, glycine) failed to do so.[14]
Branched chain amino acids (leucine, isoleucine, valine)—and in particular
leucine—show an anabolic effect by suppressing muscle protein breakdown by
as much as 40%.[44] The relationship between the availability of amino acids and
the extent of stimulation of muscle protein synthesis is curvilinear (*e.g.* linearly
regulated by the essential amino acid concentration in the blood over the
normal diurnal range), but becomes saturated at high concentrations.[45,46] This
also explains, at least partly, why it is impossible to increase muscle mass simply
by eating.
   The other major process in protein turnover, protein breakdown, also plays
an important role in the control of muscle mass. Muscle protein breakdown
decreases about 20% during feeding of a protein-rich meal, whereas muscle
protein synthesis, as already mentioned, may double.[47] Breakdown seems to be
less influenced by free amino acid levels in the blood but rather by a secondary
effect, the stimulation of insulin by components of a meal, including glucose
and amino acids. However, it appears that in adult humans, insulin does not do
what it does in growing animals (*e.g.* mice, rats, pigs) traditionally studied in

metabolic research. Through MS-based tracer incorporation measurements, it was shown in humans that muscle protein synthesis was stimulated by supplying exogenous amino acids alone while maintaining basal blood insulin concentration at the overnight fasted level.[48] Differences in response to essential amino acids on muscle protein synthesis were shown between young and old men. Basal rates of muscle protein synthesis were indistinguishable, but the elderly showed less anabolic sensitivity and responsiveness to dietary amino acids. The common link between muscle protein synthesis and breakdown may be the concentration of free amino acids in the blood, which stimulates synthesis directly and inhibits breakdown *via* stimulation of insulin.[47]

Physical activity influences total muscle protein mass and, depending on the type of exercise (anaerobic resistance or aerobic endurance training), preferential stimulation of the synthesis of myofibrillar or mitochondrial proteins is induced.[49] To favour muscle protein accumulation even more, consumption of protein during post-exercise recovery is necessary. Exercise is capable of turning on anabolic signalling cascades and feeding potentiates this effect synergistically.[50,51] It was shown that the rise in muscle protein breakdown was attenuated and thus net balance became positive when amino acid availability increased through feeding after exercise.[52] Interestingly, increases in muscle protein synthesis after acute resistance exercise is sustained for up to 48 h, even in the absence of feeding, creating a "window of anabolic opportunity" for a greater muscle protein synthetic response through exercise.[53,54] An optimal dose of protein seems to be in the range of 15–20 g of high quality protein such as beef, egg, or milk.[55]

In conclusion, research over the last 20 years has helped to better understand mechanisms influencing protein synthesis and breakdown. It has become apparent that metabolic responses are highly regulated and that the magnitude of the response depends on multiple factors related to exercise and nutrition. Only when availability of amino acids is increased during the post-exercise period can net gain in skeletal muscle mass be achieved through exercise.

10.3.2 Sarcopenia and Protein Metabolism

Sarcopenia (Greek: *sarx* flesh and *penia* loss) denotes the gradual loss of muscle mass and strength during ageing.[56] Lean muscle mass contributes up to approximately 50% of total body weight in young adults, but declines to 25% in people reaching an age of 70–80 years.[57] Rates of whole body and muscle protein metabolism decline with age, suggesting a progressive decline in the body's remodelling processes with aging. The loss of lean muscle mass is generally compensated by a gain in fat mass. Furthermore, the loss in muscle mass is accompanied by the loss of muscle strength, a decline in functional performance, and a reduction in oxidative capacity.[58] This loss in metabolic capacity and functionality potentially leads to increased risks of developing chronic metabolic diseases. Therefore, it would be highly beneficial to prevent or at least to moderate the decline in skeletal muscle mass by nutrition and exercise in order to promote healthy ageing.

Age-related reduction in muscle mass is aggravated by different factors such as sub-optimal diet[59] and sedentary lifestyle.[60] The principal causes of skeletal muscle loss, however, must be attributed to a disruption in the regulation of skeletal muscle turnover. Although some studies have described lower basal muscle protein synthesis rates in the elderly,[61] more recent studies could not confirm these findings. In consequence, the hypothesis that basal fasting protein turnover rates do not significantly differ between young and elderly receives increasing support from the scientific community.[62] To explain the obvious loss of skeletal muscle mass in sarcopenia, studies were refocused on the elucidation and analysis of molecular mechanisms that control the protein synthetic response to the main anabolic stimuli, food and physical activity.

Cuthbertson and co-workers studied the anabolic response to dietary amino acids in young and old men.[48] Basal rates of muscle protein synthesis were indistinguishable, but the elderly showed less anabolic sensitivity and responsiveness of muscle protein synthesis to essential amino acids, possibly due to decreased intramuscular expression and/or activation (phosphorylation) of amino acid sensing and signalling proteins. Another study showed that muscle protein synthesis rate was stimulated during a euglycemic hyperinsulinemic hyperaminoacidemic clamp in young subjects, but this response was significantly lower in elderly subjects.[63] These results suggest that anabolic signal sensing and/or transduction in muscle tissue, induced by insulin and amino acids, are diminished in the elderly compared to younger men.

As discussed previously, physical activity and in particular resistance-type exercise is an important stimulus for skeletal muscle protein accretion.[49] Different stable-isotope tracer and MS-based studies performed in young and elderly individuals showed exercise-induced stimulation of skeletal muscle protein synthesis.[64–66] Although only subtle differences in protein synthetic response to resistance-type exercise between young and elderly were found, a more recent study reported for elderly subjects anabolic resistance of muscle protein synthesis to resistance exercise, with an $\sim 30\%$ lower response in elderly men.[65]

Recent work by Campbell and colleagues[67] demonstrated that total dietary protein requirements do not differ for healthy older adults compared to younger adults, and that a dietary protein allowance of 0.85 g per kg body weight and day is adequate. In earlier studies, the same group supported the hypothesis that older persons who consume adequate or moderately high amounts of dietary protein ($0.9–1.2\,\mathrm{g\,kg^{-1}\,d^{-1}}$) can use resistance-type exercise to improve body composition and oral glucose tolerance without a change in body weight. Verdijk and coworkers assessed the benefits of timed protein supplementation on the increase in muscle mass and strength during prolonged resistance-type exercise training (12 weeks) in healthy elderly men who habitually consume adequate amounts of dietary protein.[68] They concluded that timed protein supplementation immediately before and after exercise did not further increase skeletal muscle mass in healthy elderly men who consumed adequate amounts of dietary protein. Taken together, the reported data suggest that adequate daily protein intake of $\sim 0.9\,g\,kg^{-1}\,d^{-1}$ in combination with resistance-type exercise training will support an augmentation in skeletal muscle mass, strength and functional performance in elderly people, thus preventing or at least moderating the decline in skeletal muscle mass seen in sarcopenia.

10.4 Conclusions

Modern approaches to study protein metabolism, the synthesis and breakdown of proteins in the body are linked closely to progress in stable-isotope analysis by mass spectrometry. Methods for stable-isotope analysis have dramatically evolved over the last few decades. Modern GC-combustion-MS systems enable accurate and sensitive measurements of isotope enrichment whereas isotope ratio MS analysis gives ultimate precision for determination of natural abundance variations and for studies using stable-isotope labelled tracers. Both MS technologies represent cornerstones for studies of protein metabolism.

With the development of proteomic technologies for qualitative and quantitative analysis of entire proteomes, new high resolution and high mass accuracy MS instruments were also used to assess isotope enrichment in labelled proteins. Although current types of mass spectrometers used in proteomics generally might not be considered sufficiently reliable for measuring less than a few per cent of a stable-isotope labelled variant, higher enrichment levels fall readily within the analytical performance of modern high resolution instruments. This opens up new possibilities for proteome-wide turnover studies, addressing protein metabolism at individual protein level. The ideal approach should combine tracer studies (constant infusion or flooding dose approach) and proteomic techniques for protein isolation, purification and characterization. Together with GC-MS and IRMS analysis for tracer enrichment quantification, this would enable the largest coverage of protein identities with concurrent synthesis and degradation data for individual proteins. In the future, this information will become even more important as systems biology requires integration and correlation of transcriptome, proteome and metabolome data, which really should also comprise protein synthesis and breakdown information.

Detailed studies over the last decades have also shown that nutrition—in particular protein and essential amino acids such a leucine—play a key role in the anabolic signalling for the stimulation of skeletal muscle protein synthesis. The combination of resistance- or endurance-type exercise training and adequate protein intake act synergistically to enhance muscle mass and performance. Studies in the elderly population confirmed the potential to address sarcopenia (*i.e.* the gradual loss of muscle mass and function with ageing) by adapted nutrition with adequate daily protein intake and resistance-type exercise training, allowing a substantial preservation in skeletal muscle mass, strength and functional performance.

References

1. J. C. Waterlow, *Protein Turnover*, CABI, Cambridge, 2006.
2. S. L. Duggleby and J. C. Waterlow, *Br. J. Nutr.*, 2005, **94**, 141.
3. D. B. Sprinson and D. Rittenberg, *J. Biol. Chem.*, 1949, **180**, 715.
4. E. B. Fern, P. J. Garlick, M. A. McNurlan and J. C. Waterlow, *Clin. Sci. (Lond.)*, 1981, **61**, 217.
5. D. E. Matthews, K. J. Motil, D. K. Rohrbaugh, J. F. Burke, V. R. Young and D. M. Bier, *Am. J. Physiol.*, 1980, **238**, E473.
6. G. N. Thompson, P. J. Pacy, H. Merritt, G. C. Ford, M. A. Read, K. N. Cheng and D. Halliday, *Am. J. Physiol.*, 1989, **256**, E631.
7. D. B. Zilversmit, *Am. J. Med.*, 1960, **29**, 832.
8. A. J. Wagenmakers, *Proc. Nutr. Soc.*, 1999, **58**, 987.
9. D. L. Hasten, J. Pak-Loduca, K. A. Obert and K. E. Yarasheski, *Am. J. Physiol. Endocrinol. Metab.*, 2000, **278**, E620.
10. P. J. Garlick, M. A. McNurlan and V. R. Preedy, *Biochem. J.*, 1980, **192**, 719.
11. T. A. Davis and P. J. Reeds, *Curr. Opin. Clin. Nutr. Metab. Care*, 2001, **4**, 51.
12. M. J. Rennie, *Proc. Nutr. Soc.*, 1999, **58**, 935.
13. T. A. Davis, M. L. Fiorotto, H. V. Nguyen and D. G. Burrin, *Am. J. Physiol.*, 1999, **277**, E103.
14. K. Smith, N. Reynolds, S. Downie, A. Patel and M. J. Rennie, *Am. J. Physiol.*, 1998, **275**, E73.
15. B. G. Southorn, J. M. Kelly and B. W. McBride, *J. Nutr.*, 1992, **122**, 2398.
16. R. R. Wolfe and D. L. Chinkes, *Isotope Tracers in Metabolic Research: Principles and Practice of Kinetic Analysis*, Wiley-Liss, New York, 2005.
17. J. Vogt, *Eur. J. Pediatr.*, 1997, **156**(Suppl 1), S9.
18. B. Koletzko, T. Sauerwald and H. Demmelmair, *Eur. J. Pediatr.*, 1997, **156**(Suppl 1), S12.
19. J. P. Godin, L. B. Fay and G. Hopfgartner, *Mass Spectrom. Rev.*, 2007, **26**, 751.
20. F. Montigon, J. J. Boza and L. B. Fay, *Rapid Commun. Mass Spectrom.*, 2001, **15**, 116.
21. M. K. Doherty and R. J. Beynon, *Expert Rev. Proteomics*, 2006, **3**, 97.
22. D. C. Liebler, *Introduction to Proteomics*, Humana Press, Totowa, NJ, 2002.
23. A. Panchaud, M. Affolter, P. Moreillon and M. Kussmann, *J. Proteomics.*, 2008, **71**, 19.
24. R. J. Beynon, M. K. Doherty, J. M. Pratt and S. J. Gaskell, *Nat. Methods*, 2005, **2**, 587.
25. S. P. Gygi, B. Rist, S. A. Gerber, F. Turecek, M. H. Gelb and R. Aebersold, *Nat. Biotechnol.*, 1999, **17**, 994.
26. S. A. Gerber, J. Rush, O. Stemman, M. W. Kirschner and S. P. Gygi, *Proc. Natl. Acad. Sci. U. S. A.*, 2003, **100**, 6940.

27. S. P. Gygi, Y. Rochon, B. R. Franza and R. Aebersold, *Mol. Cell Biol.*, 1999, **19**, 1720.
28. J. M. Pratt, J. Petty, I. Riba-Garcia, D. H. Robertson, S. J. Gaskell, S. G. Oliver and R. J. Beynon, *Mol. Cell. Proteomics*, 2002, **1**, 579.
29. B. J. Cargile, J. L. Bundy, A. M. Grunden and J. L. Stephenson Jr., *Anal. Chem.*, 2004, **76**, 86.
30. R. J. Beynon and J. M. Pratt, *Mol. Cell. Proteomics*, 2005, **4**, 857.
31. C. C. Wu, M. J. MacCoss, K. E. Howell, D. E. Matthews and J. R. Yates III, *Anal. Chem.*, 2004, **76**, 4951.
32. M. K. Doherty, C. Whitehead, H. McCormack, S. J. Gaskell and R. J. Beynon, *Proteomics*, 2005, **5**, 522.
33. M. K. Doherty, D. E. Hammond, M. J. Clague, S. J. Gaskell and R. J. Beynon, *J. Proteome Res.*, 2009, **8**, 104.
34. A. Jaleel, V. Nehra, X. M. Persson, Y. Boirie, M. Bigelow and K. S. Nair, *Am. J. Physiol. Endocrinol. Metab.*, 2006, **291**, E190.
35. Y. Boirie, J. Fauquant, H. Rulquin, J. L. Maubois and B. Beaufrere, *J. Nutr.*, 1995, **125**, 92.
36. A. Jaleel, K. R. Short, Y. W. Asmann, K. A. Klaus, D. M. Morse, G. C. Ford and K. S. Nair, *Am. J. Physiol. Endocrinol. Metab.*, 2008, **295**, E1255.
37. S. Welle, *Human Protein Metabolism,* Springer-Verlag, New York, 1999.
38. G. M. Price, D. Halliday, P. J. Pacy, M. R. Quevedo and D. J. Millward, *Clin. Sci. (Lond.)*, 1994, **86**, 91.
39. R. R. Wolfe, *Am. J. Clin. Nutr.*, 2006, **84**, 475.
40. M. K. C. Hesselink, R. Minnaard and P. Schrauwen, *Curr. Opin. Clin. Nutr. Metab. Care*, 2006, **9**, 672.
41. E. Ravussin, S. Lillioja, T. E. Anderson, L. Christin and C. Bogardus, *J. Clin. Invest*, 1986, **78**, 1568.
42. R. A. DeFronzo, R. Gunnarsson, O. Bjorkman, M. Olsson and J. Wahren, *J. Clin. Invest*, 1985, **76**, 149.
43. M. J. Rennie, R. H. Edwards, D. Halliday, D. E. Matthews, S. L. Wolman and D. J. Millward, *Clin. Sci. (Lond.)*, 1982, **63**, 519.
44. R. J. Louard, E. J. Barrett and R. A. Gelfand, *Metabolism*, 1995, **44**, 424.
45. J. Bohe, A. Low, R. R. Wolfe and M. J. Rennie, *J. Physiol.*, 2003, **552**, 315.
46. J. Bohe, J. F. Low, R. R. Wolfe and M. J. Rennie, *J. Physiol.*, 2001, **532**, 575.
47. M. J. Rennie, *Exp. Physio.l*, 2005, **90**, 427.
48. D. Cuthbertson, K. Smith, J. Babraj, G. Leese, T. Waddell, P. Atherton, H. Wackerhage, P. M. Taylor and M. J. Rennie, *FASEB J.*, 2005, **19**, 422.
49. S. B. Wilkinson, S. M. Phillips, P. J. Atherton, R. Patel, K. E. Yarasheski, M. A. Tarnopolsky and M. J. Rennie, *J. Physiol.*, 2008, **586**, 3701.
50. R. Koopman, W. H. Saris, A. J. Wagenmakers and L. J. van Loon, *Sports Med.*, 2007, **37**, 895.
51. V. Kumar, P. Atherton, K. Smith and M. J. Rennie, *J. Appl. Physiol.*, 2009, **106**, 2026.
52. K. D. Tipton, A. A. Ferrando, S. M. Phillips, D. Doyle Jr and R. R. Wolfe, *Am. J. Physiol.*, 1999, **276**, E628.

53. N. A. Burd, J. E. Tang, D. R. Moore and S. M. Phillips, *J. Appl. Physiol.*, 2009, **106**, 1692.
54. S. M. Phillips, K. D. Tipton, A. Aarsland, S. E. Wolf and R. R. Wolfe, *Am. J. Physiol.*, 1997, **273**, E99.
55. D. R. Moore, M. J. Robinson, J. L. Fry, J. E. Tang, E. I. Glover, S. B. Wilkinson, T. Prior, M. A. Tarnopolsky and S. M. Phillips, *Am. J. Clin. Nutr.*, 2009, **89**, 161.
56. I. H. Rosenberg, *J. Nutr.*, 1997, **127**, 990S.
57. K. R. Short, J. L. Vittone, M. L. Bigelow, D. N. Proctor and K. S. Nair, *Am. J. Physiol. Endocrinol. Metab.*, 2004, **286**, E92.
58. R. Koopman and L. J. C. Van Loon, *J. Appl. Physiol.*, 2009, **106**, 2040.
59. W. W. Campbell, T. A. Trappe, R. R. Wolfe and W. J. Evans, *J. Gerontol. A, Biol. Sci. Med. Sci.*, 2001, **56**, M373.
60. K. S. Nair, *Am. J. Clin. Nutr.*, 2005, **81**, 953.
61. S. Welle, C. Thornton, R. Jozefowicz and M. Statt, *Am. J. Physiol.*, 1993, **264**, E693.
62. E. Volpi, M. Sheffield-Moore, B. B. Rasmussen and R. R. Wolfe, *J. Am. Med. Assoc.*, 2001, **286**, 1206.
63. C. Guillet, M. Prod'homme, M. Balage, P. Gachon, C. Giraudet, L. Morin, J. Grizard and Y. Boirie, *FASEB J.*, 2004, **18**, 1586.
64. M. J. Drummond, H. C. Dreyer, B. Pennings, C. S. Fry, S. Dhanani, E. L. Dillon, M. Sheffield-Moore, E. Volpi and B. B. Rasmussen, *J. Appl. Physiol.*, 2008, **104**, 1452.
65. V. Kumar, A. Selby, D. Rankin, R. Patel, P. Atherton, W. Hildebrandt, J. Williams, K. Smith, O. Seynnes, N. Hiscock and M. J. Rennie, *J. Physiol.*, 2009, **587**, 211.
66. M. Sheffield-Moore, C. W. Yeckel, E. Volpi, S. E. Wolf, B. Morio, D. L. Chinkes, D. Paddon-Jones and R. R. Wolfe, *Am. J. Physiol. Endocrinol. Metab.*, 2004, **287**, E513.
67. W. W. Campbell, C. A. Johnson, G. P. McCabe and N. S. Carnell, *Am. J. Clin. Nutr.*, 2008, **88**, 1322.
68. L. B. Verdijk, R. A. Jonkers, B. G. Gleeson, M. Beelen, K. Meijer, H. H. Savelberg, W. K. Wodzig, P. Dendale and L. J. van Loon, *Am. J. Clin. Nutr.*, 2009, **89**, 608.

Section 4
Conclusion

CHAPTER 11
Conclusion

LAURENT B. FAY[a] AND MARTIN KUSSMANN[a,b]

[a] Nestlé Research Centre, Vers-chez-les-Blanc, PO Box 44, CH-1000, Lausanne 26, Switzerland; [b] Faculty of Science, Aarhus University, Ny Munkegade, Building 1521, DK-8000, Aarhus C, Denmark

Nutrition science evolved over the 20th century from basic, experimental feeding and intervention trials to today's mechanistic investigations documenting both beneficial and detrimental effects of food on human health at the molecular level. The nutritional knowledge generated, coupled with the tremendous improvement of agricultural practices, resulted in a worldwide food production that has grown to exceed population nutritional needs. However, this revolution produced also counterproductive effects with huge worldwide disparities such as one billion people being undernourished at the same time as 1.6 billion people being overweight. Between those two extremes, there are many health and disease conditions that can be either positively or negatively influenced by diet.

Undeniably, in today's world, most of the chronic diseases have a nutritional component, for example:

- regular intake of foods with saturated fats such as meat increases the risk of coronary heart disease;
- frying creatinine- and protein-containing foods such as meats generates genotoxic carcinogens (for instance heterocyclic aromatic amines); and
- high salt intake is associated with high blood pressure and stomach cancer.

RSC Food Analysis Monographs No. 9
Mass Spectrometry and Nutrition Research
Edited by Laurent B. Fay and Martin Kussmann
© The Royal Society of Chemistry 2010
Published by the Royal Society of Chemistry, www.rsc.org

On the other hand, epidemiological evidence suggests that increased fruit, vegetable and milk consumption affords significant protection against many chronic diseases. These dietary food groups contain numerous compounds that possess health-promoting activities in addition to their complement of essential nutrients. Moreover, mono-unsaturated oils such as olive or canola oil are low-risk fats; vegetables, fruits, and soy-containing products are rich in antioxidants that are essential to lower disease risk deriving from oxidative stress; and green and black teas as well as cocoa and dark chocolate are excellent sources of antioxidants because of their high content in polyphenols.

Modern nutrition research targets health promotion, disease prevention, performance improvement and risk assessment. It aims at the identification of potentially active food constituents, the demonstration of their bioavailability and efficacy, and ultimately the understanding of their mechanisms of action. Intensive efforts are dedicated to the identification and quantification of bioactive molecules in plants (*e.g.* phytochemicals) and animal derived foods (*e.g.* milk oligosaccharides, proteins and peptides). As many of these bioactive compounds represent only minor constituents in a highly complex matrix, mass spectrometry has developed into the key analytical platform in any nutrition research laboratory because of its specificity and sensitivity, especially in combination with high-performance chromatography.

In this book, we describe how mass spectrometry has developed into a tool to assess food composition and health beneficial aspects of food. To set the stage, we outline the various mass spectrometric technologies available today. Then, the first part of the book addresses the determination of some components of our diet that play a documented role in the maintenance of human health, namely, oligosaccharides, peptides, proteins and lipids. We also dedicate a chapter to naturally occurring phytonutrients, as some have received considerable attention due to their health beneficial properties. The second part of the book covers health-related issues such as cardiovascular disease, immunity and allergy, and protein turnover—in all of which specific food components and diet in general play a pivotal role.

The food components that provide health benefits beyond basic nutrition, including the prevention and management of disease, are coined "nutraceutical". This word is a contraction of *nutrient* and *pharmaceutical*. However, compared to drugs, the concentration of these active compounds is much lower. Therefore, chronic doses are likely to be the basis of any effect. As it is the combined effects of many such compounds in foods that promote health and provide benefits (rather than one of any single compound), there is still a considerable lack of knowledge concerning the efficacy of some dietary bioactive molecules. A logical extension of such conventional wisdom is the development and study of functional foods as compared to individual bioactive compounds.

When studying the relationship between exposure to bioactive compounds and—even more so—to whole foods on one hand and the health outcomes derived from complex biochemical pathways on the other hand, it is unlikely that these can be described by individual biomarkers. Rather, a combination and integration of appropriate biomarkers to describe susceptibility, exposure and effect will be required to understand the link between diet and health.

Mass spectrometry is ideally suited for studying the bioavailability and bioefficacy of bioactive compounds and even combinations thereof in food and biological samples, as well as for investigating related health outcomes. Nevertheless, an intelligent combination and integration of complementary analytical platforms will be required to further advance our understanding of the complex interactions and activities of combinations of bioactive compounds as present, for example, in functional foods. Therefore, further mass spectrometric developments in both data generation and processing (*i.e.* at hardware and software level) are needed to enable this integration. Bioinformatics increasingly plays a major role in data handling, processing and integration.

This widespread need for mass spectrometric applications in nutrition research supports the further distribution of MS platforms in food research laboratories. A remaining major barrier for broader application is the elevated cost of the instrumentation. The availability of instruments with adapted performances, from table-top to high-end devices, further improved user-friendliness and high-throughput potential should be major development areas in mass spectrometric instrumentation for food and nutrition research.

With the aim of enhancing selectivity and thus minimizing interferences, new mass analyzers with very high resolution and mass accuracy, available at moderate prices, have been developed and introduced. A typical example of this trend was the introduction of quadruple/time-of-flight (QqToF) instruments 10–15 years ago, of high-resolution triple-quadrupoles five years ago and the very recent introduction of Orbitraps. Emphasis has also been put on high mass accuracy during this development, *e.g.* for Q-ToF- (5–10 ppm), Orbitrap (<5 ppm) and Fourier transform ion cyclotron resonance (FTICR) machines (sub ppm). Although the sensitivity of current mass spectrometers is already outstanding, developments of ion guide and focusing devices are in progress to enable even higher sensitivity and allow for an even simpler sample preparation. The ultimate sample preparation could consist of only sample dilution (if necessary) in order to minimize introduced bias. Last, but certainly not least, hybrid mass spectrometers with sophisticated scan functions are increasingly being deployed, because these devices combine the advantages of different mass spectrometer types in one instrument (*e.g.* Q-ToF, Q-trap, Paul trap-Orbitrap, ion mobility-triple quadrupole).

Mass spectrometry is one of the most versatile identification and quantification techniques at the disposal of nutritionists today and, even more importantly, it is among the very few analytical tools that can deliver information-rich data. Its potential as a nutritional knowledge provider is constantly growing because of both the push from instrument manufacturers and the pull of researchers aiming to transform nutrition into a mechanism-based discipline.

Subject Index

References to figures are given in italic type; references to tables are given in bold type.